Statistical Plasma Physics

Volume I: Basic Principles

Statistical Plasma Physics

Volume I: Basic Principles

Setsuo Ichimaru
University of Tokyo

ADDISON-WESLEY PUBLISHING COMPANY
The Advanced Book Program
Redwood City, California • Menlo Park, California
Reading, Massachusetts • New York • Don Mills, Ontario
Wokingham, United Kingdom • Amsterdam • Bonn
Sydney • Singapore • Tokyo • Madrid • San Juan

Mathematics and Physics Editor: *Barbara Holland*
Production Manager: *Pam Suwinsky*
Production Assistant: *Karl Matsumoto*
Editorial Assistant: *Diana Tejo*
Electronic Composition: *Professional Book Center*

Library of Congress Cataloging-in-Publication Data

Ichimaru, Setsuo
Statistical plasma physics / Setsuo Ichimaru
p. cm.
Includes bibliographical references and index.
Contents: v. 1. Basic principles.
1. Plasma (ionized gases)--Statistical methods. 2. Statistical physics. I. Title
QCC718.S78127 1991 530.4'4072--dc20 91-40365
ISBN 0-201-55490-9 (v. 1)

1 2 3 4 5 6 7 8 9 10—MA—95 94 93 92

STATISTICAL PLASMA PHYSICS

VOLUME I: BASIC PRINCIPLES

SETSUO ICHIMARU

The following are corrigenda and a revised version of the (Subject) Index for *Statistical Plasma Physics: Basic Principles.* The author wishes to express his gratitude to Dr. Shuji Ogata of the University of Tokyo for assistance in preparing these corrigenda, and to Mr. David Miller and Mr. Jack Repcheck of Addison-Wesley Publishing Company for cooperation in producing this additional material.

CORRIGENDA

p. 36, the l.h.s. of Eq. (2.30a) should read:
$F_1(1)$

p. 92, in the 2nd line of Sec. 3.5E, "internal" should read "interaction".

p. 92, in the 4th line of Sec. 3.5E, "internal" should read "excess internal".

p. 96, in the 3rd line after Eq. (3.198), "in Chapter 11" should read "[Ichimaru, Iyetomi & Tanaka 1987]".

p. 101, the l.h.s. of the first equation in Problem 3.22 should read:
$v(k)\chi_0^c(k, z_l)$

p. 101, the first integral on the r.h.s. of the second equation in Problem 3.22 should have the domain specified as $p \leq 1$.

p. 110, the middle term of Eq. (4.23) should be multiplied by 2.

p. 146, in the 1st line, "$\omega^2 \lesssim \omega_p^2$" should read "$\omega^2 \gtrsim \omega_p^2$".

p. 201, in Fig. 6.11 (two places), "q" should read "Z".

p. 202, in the 1st line above Eq. (6.58), "kinetic" should read "total".

p. 217, on the l.h.s. of Eq. (6.103), "v" should read "$\mathbf{v}$".

p. 217, Eq. (6.106) should read:
$\omega^2 - \omega_e^* \omega - s^2 k_z^2 = 0$

p. 223, in the denominator on the r.h.s. of Eq. (6.120), "β_i" should read "Ω_i".

p. 234, the r.h.s. of Eq. (7.18) should read:

$$\mathbf{v} \cdot \frac{\partial}{\partial \mathbf{r}} + Z_\sigma e \left[\langle \mathbf{E} \rangle + \frac{\mathbf{v}}{c} \times \langle \mathbf{B} \rangle \right] \cdot \frac{\partial}{\partial \mathbf{p}} ,$$

p. 240, in Eq. (7.43), the summation should be performed with respect to both σ and σ'.

p. 243, the 2nd line of Eq. (7.46) should read:

$$-\sum_{\sigma'} \left[\frac{2\pi Z_{\sigma'} e}{(\mathbf{k} \cdot \mathbf{v})^2} \right]^2 \left[\mathbf{k} \cdot \frac{\partial F_\sigma(\mathbf{p})}{\partial \mathbf{p}} \right] \int d\mathbf{p}' F_{\sigma'}(\mathbf{p}') \delta(\mathbf{k} \cdot \mathbf{v} - \mathbf{k} \cdot \mathbf{v}') |\mathbf{v} \cdot \mathbf{Z}(\mathbf{k}, \mathbf{k} \cdot \mathbf{v}) \cdot \mathbf{v}'|^2 \Bigg\} .$$

p, 243, in the 1st line of the 2nd paragraph, "Eq. (7.65)" should read "Eq. (7.46)".

p. 250, in the last curly bracket of the 1st line of Eq. (7.77), "Z_2^{2v}" should read "$Z_2^2 v$".

p. 255, in the 2nd line, "Eq. (7.94)" should read "Eq. (7.85)".

p. 257, in Eq. (7.102) (two places), "q" should read "$\mathbf{q}$".

p. 258, in the line between Eqs. (7.105) and (7.106), "Eq. (7.103)" should read "Eq (7.102)".
p. 261, in the 12th line from the bottom, "Eq. (7.149)" should read "Eq. (7.121)".
p. 263, on the r.h.s. of Eq. (7.128) and on the l.h.s. of Eq. (7.130), terms separated by a comma should be braced by a curly bracket. For example, the l.h.s. of Eq. (7.130) should read: $\{A, B\}$
p. 264, on the r.h.s. of Eq. (7.134), terms separated by a comma inside the angular bracket $\langle\,,\,\rangle$ should be braced by a curly bracket.
p. 290, Eq. (7.240) should read:

$$\mathbf{B}(z) = \delta B[\hat{\mathbf{x}}\cos(qz) + \hat{\mathbf{y}}\sin(qz)]$$

p. 315, Sec. 8.4C should start with as follows:

C. Calculation of the First Moment

As Eq. (8.75) implies, the first moment $\langle \Delta X_1 \rangle$ is proportional to the average force on a particle "1" in the y direction due to scattering in the fields produced by the particles "2". Such a force arises from two separate sources: one due to the field fluctuations and another due to the polarization. These may be calculated, respectively, as

$$\frac{\langle \Delta X_1 \rangle_{\rm f}}{\tau} = i\frac{e^2}{2B^2}\sum_{\mathbf{k}}\int_{-\infty}^{\infty} d\omega \int_{-\infty}^{\infty} dt \frac{k_y}{k^2}\mathbf{k}\cdot\mathsf{H}(t)\cdot\mathbf{k}\langle|E^2|(\mathbf{k},\omega)\rangle_2 \exp\{i[\mathbf{k}\cdot\Delta\mathbf{r}(t) - \omega t]\} \quad (8.81)$$

$$\frac{\langle \Delta X_1 \rangle_{\rm p}}{\tau} = -\frac{2Z_1 ec}{B}\sum_{\mathbf{k}}\int_{-\infty}^{\infty} d\omega \int_{-\infty}^{\infty} dt \frac{k_y}{k^2}\frac{\mathrm{Im}\epsilon(\mathbf{k},\omega)_2}{|\epsilon(\mathbf{k},\omega)|^2}\exp\{i[\mathbf{k}\cdot\Delta\mathbf{r}(t) - \omega t]\} \quad (8.82)$$

where $\mathrm{Im}\epsilon(\mathbf{k},\omega)_2$ denotes the imaginary part of that portion of $\epsilon(\mathbf{k},\omega)$ which arises from the polarizability of the particles "2". For a Maxwellian plasma,

$$\mathrm{Im}\epsilon(\mathbf{k},\omega)_2 = \frac{4\pi^2(Z_2 e)^2\omega}{T_2 k^2} S_2^{(0)}(\mathbf{k},\omega) \quad (8.83)$$

where $S_2^{(0)}(\mathbf{k},\omega)$ is given by Eq. (8.80).

In a homogeneous system, Eqs. (8.81) and (8.82) vanish identically because of symmetry. The leading terms should therefore be proportional to the density gradient $\partial n_2/\partial X$ of the particles "2". Physically, such a density gradient acts in a number of ways to produce nonvanishing contributions in $\langle \Delta X_1 \rangle$: For instance, it gives rise to a net drift motion of the particles "2" in the y direction with the velocity (see Eq. 6.50)

p. 332, in the 4th line from Eq. (9.27) and in the 2nd line of the 4th paragraph, "$|\gamma_k/\omega_k|\Delta^0$" should read "$|\gamma_k/\omega_k| \sim \Delta^0$".
p. 345, in Eq. (9.71), "$\sim$" should be inserted after $\langle \delta n^2 \rangle$.
p. 345, in Eq. (9.70c), "$\sim$" should be inserted after ρ_E.
p. 349, the 2nd line above Sec. 9.3 should read:
mately zero) merged much faster than those of parallel helicity (where the field

Frontiers in Physics

DAVID PINES/Editor

Volumes of the Series published from 1961 to 1973 are not officially numbered. The parenthetical numbers shown are designed to aid librarians and bibliographers to check the completeness of their holdings.

Titles published in this series prior to 1987 appear under either the W. A. Benjamin or the Benjamin/Cummings imprint; titles published since 1986 appear under the Addison-Wesley imprint.

(1)	N. Bloembergen	Nuclear Magnetic Relaxation: A Reprint Volume, 1961
(2)	G. F. Chew	S-Matrix Theory of Strong Interactions: A Lecture Note and Reprint Volume, 1961
(3)	R. P. Feynman	Quantum Electrodynamics: A Lecture Note and Reprint Volume, 1961
(4)	R. P. Feynman	The Theory of Fundamental Processes: A Lecture Note Volume, 1961
(5)	L. Van Hove, N. M. Hugenholtz and L. P. Howland	Problem in Quantum Theory of Many-Particle Systems: A Lecture Note and Reprint Volume, 1961
(6)	D. Pines	The Many-Body Problem: A Lecture Note and Reprint Volume, 1961
(7)	H. Frauenfelder	The Mössbauer Effect: A Review—With a Collection of Reprints, 1962
(8)	L. P. Kadanoff G. Baym	Quantum Statistical Mechanics: Green's Function Methods in Equilibrium and Nonequilibrium Problems, 1962
(9)	G. E. Pake	Paramagnetic Resonance: An Introductory Monograph, 1962 [cr. (42)—2nd edition]
(10)	P. W. Anderson	Concepts in Solids: Lectures on the Theory of Solids, 1963
(11)	S. C. Frautschi	Regge Poles and S-Matrix Theory, 1963
(12)	R. Hofstadter	Electron Scattering and Nuclear and Nucleon Structure: A Collection of Reprints with an Introduction, 1963
(13)	A. M. Lane	Nuclear Theory: Pairing Force Correlations to Collective Motion, 1964
(14)	R. Omnès M. Froissart	Mandelstam Theory and Regge Poles: An Introduction for Experimentalists, 1963
(15)	E. J. Squires	Complex Angular Momenta and Particle Physics: A Lecture Note and Reprint Volume, 1963
(16)	H. L. Frisch J. L. Lebowitz	The Equilibrium Theory of Classical Fluids: A Lecture Note and Reprint Volume, 1964
(17)	M. Gell-Mann Y. Ne'eman	The Eightfold Way (A Review—With a Collection of Reprints), 1964
(18)	M. Jacob G. F. Chew	Strong-Interaction Physics: A Lecture Note Volume, 1964
(19)	P. Nozières	Theory of Interacting Fermi Systems, 1964
(20)	J. R. Schrieffer	Theory of Superconductivity, 1964 (revised 3rd printing, 1983)
(21)	N. Bloembergen	Nonlinear Optics: A Lecture Note and Reprint Volume, 1965
(22)	R. Brout	Phase Transitions, 1965
(23)	I. M. Khalatnikov	An Introduction to the Theory of Superfluidity, 1965
(24)	P. G. deGennes	Superconductivity of Metals and Alloys, 1966
(25)	W. A. Harrison	Pseudopotentials in the Theory of Metals, 1966
(26)	V. Barger D. Cline	Phenomenological Theories of High Energy Scattering: An Experimental Evaluation, 1967
(27)	P. Choquàrd	The Anharmonic Crystal, 1967

(28)	T. Loucks	Augmented Plane Wave Method: A Guide to Performing Electronic Structure Calculations—A Lecture Note and Reprint Volume, 1967
(29)	Y. Ne'eman	Algebraic Theory of Particle Physics: Hadron Dynamics in Terms of Unitary Spin Current, 1967
(30)	S. L. Adler R. F. Dashen	Current Algebras and Applications to Particle Physics, 1968
(31)	A. B. Migdal	Nuclear Theory: The Quasiparticle Method, 1968
(32)	J. J. J. Kokkedee	The Quark Model, 1969
(33)	A. B. Migdal V. Krainov	Approximation Methods in Quantum Mechanics, 1969
(34)	R. Z. Sagdeev and A. A. Galeev	Nonlinear Plasma Theory, 1969
(35)	J. Schwinger	Quantum Kinematics and Dynamics, 1970
(36)	R. P. Feynman	Statistical Mechanics: A Set of Lectures, 1972
(37)	R. P. Feynman	Photon-Hadron Interactions, 1972
(38)	E. R. Caianiello	Combinatorics and Renormalization in Quantum Field Theory, 1973
(39)	G. B. Field, H. Arp, and J. N. Bahcall	The Redshift Controversy, 1973
(40)	D. Horn F. Zachariasen	Hadron Physics at Very High Energies, 1973
(41)	S. Ichimaru	Basic Principles of Plasma Physics: A Statistical Approach, 1973 (2nd printing, with revisions, 1980)
(42)	G. E. Pake T. L. Estle	The Physical Principles of Electron Paramagnetic Resonance, 2nd Edition, completely revised, enlarged, and reset, 1973 [cf. (9)—1st edition]

Volumes published from 1974 onward are being numbered as an integral part of the bibliography.

43	R. C. Davidson	Theory of Nonneutral Plasmas, 1974
44	S. Doniach E. H. Sondheimer	Green's Functions for Solid State Physicists, 1974
45	P. H. Frampton	Dual Resonance Models, 1974
46	S. K. Ma	Modern Theory of Critical Phenomena, 1976
47	D. Forster	Hydrodynamic Fluctuations, Broken Symmetry, and Correlation Functions, 1975
48	A. B. Migdal	Qualitative Methods in Quantum Theory, 1977
49	S. W. Lovesey	Condensed Matter Physics: Dynamic Correlations, 1980
50	L. D. Faddeev A. A. Slavnov	Gauge Fields: Introduction to Quantum Theory, 1980
51	P. Ramond	Field Theory: A Modern Primer, 1981 [cf. 74—2nd ed.]
52	R. A. Broglia A. Winther	Heavy Ion Reactions: Lecture Notes Vol. I, Elastic and Inelastic Reactions, 1981
53	R. A. Broglia A. Winther	Heavy Ion Reactions: Lecture Notes Vol. II, 1990
54	H. Georgi	Lie Algebras in Particle Physics: From Isospin to Unified Theories, 1982
55	P. W. Anderson	Basic Notions of Condensed Matter Physics, 1983
56	C. Quigg	Gauge Theories of the Strong, Weak, and Electromagnetic Interactions, 1983
57	S. I. Pekar	Crystal Optics and Additional Light Waves, 1983

58	S. J. Gates M. T. Grisaru M. Roček W. Siegel	Superspace *or* One Thousand and One Lessons in Supersymmetry, 1983
59	R. N. Cahn	Semi-Simple Lie Algebras and Their Representations, 1984
60	G. G. Ross	Grand Unified Theories, 1984
61	S. W. Lovesey	Condensed Matter Physics: Dynamic Correlations, 2nd Edition, 1986
62	P. H. Frampton	Gauge Field Theories, 1986
63	J. I. Katz	High Energy Astrophysics, 1987
64	T. J. Ferbel	Experimental Techniques in High Energy Physics, 1987
65	T. Applequist A. Chodos P. G. O. Freund	Modern Kaluza-Klein Theories, 1987
66	G. Parisi	Statistical Field Theory, 1988
67	R. C. Richardson E. N. Smith	Techniques in Low-Temperature Condensed Matter Physics, 1988
68	J. W. Negele H. Orland	Quantum Many-Particle Systems, 1987
69	E. W. Kolb M. S. Turner	The Early Universe, 1990
70	E. W. Kolb M. S. Turner	The Early Universe: Reprints, 1988
71	V. Barger R. J. N. Phillips	Collider Physics, 1987
72	T. Tajima	Computational Plasma Physics, 1989
73	W. Kruer	The Physics of Laser Plasma Interactions, 1988
74	P. Ramond	Field Theory: A Modern Primer 2nd edition, 1989 [cf. 51—1st edition]
75	B. F. Hatfield	Quantum Field Theory of Point Particles and Strings, 1989
76	P. Sokolsky	Introduction to Ultrahigh Energy Cosmic Ray Physics, 1989
77	R. Field	Applications of Perturbative QCD, 1989
80	J. F. Gunion H. E. Haber G. Kane S. Dawson	The Higgs Hunter's Guide, 1990
81	R. C. Davidson	Physics of Nonneutral Plasmas, 1990
82	E. Fradkin	Field Theories of Condensed Matter Systems, 1991
83	L. D. Faddeev A. A. Slavnov	Gauge Fields, 1990
84	R. Broglia A. Winther	Heavy Ion Reactions, Parts I and II, 1990
85	N. Goldenfeld	Lectures on Phase Transitions and the Renormalization Group, 1992
86	R. D. Hazeltine J. D. Meiss	Plasma Confinement, 1992
87	S. Ichimaru	Statistical Plasma Physics, Volume I: Basic Principles, 1992
88	S. Ichimaru	Statistical Plasma Physics, Volume II: Condensed Plasmas, 1992

Editor's Foreword

The problems of communicating in a coherent fashion recent developments in the most exciting and active fields of physics continues to be with us. The enormous growth in the number of physicists has tended to make the familiar channels of communication considerably less effective. It has become increasingly difficult for experts in a given field to keep up with the current literature; the novice can only be confused. What is needed is both a consistent account of a field and the presentation of a definite "point of view" concerning it. Formal monographs cannot meet such a need in a rapidly developing field, while the review article seems to have fallen into disfavor. Indeed it would seem that the people most actively engaged in developing a given field are the people least likely to write at length about it.

FRONTIERS IN PHYSICS was conceived in 1961 in an effort to improve the situation in several ways. Leading physicists frequently give a series of lectures, a graduate seminar, or a graduate course in their special fields of interest. Such lectures serve to summarize the present status of a rapidly developing field and may well constitute the only coherent account available at the time. One of the principal purposes of the FRONTIERS IN PHYSICS Series is to make notes on such lectures available to the physics community.

As FRONTIERS IN PHYSICS has evolved, a second category of book, the informal text/monograph, an intermediate step between lecture notes and formal texts or monographs, has played an increasingly important role in the series. In an informal text or monograph, an author has reworked his or her lecture notes to the point at which the manuscript represents a coherent summation of a newly-developed field complete with references and problems, suitable for either classroom teaching or individual study.

Setsuo Ichimaru's introductory text/monograph, *Basic Principles of Plasma Physics*, published in 1973, represented an early, and highly successful example of such a book. In it, Professor Ichimaru, who has made a number of important contributions to fundamental plasma theory and to its applications in astrophysics, condensed matter physics, atomic, and nuclear physics, presented a coherent and self-contained account of plasma theory, an account in which plasma theory is viewed not as an isolated topic, but as an integral part of statistical physics. During the succeeding eighteen years, countless students and faculty alike have been introduced to statistical plasma physics by Ichimaru's volume. Over this same period, the subfield of statistical plasma physics has continued to grow and flourish. A revision of *Basic Principles of Plasma Physics* thus seemed both timely and appropriate, and I approached Professor Ichimaru a few years ago to prepare such a volume. What has emerged is a thoroughly revised and expanded version, which we have decided to publish in two volumes, with the title *Statistical Plasma Physics*, chosen to reflect the breadth and depth of the author's treatment.

In this first volume, which is self-contained, the basic principles are introduced for equilibrium situations and then applied to nonequilibrium conditions in which transient processes, instabilities, and turbulence may play a role. The second volume will be devoted to topics in the fields of condensed plasma physics, with applications to subfields which range from condensed matter physics to astrophysics. It gives me great pleasure to welcome Professor Ichimaru once more to the ranks of authors for FRONTIERS IN PHYSICS.

David Pines

Urbana, Illinois
November, 1991

Preface

Plasma physics is concerned with the equilibrium and nonequilibrium properties of a statistical system containing many charged particles. The forces of interaction between the particles are electromagnetic, extending themselves over wide ranges. The system is characterized by an enormous number of microscopic degrees of freedom arising from the motion of individual particles. Statistics thus enable us to perform a theoretical analysis of the macroscopic behavior of such a system; theories of plasmas basically involve the many-body problems. The principal aim of this volume, *Statistical Plasma Physics: I. Basic Principles* is to present a coherent and self-contained account of the fundamental theories in plasma physics, not as an isolated subject, but as an integral part of the statistical physics.

The author taught these subjects to undergraduate and graduate students at both the University of Tokyo and the University of Illinois for several years. In 1973, *Basic Principles of Plasma Physics* (*BPPP*) was published as one of the volumes in the *Frontiers in Physics* series. The book represented an outgrowth of those lectures and stressed the fundamentals of statistical theory in plasma physics.

The present volume was conceived initially as a revision of *BPPP*. Thus, all the chapters, problems, and appendices in *BPPP* have been thoroughly revised and rewritten. In addition, a good deal of new material has been incorporated in the revised edition. All of these activities culminated in a book with new title and contents. The volume is intended as a textbook for an advanced undergraduate to first-year graduate course on plasma and statistical physics with 36 to 45 lecture hours.

In Chapter 1 *Introduction*, an overview of the plasma phenomenon in nature is presented, followed by discussion of various parameters characterizing plasmas, such as the Coulomb coupling parameters and the degeneracy parameters. Basic consequences arising from the Coulomb interaction, such as screening and oscillations, are then considered with emphasis on the interplay between the collective motion and the individual particlelike behavior.

In Chapter 2 *Kinetic Equations*, the structures and properties of the kinetic equations which govern the dynamics of plasmas are studied. In terms of Klimontovich's microscopic formalism and the Bogoliubov-Born-Green-Kirkwood-Yvon hierarchy equations for the multiparticle distribution functions, the Vlasov equation for the single-particle distributions is derived and elucidated. The collision term in the kinetic equation beyond the Vlasov formalism is treated.

In Chapter 3 *Plasmas as Dielectric Media*, a theoretical foundation for the linear-response formalism and the electrodynamic description of plasmas as continuous media are laid down. The linear response functions are introduced and the fluctuation-dissipation theorem outlined; combination between these establishes a rigorous description for the fluctuations and correlations in plasmas under thermodynamic equilibria. The dielectric tensor, defined subsequently, sets the macroscopic electrodynamic equations with the dispersion relations, which characterize the nature of the longitudinal and transverse excitations. The dielectric tensors for Vlasov plasmas are explicitly calculated, both in the absence and in the presence of an external magnetic field. Finally the dielectric formulation for the electron gas at finite temperatures is set forth within the framework of the random-phase approximation (RPA); strong coupling effects beyond the RPA are also described.

In Chapter 4 *Electromagnetic Properties of Vlasov Plasmas in Thermodynamic Equilibria*, the dielectric formulation of the Vlasov plasmas in equilibrium states is presented. In the first half of this chapter, the longitudinal dielectric properties are studied. Propagation and the Landau damping of various collective modes are investigated in plasmas with and without an external magnetic field. In the second half, cases of the electromagnetic-wave propagation in plasmas are treated. Topics include the cutoff at the plasma frequency, resonances at the cyclotron frequencies, and low-frequency modes of the electromagnetic-wave propagation such as helicon, the Alfvén wave, and whistler.

In Chapter 5 *Transient Processes*, the temporal and spatial propagation of small- and large-amplitude plasma waves, the plasma-wave echo, and nonlinear processes associated with propagations of large-amplitude waves are treated. The latter include a formation of solitary waves, self-trapping, and self-focusing; relevant nonlinear mechanisms are the hydrodynamic flows and the ponderomotive forces.

In Chapter 6 *Instabilities*, we investigate the conditions under which onsets of various plasma instabilities may be expected. The physical mechanisms responsible for the onset of a plasma instability may be sought in the excess of free energy due to a departure from a thermodynamic equilibrium. Hence it is meaningful to classify instabilities into two groups: instabilities in homogeneous plasmas and those in inhomogeneous plasmas. In the former cases, the kinetic energies associated with a stream or an anisotropic velocity distribution of the particles form major sources of the extra free energy. Instabilities in this category are treated in the first half of this chapter.

In the second half, those instabilities characteristic of inhomogeneous plasmas are considered. Additional sources of free energy associated with the spatial variations of physical quantities are available in these circumstances. A nonuniform system therefore has a tendency to release this extra amount of free energy and thereby to approach a uniform equilibrium state through the onset of an instability; an anomalous relaxation would take place in such a plasma. The Rayleigh-Taylor instabilities and the drift-wave instabilities are the two major instabilities treated in this category.

In Chapter 7 *Fluctuations*, microscopic formulations on the static and dynamic correlations are presented for those plasmas with weak and strong coupling, followed by elucidation of the elementary processes stemming from fluctuations. These include collisional processes, density-fluctuation excitations, transport processes, and electromagnetic radiation.

In Chapter 8 *Relaxations*, the relaxation processes for the motion of a single particle are approached by a fluctuation-theoretic method in a constant external magnetic field. The Fokker-Planck coefficients, temperature relaxations, and diffusion across the magnetic field are thereby treated.

In Chapter 9 *Plasma Turbulence*, problems arising from anomalous excitations of field fluctuations are considered. Wave-kinetic equations are developed and applied to the cases of a current-carrying unstable plasma. Anomalous resistivities and reconnection rates of the magnetic fields are discussed. Finally the turbulent magnetic viscosities in the accretion-disk plasmas are considered with an application to astrophysics.

The presentation in this book is self-contained and should be read without difficulty by those who have adequate preparation in classical mechanics, electricity and magnetism, elementary quantum mechanics and statistics. The level of treatment is elementary to intermediate. References are cited mainly for the purpose of assisting the reader in finding relevant discussions extending and supplementing the material presented here.

Volume II in this series is concerned with treatment of subjects in the fields of condensed plasma physics. The volume is intended as a graduate-level textbook with 30 to 42 lecture hours. It aims at elucidating a number of basic topics in plasma physics interfacing with condensed matter physics, atomic physics, nuclear physics, and astrophysics; as such it will be useful also for research physicists and engineers in these fields.

For the completion of the present volume of *Statistical Plasma Physics*, the author has been indebted enormously to a large number of his fellow scientists, colleagues, friends, and associates. In the Preface to *BPPP*, the author recorded the acknowledgment "I should like to express my sincere gratitude to Professor David Pines, who introduced me to this field and has given me much helpful advice and kind encouragement over the years as well as during the writing of this book. I wish to thank Professor John Bardeen and Professor Marshall N. Rosenbluth, whose guidance and help have had a deep influence on my scientific and human evolution. Finally, I wish to thank Professor Ralph O. Simmons, head of the physics department at the University of Illinois, for his hospitatity during the course of this work." During the writing of the present series, the author was deeply saddened at the death of Professor John Bardeen, and wishes here to express his sincere condolences.

The encouragement and pertinent suggestions extended by D. Pines have continued to be a constant motive for the realization of this volume. For the completion of this volume, the author has been encouraged and assisted by H. Van Horn through his considerate advice. During the course of the work, the author has received valuable advice from and has benefited from illuminating discussions with many fellow scientists, namely, R. Abe, N. Ashcroft, M. Baus, D. Baldwin, R. Balescu, C. Deutsch, H. DeWitt, D. DuBois, D. Ceperley, Ph. Choquard, R. Davidson, V.E. Fortov, Y. Furutani, M. Goldman, J.-P. Hansen, W. Horton, W. Hubbard, K. Husimi, Y.-H. Ichikawa, A. Isihara, B. Jancovici, R. Kalia, G. Kalman, A. Kaufman, T. Kihara, W. Kohn, W. Kraeft, R. Kubo, J. Malmberg, T. O'Neil, K. Nishijima, K. Nishikawa, C. Pethick, D. Pines, M. Raether, M. Rosenbluth, N. Rostoker, K. Singwi (dec.), A. Sitenko, A. Sjölander, S. Takeno, M. Tanaka, T. Taniuti, D. ter Haar, W. Thompson, S.-t. Tsai, P. Vashishta, H. Van Horn, Y. Wada, M. Watabe, H. Wilhelmsson, I. Yakimenko, and F. Yonezawa. Past and present associates with the author's laboratory at the University of Tokyo, namely, N. Itoh, N. Iwamoto, H. Iyetomi, H. Miesenböck-Böhm, S. Nagano, A. Nakano, T. Nakano, S. Ogata, T. Tajima, K. Tago, S. Tanaka, T. Tange (dec.), H. Totsuji, K. Utsumi, T. Watari, M. Yamada, I. Yamashita, and X.-Z. Yan, assisted him through useful comments. S. Ogata has set the final manuscript in the TeX format; K. Tsuruta has provided assistance in the preparation of the drawings. Student participants in the Spring, 1991 reading class on plasma physics, namely, A. Fushimi, H. Kaneda, H. Kitamura, J. Yamamoto, and S. Yunoki, contributed to the improvement of the presentation in the original manuscript. Finally, the author's wife, Tomoko Ichimaru, rendered continual help to him in every way throughout the project. All of these are gratefully acknowledged.

Setsuo Ichimaru

Tokyo, Japan
September 1991

Contents

Statistical Plasma Physics

Volume I: Basic Principles

CHAPTER 1

Introduction

A plasma may be defined as any statistical system containing mobile charged particles. Vague as it may sound, the foregoing statement is sufficient to define what is known as plasma in physics and engineering.

We can trace the history of plasma research back to the time of Michael Faraday's study of the "dark discharges". From the middle of the last century until the earlier part of the present century, many prominent physicists engaged in research on electric discharges in gases; among them were J.J. Thomson, J.S. Townsend, and I. Langmuir. It was Tonks and Langmuir who coined the word "plasma" to mean that part of a gaseous discharge which contained almost equal densities of electrons and positive ions. They did so in connection with the oscillatory behavior observed in it, the so-called plasma oscillation.

Many significant contributions to the basic understanding of plasma phenomena have been advanced by astronomers and geophysicists. Indeed, problems in astronomy and geophysics, such as the propagation of electromagnetic radiation through the upper atmosphere, the dynamics of Earth's magnetosphere, the turbulent behavior of the ionized matter and magnetic fields near the surfaces of the Sun and stars, the dispersion and broadening of signals traveling through interstellar space, the emission mechanisms of pulsars and radio sources, the origin of cosmic rays, the physical processes in the X-ray and γ-ray sources, and the evolution and internal structures of stellar objects, are all closely related to the fundamental aspects of plasma physics.

Work on plasmas was confined to comparatively few individuals and laboratories until the early 1950s, when intensive work was begun in the United States, the United Kingdom, the Soviet Union, and other countries on the realization of controlled release of nuclear fusion energy. Although it has become clear that many years will pass before this objective is successfully achieved, the study of plasmas, which was once regarded as one of the old-fashioned fields in physics, has under this impetus become one of the central topics in science and engineering. In addition, interest in plasmas has been spurred in recent years by other possible technological applications, such as the direct conversion of heat energy into electricity by a magnetohydrodynamic means, the propulsion of space vehicles, and the development of new electronic devices.

Let us note that, according to the definition presented at the beginning of this section, plasmas may be found not only in gases but also in solids; electronic phenomena in semiconductors, semimetals, and metals can all be viewed as examples of plasma effects in solids. In the cases of the electrons in metals and semimetals, the density is so high that the degeneracy brought about by the Pauli principle must be taken into account. The degenerate conduction electrons found in such metals are accordingly called a *quantum plasma*.

1.1 PLASMAS IN NATURE

Atoms and molecules, building blocks of ordinary substances, consist of nuclei and electrons. An electron has a negative unit charge ($e = 4.803 \times 10^{-10}$cgs esu). Ions produced by removing some or all of the orbital electrons from atoms or molecules carry positive charges. A system containing many of such mobile charged particles constitutes a plasma.

When "plasma" is mentioned, many readers may associate this term with "controlled thermonuclear fusion research". This is a scheme of devising new energy sources by exploiting nuclear energy released in fusion processes between light nuclei of hydrogen isotopes such as deuterium and tritium. Since atomic nuclei are positively charged, they repel each other by the Coulomb forces. To induce nuclear fusion reactions effectively by overcoming such repulsive forces, those nuclei have to collide vigorously with each other. The minimum conditions for net production of energy by a magnetic-confinement scheme are estimated to be dense, high-temperature plasmas of 10^{14} to 10^{15} cm^{-3} and $\sim 10^8$ K held for more than a second. Stable confinement and heating of a plasma are essential problems in realization of the controlled release of such nuclear fusion energy.

Plasmas are produced in the laboratory by electric discharge. When a strong electric field is applied to an ordinary gas, charged particles in it are accelerated. The charged particles accelerated to high energies collide with atoms or molecules, and thereby ionize those neutral particles by removing orbital electrons. Those ions and electrons produced by collisional ionization are then accelerated by the electric field and ionize other neutral particles by collisions; ionization processes thus proceed like an avalanche. This is a phenomenon of the *electric discharge*. The ionized gas produced by electric discharge is a typical example of plasmas; it is found commonly in our daily life such as in the fluorescent lamp, neon tube, and arc welding.

The electron densities and temperatures of the plasmas produced by electric discharge vary widely depending on the condition of the discharge. For example, in the glow discharge at a gas pressure of ~1 Torr, one usually finds an electron density of 10^9 to 10^{11} cm^{-3} and an electron temperature of $\sim 10^4$ K.

Without using an electric discharge, a plasma state may be attained by simply raising the temperature of a neutral gas. In these circumstances, thermal energy of the particles is utilized for impact ionization and hence *thermal ionization*. Since the binding energy of an electron in a neutral molecule takes on the order of a few electron volts,* the temperature must correspondingly exceed 10^4 to 10^5 K.

As remarked earlier, *metals* or *semiconductors* contain relatively mobile, free charged particles. For instance, one finds conduction electrons in metal forming a quantum-mechanically degenerate plasma at a density of approximately $10^{23}cm^{-3}$. A semiconductor contains mobile electrons and holes, though their densities are substantially smaller than those of metals.

Above the surface of the Earth, layers of ionized gas, called the *ionosphere*, exist at altitudes of 70 to 500 km over the stratosphere. The electron density is largest in the upper part, at altitudes to 200 to 500 km (F-layer). Its value is approximately 10^6 cm^{-3}, and the electron temperature is in the vicinity of 2000 K, although those values vary depending on day or night, seasons, and solar activities.

As the influence of solar activities on the ionosphere may suggest, the Sun itself is an important source of plasmas. A flow of plasma, the solar wind, is ejected from the Sun into interplanetary space. Wind velocity near the Earth is about 250 to 700 km/s; the number density of plasma particles is 1 to 10 cm^{-3}, and the electron temperature is $\sim 10^5$ K. The solar wind collides with the magnetic field of the Earth at about 10 to $13R_E$ ($R_E = 6400$ km: the radius of the

---

* Throughout this volume, we express temperature in energy units unless otherwise specified. Conversion may be accomplished via

$$1 \text{ eV} = 1.6021 \times 10^{-12} \text{ erg} = 1.1605 \times 10^4 \text{ K}.$$

Earth), and thereby produces the boundary surface of the domain called the *magnetosphere*, within which Earth's magnetic fields are more or less confined. Charged particles emitted by the Sun cannot cross Earth's magnetic lines of force; they form a bow shock. Behind the Earth, the various strands of solar wind join up again, and this region is called the magnetotail. Reconnection of magnetic fields is related to the dynamical processes in the magnetosphere.

As for the Sun, it has a region containing solar atmosphere called the *chromosphere* (~2000 km in depth). It is a plasma of an average electron density 10^{10} to 10^{11} cm^{-3} and temperature ~6000 K. Above the chromosphere there extends a large region of corona, whose plasma parameters such as the electron density depend considerably on the distance from the solar surface. At 10^5 km, the density and temperature are estimated approximately as 10^8 cm^{-3} and 10^6 K; at 10^6 km, they are $\sim 10^6$ cm^{-3} and $\sim 10^6$ K, respectively.

Interiors of *main sequence stars* such as the Sun are plasmas constituting mostly of hydrogen. The central part of the Sun has a pressure of $\sim 10^5$ Mbar, a mass density of $\sim 10^2$ g/cm^3, and a temperature of $\sim 10^7$ K. Dense-plasma effects [e.g., Ichimaru 1982, 1990] such as Coulomb coupling between ions and quantum-mechanical polarization effects of electrons begin to play crucial parts in the determination of atomic states for those "impurities" starting with helium. Opacities and other transport coefficients depend sensitively on the electronic and atomic states.

The material inside a *Jovian planet* (Jupiter, Saturn, Uranus, ...) offers an important subject of study in dense-plasma physics. Here one considers a hydrogen plasma with a few percent admixture of helium at the mass density 1 to 10 g/cm^3 and $T \approx 10^4$ K; the pressure ranges as high as 50 Mbar. Since these temperatures are much lower than those of the solar interior, the dense-plasma effects are much more pronounced here than in the Sun (cf. Sec. 1.2). The Pioneer and Voyager satellites transmitted to us valuable data on Jovian planets, such as the excess infrared-radiation spectra and the harmonics of the gravitational-field distributions. To account for these observational data, thermodynamic and transport properties of dense Jovian matter need to be clarified [Hubbard, W.B. 1980].

The interior of a *white dwarf* [e.g., Shapiro & Teukolsky 1983], one of the final stages of stellar evolution, consists of dense material with mass density 10^5 to 10^9g/cm^3, corresponding to the electron density 10^{25} to 10^{32} cm^{-3}, and $T = 10^7$ to 10^8 K. Since the density is so high, it becomes an essential problem to analyze the possibility of a freezing transition and the phase properties of the system. For the progenitor of a Type I supernova, one often assumes a white dwarf with an interior consisting of a carbon-oxygen mixture [Starrfield et al. 1972]. Physical problems in such a binary-ionic mixture include assessment of the possibility of a chemical separation at the freezing transition and the resulting phase diagrams [Iyetomi, Ogata & Ichimaru 1989]; those are related to the cooling rate and a detailed mechanism of the supernova explosion [e.g., Van Horn 1990].

The *neutron star* [e.g., Shapiro & Teukolsky 1983], another final stage of stellar evolution, is a highly condensed material corresponding approximately to a compression of a solar mass ($\approx 2 \times 10^{33}$ g) into a sphere with radius ~10 km. According to a theoretical model study, it has a crust with a thickness of several hundred meters and a mass density in the range 10^4 to 10^7 g/cm^3, consisting mostly of iron. Here it is again an essential problem to analyze the phase properties of the system with inclusion of the possibilities of freezing transitions. The state of matter and the transport properties in the crustal matter are considered to form those physical elements which crucially control the cooling rate of a neutron star [Gudmundsson, Pethick & Epstein 1982].

Finally we mention the interstellar space containing those stars; although dilute, it also is a good example of a plasma. Typical electron densities range from 10^{-2} to 10^1 cm^{-3} and the temperature ~10^4 K. Furthermore we find various phenomena related to plasmas in those conspicuous astrophysical phenomena, such as the supernova remnants, pulsating stars (pulsars), compact X-ray sources (X-ray stars), and the active galactic nuclei (e.g., quasars).

As a summary of this section, we show a schematic presentation of the parameter domains for those various plasmas in Fig. 1.1.

1.2 COULOMB COUPLING PARAMETERS

As we have overviewed in the previous section, the plasmas found in nature or in the laboratory are characterized by temperature and density parameters over a wide range of variation. It is therefore convenient to introduce dimensionless coupling constants, called *Coulomb coupling parameters*, which characterize strengths of the particle interactions in those various charged-particle systems. We shall also consider parameters describing the extents to which the quantum-mechanical effects are involved in a plasma. These considerations will assist in locating the ranges of validity for the classical versus quantum-theoretical treatments of plasmas.

In many instances, we introduce and consider a spatially homogeneous *one-component plasma* (OCP) for simplicity. It is a system consisting of a single species of charged particles embedded in a uniform background of neutralizing charges. Such an OCP is a substantially idealized model of a real plasma; some plasmas in nature, as we shall see, do indeed satisfy the conditions for such idealization. Wherever appropriate, we shall extend our consid-

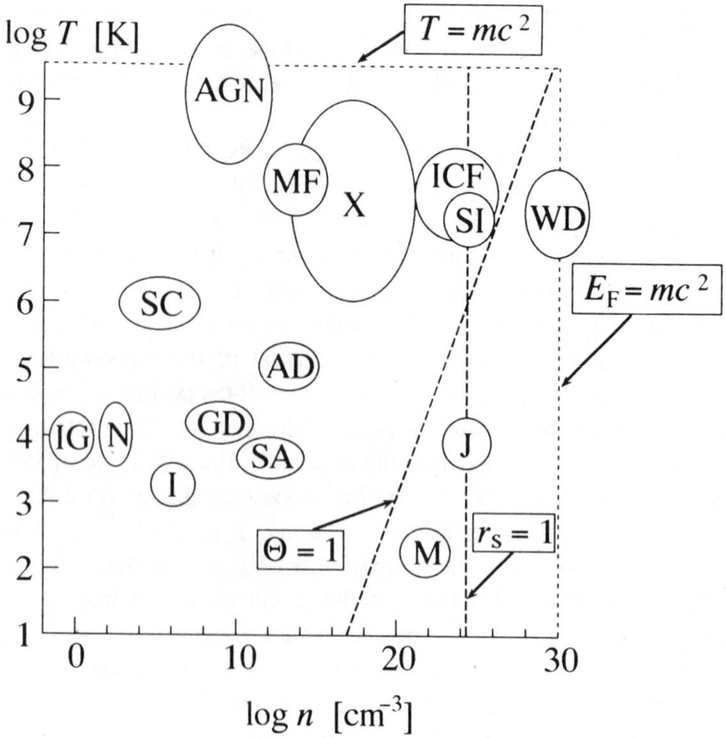

Figure 1.1 Typical values of the electron density n and the temperature T for various plasmas: interstellar gas (IG), gaseous nebula (N), ionosphere (I), glow discharge (GD), solar atomosphere (SA), arc discharge (AD), solar corona (SC), active galactic nucleus (AGN), magnetic fusion (MF), X-ray star (X), inertial confinement fusion (ICF), solar interior (SI), metal (M), Jovian interior (J), and white dwarf (WD).

erations also to plasmas consisting of multiple species of charged particles (multicomponent plasmas).

Let us define a coupling constant of a plasma as a ratio of an average Coulomb-interaction energy to an average kinetic energy. We are thus concerned with the strength of coupling due to Coulomb interaction. Those plasmas with values of the coupling constant much smaller than unity may be called *weakly coupled plasmas*; those with the coupling constant around or greater than unity, *strongly coupled plasmas*.

We first estimate the average kinetic energy. When the charged-particle system under consideration obeys the classical statistics, the velocity distribution of the individual particles is given by the Maxwellian:

$$f_M(\mathbf{v}) = \left(\frac{m}{2\pi T}\right)^{3/2} \exp\left(-\frac{mv^2}{2T}\right) \tag{1.1}$$

where m denotes the mass of a particle. The average value of the kinetic energy per particle is then calculated as $(3/2)T$, which we take as T in an order-of-magnitude estimate.

For a high-density electron system such as conduction electrons in metal, one uses the Fermi energy

$$E_F = \frac{\hbar^2 (3\pi^2 n)^{2/3}}{2m} \tag{1.2}$$

instead of the classical evaluation mentioned above. Here $\hbar = 1.0546 \times 10^{-27}$ erg · s is the Planck constant divided by 2π, $m = 9.1095 \times 10^{-28}$g is the electron mass, and n denotes the number density of the electrons. Since Eq. (1.2) is an increasing function of n, $E_F >> T$ may be realized in a high-density electron system. A degeneracy parameter of the electrons is thus defined by

$$\Theta = \frac{T}{E_F}. \tag{1.3}$$

When $\Theta << 1$, the Fermi degeneracy brought about by the quantum statistics becomes more important than the effects of thermal motion represented by T. In these circumstances, it is relevant to use the Fermi energy as an estimate of the kinetic energy.

For an OCP containing n particles with electric charge Ze in a unit volume, the Coulomb energy per particle may likewise be estimated as $(Ze)^2/a$, where

$$a = \left(\frac{4\pi n}{3}\right)^{-1/3} \tag{1.4}$$

is the radius of a sphere with the characteristic volume $1/n$. This radius is usually referred to as the *ion-sphere radius* or the *Wigner-Seitz radius*.

The Coulomb coupling parameter for the degenerate electron system is

$$\frac{e^2}{aE_F} = 0.543 r_s , \tag{1.5}$$

where

$$r_s = \left(\frac{3}{4\pi n}\right)^{1/3} \frac{me^2}{\hbar^2} \tag{1.6}$$

is the Wigner-Seitz radius for electrons measured in units of the Bohr radius, $a_B = \hbar^2/me^2 = 5.292 \times 10^{-9}$cm. As Eq. (1.5) indicates, r_s is equivalent to the Coulomb coupling parameter for a degenerate electron system; $r_s = 1$ corresponds to $n = 1.6 \times 10^{24}\text{cm}^{-3}$.

The Coulomb coupling parameter for a plasma obeying the classical statistics is

$$\Gamma = \frac{(Ze)^2}{aT} = 2.69 \times 10^{-5} Z^2 \left(\frac{n}{10^{12}\text{cm}^{-3}}\right)^{1/3} \left(\frac{T}{10^6\text{K}}\right)^{-1} . \tag{1.7}$$

This expression indicates that for $Z=1$ and $T=10^6$K, the density n must become as high as $\sim 10^{26}$ cm^{-3} to make $\Gamma \approx 1$.

Most of the classical plasmas that we encounter, however, are characterized by $\Gamma \ll 1$. For example, we may assume $n \approx 10^{11}\text{cm}^{-3}$, $T \approx 10^4$K for a gaseous-discharge plasma, $n \approx 10^{16}\text{cm}^{-3}$, $T \approx 10^8$K for a plasma in a controlled thermonuclear-fusion experiment, and $n \approx 10^6\text{cm}^{-3}$, $T \approx 10^6$K for a plasma in the solar corona. For those plasmas, respectively, we find $\Gamma = 10^{-3}$, 10^{-5}, and 10^{-7}. They are thus weakly coupled plasmas; their thermodynamic properties, for instance, are analogous to those of an ideal gas.

A typical example of a strongly coupled classical plasma may be seen in the system of ions inside a highly evolved star. The interior of such a star is in a compressed, high-density state. The Fermi energy of the electron system takes on a value much greater than the binding energy of an electron around an atomic nucleus; all the atoms are thus in ionized states (*pressure ionization*). The electron system constitutes a weakly coupled ($r_s \ll 1$), degenerate plasma with an immensely large Fermi energy ($E_F \approx mc^2$). It makes an ideal neutralizing background of negative charges to the ion system; since $r_s \ll 1$, the polarizability of the background may be negligible. Those atomic nuclei stripped of the electrons form an ion plasma obeying the classical statistics; their de Broglie wavelengths are much smaller on the average than the interparticle spacing, that is,

$$Y \equiv \frac{\hbar}{a\sqrt{MT}} \ll 1 , \tag{1.8}$$

where M is the ionic mass. In the interior of a highly evolved star, the coupling parameter Γ for such an ion plasma is usually greater than unity; in a white dwarf one estimates that $\Gamma = 10$ to 200 [e.g., Schatzman 1958; Van Horn 1971].

An example of a strongly coupled plasma closer to the Earth may be offered by the Jovian interior, which consists mostly of hydrogen and a small fraction of helium. Its electron system is characterized by $r_s = 0.6$ to 1; its ion system, $\Gamma = 20$ to 50. The Jovian interior thus appears to make a complex system of strongly coupled classical plasma immersed in a polarizable background of degenerate electrons. In addition, the ion plasma may have to be looked upon as a binary-ionic mixture of hydrogen and helium.

As an example of a strongly coupled plasma in the laboratory, we can think of plasmas produced by shock compression [Mostovych, Kearney & Stamper 1990; Fortov et al. 1990]. Other examples of strongly-coupled laboratory plasmas include liquid and/or ultrahigh pressure metals, superionic conductors, and cryogenic non-neutral plasmas contained in electromagnetic traps, such as the laser-cooled pure ion plasmas [Bollinger et al. 1990] and the strongly magnetized pure electron plasmas [O'Neil et al. 1990]. Strongly coupled plasmas may also be found in low dimensions, such as the two-dimensional systems of electrons and holes in the interface states of semiconductor-insulators [Ando, Fowler & Stern 1982], semiconductor heterostructures [Ando 1990], and the electrons in the surface states of dielectrics [Cole 1974; Grimes 1978].

1.3 ELECTROSTATIC CONSEQUENCES OF THE COULOMB INTERACTION

A plasma is a collection of charged particles. The Coulomb force with which the charged particles interact is known to be long-ranged. Consequently the physical properties of a plasma exhibit remarkable differences from those of an ordinary neutral gas.

Most of the salient features in plasmas can be understood by investigating the behavior of the OCP, introduced in the preceding section. In this section, we shall thus adopt such an OCP model to study the basic consequences of the Coulomb interaction in the static properties of the plasma; the study will be extended also to cases with various charged species.

We begin this section with a consideration of the Coulomb cross sections for transfer of momentum and energy. This example is intended to illustrate how a naive substitution of Coulomb potential in the calculation of plasma properties would lead to a false prediction. It will at the same time point to the necessity of taking into account the organized or collective behavior of many charged particles brought about by the long-ranged Coulomb forces; the plasma thus exhibits a mediumlike character. The ion-sphere model is introduced as a useful concept replacing the Debye screening in a strongly coupled plasma. Frequencies of Coulomb collisions resulting from those considerations will be approximately estimated.

Dynamic processes in density fluctuations, including plasma oscillation, will be considered in the next section.

A. Cross Sections for the Coulomb Scattering

Suppose that charged particles with masses m_1 and m_2 and electric charges Z_1e and Z_2e scatter each other by the Coulomb force with the relative velocity, $v = |\mathbf{v}_1 - \mathbf{v}_2|$, and impact parameter b. In the frame of reference comoving at the center-of-mass velocity,

$$\mathbf{V} = \frac{m_1\mathbf{v}_1 + m_2\mathbf{v}_2}{m_1 + m_2}, \tag{1.9}$$

the angle χ of scattering (cf., Fig. 1.2) is related to the impact parameter via

$$\cot\left(\frac{\chi}{2}\right) = \frac{b\mu v^2}{Z_1Z_2e^2} \tag{1.10}$$

where $\mu = m_1m_2/(m_1 + m_2)$ is the reduced mass. The differential cross section dQ for scattering into an infinitesimal solid angle do around a scattering angle is given by the Rutherford formula

$$\frac{dQ}{do} = \left[\frac{Z_1Z_2e^2}{2\mu v^2\sin^2(\chi/2)}\right]^2. \tag{1.11}$$

The cross section Q_m for momentum transfer can then be calculated by integrating (1.11) over solid angles with a weighting function $(1 - \cos\chi)$, which represents the fractional change of momentum on scattering. If for the moment a finite angle $\chi_{\min}$ is chosen for the lower limit of the χ integration, we calculate

$$\begin{aligned} Q_m &= \int_{\chi_{\min}}^{\pi} d\chi\, 2\pi \sin\chi (1 - \cos\chi)\left[\frac{Z_1Z_2e^2}{2\mu v^2 \sin^2(\chi/2)}\right]^2 \\ &= 4\pi\left(\frac{Z_1Z_2e^2}{\mu v^2}\right)^2 \ln\left[\frac{1}{\sin(\chi_{\min}/2)}\right]. \end{aligned} \tag{1.12}$$

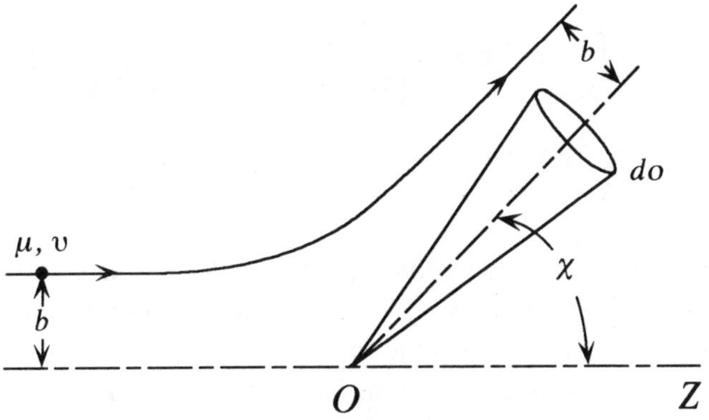

Figure 1.2 Coulomb scattering.

When $\chi_{\min} << 1$, Q_m becomes proportional to $\ln(2/\chi_{\min})$. The logarithmic factor, whose appearance is typical of the Coulomb interaction, is called a *Coulomb logarithm*. The cross section is seen to diverge logarithmically as $\chi_{\min}$ approaches zero.

This anomaly points to an inadequacy of the foregoing treatment when it is applied directly to the charged particles in the plasma. The cross section for the momentum transfer is proportional to the electric resistivity of the plasma; experiments tell us that ordinary plasmas are characterized by finite values of resistivity, not infinite ones. The origin of the logarithmic divergence mentioned above can be traced to those scattering acts which take place with large impact parameters and thus with small scattering angles. Charged particles in the plasma located at long distances from a scattering center will undoubtedly be influenced by many other scattering centers with similar strengths of interaction. Hence a simple picture of binary Coulomb scattering cannot correctly describe the behavior of charged particles interacting at large distances in the plasma.

The cross section Q_E for energy transfer can be treated analogously. The amount of energy exchanged at a collision with the transfer of momentum $\Delta\mathbf{p}$ is given by (Problem 1.5)

$$\Delta w = \mathbf{V} \cdot \Delta\mathbf{p} \tag{1.13}$$

so that

$$Q_E = \int_{\chi_{\min}}^{\pi} d\chi \, 2\pi \sin\chi \left[\frac{\Delta w}{(\mu/2)v^2}\right]\left[\frac{Z_1 Z_2 e^2}{2\mu v^2 \sin^2(\chi/2)}\right]^2$$

$$\approx \left\langle \frac{2V}{v} \right\rangle Q_m \,. \tag{1.14}$$

Through Q_m, a Coulomb logarithm appears in the cross section for energy transfer as well. The forefactor, $\langle 2V/v \rangle$, introduces an additional dependence of the cross section on the mass ratio, m_1/m_2.

The appearance of such a Coulomb anomaly is not limited to the cases of transport coefficients alone. Another notorious example is found in a calculation of the ground state energy of a degenerate electron gas [e.g., Pines 1963]. Regarding Coulomb interactions as a perturbation, we find that a divergence of similar character starts to appear from the second order of the perturbation calculations. The origin of this divergence is again traced to those scattering acts involving small momentum transfers.

B. Debye Screening

Having thus recognized a failure of a simple picture of binary collisions in a plasma, we now take up the problem of determining an effective interaction between charged particles in a plasma. We shall thus calculate the potential field around a Coulomb scattering center in an OCP, by taking explicit account of the statistical distribution of other charged particles. This calculation will lead to the notion of *Debye screening*, an illuminating example of the cooperative phenomena. The force range of a charged particle is now confined within a certain characteristic length determined by the density and temperature of the plasma. A number of immediate corollaries from such a calculation will then be discussed: The logarithmic divergence of the Coulomb cross section can be cured by cutting off the interaction at that characteristic length; a plasma tends to stay neutral when viewed on a macroscopic scale.

Consider a point charge $Z_0 e$ located at the origin ($\mathbf{r} = \mathbf{0}$): In vacuum it produces a potential field

$$\Phi_0(\mathbf{r}) = \frac{Z_0 e}{r} \,. \tag{1.15}$$

In the plasma such a potential field acts to disturb the spatial distribution of charged particles. The space-charge field so induced around the point charge in turn produces an extra potential field, which should be added to the original potential $\Phi_0(\mathbf{r})$; a new effective potential $\Phi(\mathbf{r})$ is thus obtained as a summation of the two. The space-charge distribution in the plasma is determined, not from the bare potential $\Phi_0(\mathbf{r})$, but from the total potential field in a self-consistent fashion. A calculation along these lines was originally carried out by Debye and Hückel [1923] in connection with the theory of strong electrolytes.

The Poisson equation for the total field is thus written as

$$\nabla^2\Phi(\mathbf{r}) = -4\pi[Z_0 e\delta(\mathbf{r}) + Ze\langle\delta\rho(\mathbf{r})\rangle] , \tag{1.16}$$

where $\delta(\mathbf{r})$ is the three-dimensional delta function. The average density variation $\langle\delta\rho(\mathbf{r})\rangle$ in the OCP is calculated by applying the usual Boltzmann statistics:

$$\langle\delta\rho(\mathbf{r})\rangle = n\exp\left[-\frac{Ze\Phi(\mathbf{r})}{T}\right] - n , \tag{1.17}$$

where n refers to the uniform, average number density of the OCP particles. The exponent, $-Ze\Phi(\mathbf{r})/T$, in Eq. (1.17) represents the ratio of the potential energy $-Ze\Phi(\mathbf{r})$ per particle to the average kinetic energy T. When $\Gamma << 1$ as in the ordinary gaseous plasmas, this ratio may be regarded as sufficiently small on the average, so that we may expand Eq. (1.17) as*

$$\langle\delta\rho(\mathbf{r})\rangle = -\frac{Zen\Phi(\mathbf{r})}{T} . \tag{1.18}$$

Substitution of this in Eq. (1.16) yields a differential equation for determination of the effective potential $\Phi(\mathbf{r})$:

$$-\nabla^2\Phi(\mathbf{r}) + \frac{4\pi n(Ze)^2}{T}\Phi(\mathbf{r}) = 4\pi Z_0 e\delta(\mathbf{r}) . \tag{1.19}$$

The solution to this equation with the boundary condition that $\Phi(\mathbf{r})$ vanishes at infinity is

$$\Phi(\mathbf{r}) = \frac{Z_0 e}{r}\exp\left(-\frac{r}{\lambda_D}\right). \tag{1.20}$$

The parameter

$$\lambda_D = \sqrt{\frac{T}{4\pi n(Ze)^2}} \tag{1.21}$$

* In the vicinity of $r = 0$, this expansion is inapplicable even with $\Gamma << 1$; strong-coupling effects persist there.

introduced here is called the *Debye length*; it is an important characteristic length for the weakly coupled plasmas. Putting numbers for the known physical parameters in Eq. (1.21), we find the following convenient numerical formula for the Debye length:

$$\lambda_{\mathrm{D}} = 6.9 \frac{\sqrt{T/n}}{Z} \text{ (cm)} \tag{1.22}$$

where T and n are to be measured in units of K and cm^{-3}. For an OCP with $Z = 1$, $n = 10^{10}$, and $T = 10^4$, λ_{D} is computed to be approximately 7×10^{-3} cm; the Debye length usually takes on a value smaller than a macroscopic scale of distance.

The meaning of Eq. (1.20) is clear by comparison with Eq. (1.15). For distances smaller than λ_{D}, the effective potential is virtually identical to the bare Coulomb potential, Eq. (1.15). For distances larger than λ_{D}, the potential field decreases exponentially; for $r >> \lambda_{\mathrm{D}}$, one can take $\Phi(\mathbf{r}) \approx 0$. In other words, the potential field around a point charge is effectively screened out by the induced space-charge field in the OCP for distances greater than the Debye length.

Let us consider the number of plasma particles involved in such an act of Debye screening. For such a purpose, we define the *Debye number*, N_{D}, the average number of particles contained in a sphere with a radius λ_{D}:

$$N_{\mathrm{D}} = \left(\frac{4\pi n}{3}\right)\lambda_{\mathrm{D}}^3 = (3\Gamma)^{-3/2}$$

$$= 1.38 \times 10^6 Z^{-3} \left(\frac{n}{10^{12}\mathrm{cm}^{-3}}\right)^{-1/2} \left(\frac{T}{10^6\mathrm{K}}\right)^{3/2} . \tag{1.23}$$

As we find here, the Debye number takes on an extremely large value for a weakly coupled plasma ($\Gamma << 1$). On the other hand, the total number of plasma particles necessary for such a screening is $|Z_0/Z|$, which is on the order of unity. Consequently, for realization of Debye screening, so many ($\sim N_{\mathrm{D}}$) charged particles act cooperatively by an extremely small amount each ($\sim N_{\mathrm{D}}^{-1}$), to erase the field produced by the external particle. It is in this sense that the Debye screening makes an illuminating example of the cooperative phenomena brought about by the long-range Coulomb interaction in the plasma.

The calculation of the Debye screening can be easily extended to cases that involve a variety of mobile charged components. If we use a suffix σ to distinguish between various constituents of the plasma and let $Z_\sigma e$ and n_σ denote the electric charge and the number density of a given species of particles, the Debye length of such a plasma is now expressed as

$$\lambda_{\mathrm{D}} = \sqrt{\frac{T}{\sum_\sigma 4\pi n_\sigma (Z_\sigma e)^2}} \, . \tag{1.24}$$

Based on the idea of Debye screening described above, let us return to the task of evaluating the Coulomb cross sections in the plasma. The field of a point charge is now effectively screened out at distances larger than the Debye length. In the calculation of the Coulomb cross section in Eq. (1.12), rather than letting the lower limit of the χ integration approach zero, we may hold it at a finite angle $\chi_{\min}$ corresponding to a maximum impact parameter $b_{\max}$ equal to the Debye length; we thus avoid the logarithmic divergence. In weakly coupled plasmas, the Coulomb logarithm determined in this way assumes a large number (typically, 15). Hence in the actual evaluations, whether the impact parameter should be chosen equal to the Debye length or some multiple of it is immaterial because such an ambiguity affects only the argument in the logarithm. Within such a logarithmic accuracy, we can therefore assume $Z_1 = Z_2 = Z$ and a statistical average, $\langle \mu v^2 \rangle = 3T$, in the OCP to estimate the logarithm in the final expression of Eq. (1.12). The Coulomb logarithm thus takes on a magnitude

$$\ln \Lambda = \ln (9N_{\mathrm{D}}) \, . \tag{1.25}$$

The frequency of Coulomb collisions associated with transfer of momentum between particles may be formulated as

$$\nu_m(v) = nQ_m v \, . \tag{1.26}$$

Substitution of Eqs. (1.12) and (1.25) in Eq. (1.26) yields

$$\nu_m(v) = 4\pi n \left[\frac{(Ze)^4}{\mu^2 v^3} \right] \ln(9N_{\mathrm{D}}) \, . \tag{1.27}$$

Using a Maxwellian average

$$\langle v^3 \rangle = 8\sqrt{\frac{2}{\pi}} \left(\frac{T}{\mu} \right)^{3/2} ,$$

we find the average collision frequency approximately as

$$V_m = \left(\frac{\pi}{2}\right)^{3/2} \left[\frac{n(Ze)^4}{\sqrt{\mu}T^{3/2}}\right] \ln(9N_D) . \tag{1.28}$$

The analysis just presented has been concerned with the screening of the potential field around a point charge in a plasma. Similar arguments are also applicable to the space-charge distributions produced by spontaneous fluctuations. Suppose that the space-charge neutrality of the plasma is destroyed locally by some mechanism (e.g., by thermal agitation) and that a local distribution of space charge appears; other charged particles in the plasma then act to neutralize such a spontaneous fluctuation of space charges. We may then argue that it is energetically quite unfavorable for a space-charge distribution to appear over a distance larger than the Debye length. Since any plasma contains a collection of randomly moving charged particles, we see that the *macroscopic neutrality* is maintained by such a statistical mechanism.

Finally, let us evaluate the internal energy U per unit volume of a weakly coupled plasma. In the Debye-Hückel approximation, the interaction-energy density is calculated as

$$U_{\text{int}} = \frac{Zen}{2} [\Phi(0) - \Phi_0(0)] \tag{1.29}$$

where we set $Z_0 = Z$. Taking account of additional terms arising from the cluster-expansion calculations [Abe 1959], one finds

$$\frac{U}{nT} = \frac{3}{2} - \frac{\sqrt{3}}{2}\Gamma^{3/2} - 3\Gamma^3\left[\frac{3}{8}\ln(3\Gamma) + \frac{\gamma}{2} - \frac{1}{3}\right] \tag{1.30}$$

where $\gamma = 0.57721 \cdots$ is Euler's constant.

C. The Ion-Sphere Model

Thus far we have considered a problem of the effective potential in a weakly coupled plasma and arrived at the concept of Debye screening. As we move into the strong-coupling regime ($\Gamma \geq 1$), the Debye number becomes smaller than unity; the concept of Debye screening as a cooperative phenomenon is no longer applicable. The probability of finding other charged particles in a sphere with the radius λ_D almost vanishes. A charged particle creates a sizable domain around itself where other particles, having been repelled, are unlikely to be found, a sort of a territorial domain of its own influence which may be looked upon as a *Coulomb hole*.

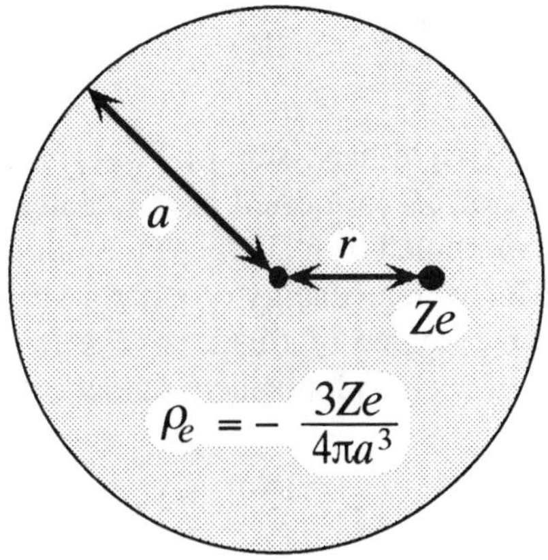

Figure 1.3 Ion-sphere model; ρ_e refers to the average charge density in the sphere.

To understand salient features of such a strongly coupled plasma, it is instructive to introduce an *ion-sphere model* [Salpeter 1954], equivalent to the Wigner-Seitz sphere used in solid-state physics [e.g., Pines 1963]. As Fig. 1.3 illustrates, one considers a charged particle Ze and its surrounding neutralizing charge sphere, whose total electric charge is just enough to cancel the charge Ze. This sphere thus represents the territorial domain of influence for the charge Ze. Its radius is a of Eq. (1.4) and the charge density is $-3Ze/4\pi a^3$.

The ion sphere thus consists of a single ion and its surrounding negative-charge sphere. The electrostatic energy E_{IS} associated with the ion sphere is then calculated in the following way: First the electrostatic potential produced by the negative-charge sphere at r ($\leq a$) is

$$\Phi_{IS}(r) = -\frac{3Ze}{4\pi a^3}\int_{r'\leq a} d\mathbf{r}' \frac{1}{|\mathbf{r}-\mathbf{r}'|} = -\frac{3Ze}{2a} + \frac{Ze}{2a}\left(\frac{r}{a}\right)^2 . \tag{1.31}$$

The electrostatic energy of the negative-charge itself is then

$$-\frac{1}{2}\int_{r\leq a} d\mathbf{r}\, \frac{3Ze}{4\pi a^3}\Phi_{IS}(r) = \frac{3}{5}\frac{(Ze)^2}{a} . \tag{1.32}$$

Summation of this energy and $Ze\Phi_{IS}(r)$ yields

$$\frac{E_{\text{IS}}}{T} = -0.9\Gamma + \frac{\Gamma}{2}\left(\frac{r}{a}\right)^2 . \tag{1.33}$$

The first term on the right-hand side of Eq. (1.33) represents the electrostatic energy when the ion Ze is located at the center of the ion sphere. The second term, being proportional to r^2, induces a motion of the harmonic-oscillator type to the ion. The average energy of a harmonic oscillator including the kinetic energy is T per a degree of freedom. The density of internal energy calculated on the basis of such an ion-sphere model is thus

$$\frac{U}{nT} = -0.9\Gamma + 3 . \tag{1.34}$$

This estimate, in fact, takes on a value close to computer simulation results in the domain of $\Gamma >> 1$ [Ichimaru 1982].

The ion-sphere model has played an important part in elucidating the properties of the strongly coupled plasma. Here we consider how the ion-sphere model may be applied to the calculation of the Coulomb cross section, Eq. (1.12), and the collision frequency, Eq. (1.26).

According to the ion-sphere model of Fig. 1.3, the range of the electrostatic potential of the charged particle Ze is confined within a distance $\alpha_1 a$, where α_1 is a correction factor of order unity accounting for uncertainty involved in a strict enforcement of the ion-sphere model. Putting $Z_1 = Z_2 = Z$ in Eq. (1.10), we determine the scattering angle $\chi_{\min}$ corresponding to the impact parameter at the distance $\alpha_1 a$,

$$\cot\left(\frac{\chi_{\min}}{2}\right) = \frac{\alpha_1 a \mu v^2}{(Ze)^2} . \tag{1.35}$$

Equation (1.12) then becomes

$$Q_m = 2\pi\left[\frac{(Ze)^2}{\mu v^2}\right]^2 \ln\left\{1 + \left[\frac{\alpha_1 a \mu v^2}{(Ze)^2}\right]^2\right\} . \tag{1.36}$$

For a strongly coupled plasma ($\Gamma \geq 1$), $\alpha_1 a \mu v^2/(Ze)^2 < 1$ holds for a majority of particles. Retaining only the first term after expanding the logarithmic term in Eq. (1.36), we obtain

$$Q_m \approx 2\pi(\alpha_1 a)^2 , \tag{1.37}$$

and Eq. (1.28) yields

$$v_m \approx 4\sqrt{2}\pi n(\alpha_1 a)^2\sqrt{\frac{T}{\mu}}. \tag{1.38}$$

As we would anticipate from the construction of the ion-sphere model, the scattering cross-section is proportional to the cross section πa^2 of the ion sphere.

Equation (1.37) thus takes a form easy to interpret qualitatively. Quantitatively, however, it depends strongly on an estimation of the undetermined parameter α_1. A theoretical determination of α_1 is a complex problem that calls for a detailed knowledge of interparticle correlations in the strongly coupled plasma, which we shall study in some detail later in Chapter 7.

The Wigner-Seitz sphere model played the central part in Wigner's theory of low-density electron solid [Wigner 1934,1938]. It is the *Wigner transition* in a quantum electron gas. The classical counter part to such a Wigner transition takes place when the Coulomb coupling parameter Γ exceeds a certain critical value, $\Gamma_m \approx 180$, in an OCP; *Coulomb solids* are thus formed. Freezing transitions and the properties of the resulting Coulomb solids will be studied in Vol. II.

1.4 DENSITY FLUCTUATION EXCITATIONS

In the preceding section, we considered the static properties of plasmas such as screening, which did not involve a time-dependent behavior of density variations in the plasma. The long-range Coulomb interaction produces not only such a static cooperative effect but also the *plasma oscillation*, viewed as a dynamic cooperative phenomenon or a collective motion.

In this section, we begin with the equation of motion describing the temporal behavior of density fluctuations in plasma. We shall thus find that plasma is characterized by a fascinating interplay between its mediumlike character due to the long-range nature of the Coulomb interaction and the individual-particlelike behavior. The latter arises from a simple fact that plasma consists of an extremely large number of distinct particles like an ordinary gas. In these connections, there will be introduced the concept of a fluid limit which singles out the mediumlike character, and the dimensionless discreteness parameter Δ representing the individual-particlelike behavior.

A. Plasma Oscillation

Let $\mathbf{r}_j(t)$ be the spatial coordinates at time t of a jth particle in n charged particles contained in a unit volume. In terms of the three-dimensional delta-function introduced in Eq. (1.16), the density distribution arising from the jth particle is $\delta[\mathbf{r}-\mathbf{r}_j(t)]$. Summing this for all the n particles, we express the particle-density distribution as

$$\rho(\mathbf{r},t)=\sum_j \delta[\mathbf{r}-\mathbf{r}_j(t)]\,. \tag{1.39}$$

With the periodic boundary conditions appropriate to the unit volume, the Fourier component* of the density fluctuations with wave vector $\mathbf{k}$ is calculated as

$$\rho_{\mathbf{k}}(t)=\int_{\text{unit volume}} d\mathbf{r}\rho(\mathbf{r},t)\exp(-i\mathbf{k}\cdot\mathbf{r})=\sum_j \exp[-i\mathbf{k}\cdot\mathbf{r}_j(t)]\,. \tag{1.40}$$

An equation of motion may be obtained by differentiating this density fluctuation twice with respect to time. Since $\dot{\mathbf{r}}_j=\mathbf{v}_j$ represents the velocity of the jth particle, we find

$$\ddot{\rho}_{\mathbf{k}}=-\sum_j\left[(\mathbf{k}\cdot\mathbf{v}_j)^2+i\mathbf{k}\cdot\dot{\mathbf{v}}_j\right]\exp(-i\mathbf{k}\cdot\mathbf{r}_j)\,. \tag{1.41}$$

The acceleration $\dot{\mathbf{v}}_j$ of the jth particle can be expressed in terms of the force exerted on it by all the other charged particles and the neutralizing background charges,

$$\dot{\mathbf{v}}_j=-\frac{1}{m}\frac{\partial}{\partial\mathbf{r}_j}\sum_{l(\neq j)}\frac{(Ze)^2}{|\mathbf{r}_j-\mathbf{r}_l|}$$
$$+\,(\text{acceleration by the background charges})\,. \tag{1.42}$$

---

* Throughout this volume, unless specified otherwise, we use the following convention on the Fourier expansions and transformations: With the periodic boundary conditions appropriate to a cube of volume L^3, a function $F(\mathbf{r})$ may be expanded as

$$F(\mathbf{r})=\sum_{\mathbf{k}} F_{\mathbf{k}}\exp(i\mathbf{k}\cdot\mathbf{r})\rightarrow L^3\int\frac{d\mathbf{k}}{(2\pi)^3}F_{\mathbf{k}}\exp(i\mathbf{k}\cdot\mathbf{r})\,.$$

The Fourier components are thus calculated as

$$F_{\mathbf{k}}=L^{-3}\int_{L^3} d\mathbf{r}F(\mathbf{r})\exp(-i\mathbf{k}\cdot\mathbf{r})\,.$$

For a translationally invariant, infinite plasma (without boundary), the choice of the volume is practically unrestricted. For simplicity, we may sometimes choose $L^3=1$, which leads to a representation by the Fourier transformation through a passage to a homogeneous case.

Here, m denotes the mass of a plasma particle and $\partial/\partial\mathbf{r}_j$ means a vectorial differential operation—gradient—with respect to the $\mathbf{r}_j$ coordinates.

Fourier expansion of the Coulomb potential in the cube of a unit volume is given by

$$\frac{Ze}{r} = \sum_{\mathbf{k}} \frac{4\pi Ze}{k^2} \exp(i\mathbf{k}\cdot\mathbf{r}) . \tag{1.43}$$

This expression is substituted in Eq. (1.42), where we note that the component with $\mathbf{k} = \mathbf{0}$ cancels exactly with the acceleration due to the neutralizing charges; we thus obtain

$$\dot{\mathbf{v}}_j = -i\frac{4\pi(Ze)^2}{m} \sum_{\mathbf{q}}{}' \frac{\mathbf{q}}{q^2} \rho_{\mathbf{q}} \exp(i\mathbf{q}\cdot\mathbf{r}_j) . \tag{1.44}$$

Here $\sum_{\mathbf{q}}{}'$ implies that the $\mathbf{q} = \mathbf{0}$ term is to be excluded in the summation.

Substituting Eq. (1.44) in Eq. (1.41), we find

$$\ddot{\rho}_{\mathbf{k}} = -\sum_j (\mathbf{k}\cdot\mathbf{v}_j)^2 \exp(-i\mathbf{k}\cdot\mathbf{r}_j) - \frac{4\pi(Ze)^2}{m} \sum_{\mathbf{q}}{}' \frac{\mathbf{k}\cdot\mathbf{q}}{q^2} \rho_{\mathbf{k}-\mathbf{q}}\rho_{\mathbf{q}} . \tag{1.45}$$

The first term on the right-hand side represents the effect of translational motion of the individual particles; the second term stems from the Coulomb interaction. We particularly single out the $\mathbf{q} = \mathbf{k}$ term in the interaction term; since $\rho_0 = n$, Eq. (1.45) is rewritten as

$$\ddot{\rho}_{\mathbf{k}} + \omega_p^2 \rho_{\mathbf{k}} = -\sum_j (\mathbf{k}\cdot\mathbf{v}_j)^2 \exp(-i\mathbf{k}\cdot\mathbf{r}_j)$$

$$- \frac{4\pi(Ze)^2}{m} \sum_{\mathbf{q}(\neq\mathbf{k})}{}' \frac{\mathbf{k}\cdot\mathbf{q}}{q^2} \rho_{\mathbf{k}-\mathbf{q}}\rho_{\mathbf{q}} . \tag{1.46}$$

Here another important parameter,

$$\omega_p \equiv \sqrt{\frac{4\pi n(Ze)^2}{m}} , \tag{1.47}$$

called the *plasma frequency*, arises; for an electron plasma it is computed as

$$\omega_p = 5.64 \times 10^{10} \sqrt{\frac{n}{10^{12}\text{cm}^{-3}}} \ (\text{s}^{-1}) . \tag{1.48}$$

For a multicomponent plasma, one finds

$$\omega_p = \sqrt{\sum_{\sigma} \frac{4\pi n_\sigma (Z_\sigma e)^2}{m_\sigma}} \, . \tag{1.49}$$

If the right-hand side of Eq. (1.46) is neglected altogether, the temporal variation of the density fluctuation with wave vector **k** obeys an equation of motion for a harmonic oscillator with frequency ω_p [Tonks and Langmuir 1929]. As we note in Eq. (1.40), $\rho_{\mathbf{k}}$ contains the coordinates of extremely numerous charged particles; those may be regarded as *collective variables.* We here observe a possibility that such collective variables exhibit an orderly oscillatory behavior, a *collective motion,* under the action of the long-range Coulomb forces. In the next subsection we investigate the conditions under which the density fluctuations behave collectively in a plasma.

B. Collective Motion and Individual-Particlelike Behavior

Carrying out Maxwellian average of the velocity variables contained in the first term on the right-hand side of Eq. (1.46), we obtain

$$\left\langle \sum_j (\mathbf{k}\cdot\mathbf{v}_j)^2 \exp(-i\mathbf{k}\cdot\mathbf{r}_j) \right\rangle \approx k^2 \left(\frac{T}{m}\right) \rho_{\mathbf{k}} \, . \tag{1.50}$$

The second term in Eq. (1.46) represents a nonlinear term consisting of the products of density fluctuations. On the complex-number plane, the Fourier component of density fluctuations is expressed as a summation of unit vectors over the total number of particles; the phase angle of each vector is given by $-\mathbf{k}\cdot\mathbf{r}_j$. For a uniform plasma, such phase angles distribute randomly as long as $\mathbf{k} \neq \mathbf{0}$; hence the expectation value $\langle \rho_{\mathbf{k}} \rangle = 0$.

Bohm and Pines [1951,1953; Pines & Bohm 1952], in their theory of collective variables in many-electron systems, considered a possibility that an analogous situation may become applicable to the second term on the right-hand side of Eq. (1.46) consisting of products of density fluctuations. They thereby introduced an approximation, called the *random-phase approximation* (RPA), in which those product terms are ignored. Even if the phases $-\mathbf{k}\cdot\mathbf{r}_j$ are distributed randomly, however, the effects arising from the products of fluctuations $\rho_{\mathbf{k}-\mathbf{q}}\,\rho_{\mathbf{q}}$ or the mode-coupling terms cannot generally be neglected. Setting the origin of the nomenclature aside, we may interpret the RPA as an approximation applicable to the cases of weak fluctuations whereby the equation of motion may be linearized with respect to the fluctuations. The RPA thus provides an accurate description of a weakly coupled plasma near ther-

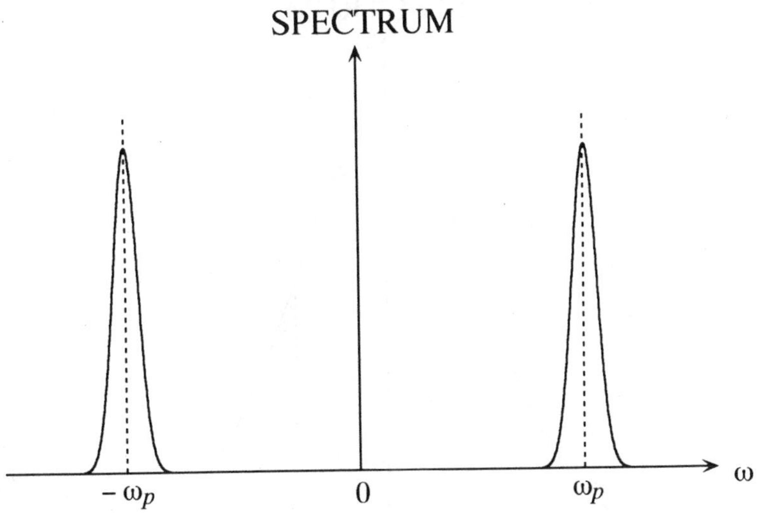

Figure 1.4 Frequency spectrum of density fluctuations in the long-wavelength regime—collective motions.

modynamic equilibrium. In these circumstances we may ignore the second term on the right-hand side of Eq. (1.46), since they are nonlinear terms of fluctuations.

Let us thus adopt such an RPA. The condition that the density fluctuations behave collectively and produce plasma oscillations may be derived from a comparison between the second term on the left-hand side of Eq. (1.46) and Eq. (1.50) as

$$k^2 << k_D^2 . \tag{1.51}$$

The parameter

$$k_D = \sqrt{\frac{4\pi n(Ze)^2}{T}} \tag{1.52}$$

is the inverse of the Debye length defined in Eq. (1.21) (or in Eq. 1.24 for a multicomponent plasma); we shall call it the *Debye wave number.*

In the long-wavelength regime satisfying Eq. (1.51), the density fluctuations in the plasma behave collectively and oscillate at a frequency in the vicinity of ω_p. In such a regime the frequency spectrum of fluctuations may look as shown schematically in Fig. 1.4.

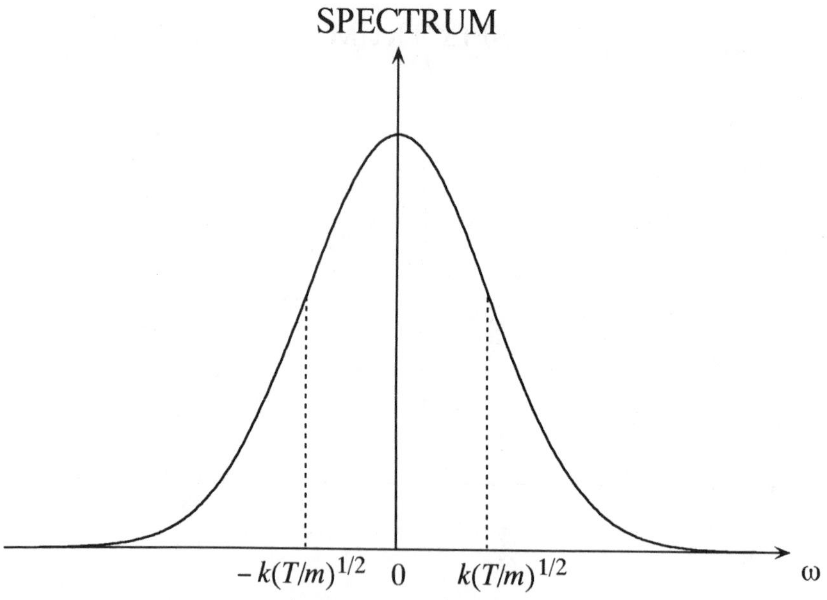

Figure 1.5 Frequency spectrum of density fluctuations in the short-wavelength regime—individual-particlelike fluctuations.

In the short-wavelength regime such that $k^2 \gg k_D^2$, however, the first term on the right-hand side of Eq. (1.46) is the major contribution and the equation of motion becomes

$$\ddot{\rho}_{\mathbf{k}} = -\sum_j (\mathbf{k} \cdot \mathbf{v}_j)^2 \exp(-i\mathbf{k} \cdot \mathbf{r}_j) \, . \tag{1.53}$$

This is nothing but the equation of motion for fluctuations in the system where the Coulomb interaction is neglected altogether and individual particles are assumed to move on rectilinear trajectories without changing velocities $\mathbf{v}_j$. Consequently, the fluctuations in this case may be looked upon as a superposition of the random motion of discrete individual particles. The frequency spectrum of fluctuations now exhibits a characteristic thermal spread as shown in Fig. 1.5.

The individual-particlelike spectrum of Fig. 1.5 is a characteristic feature of a neutral-particle system of an ideal-gas type. In a plasma, such an individual-particlelike behavior in the short-wavelength domain coexists with the collective motion, that is, the plasma oscillation in the long-wavelength regime; the latter represents the principal effect of the Coulomb interaction. Such a coexistence of the collective and individual-particlelike behaviors adds a variety of interesting features to the properties of plasma, as we shall study in subsequent chapters.

C. Discreteness Parameters and Fluid Limit

The Debye length (1.21) and the plasma frequency (1.47) can be computed when the values of the four parameters, m, Ze, $1/n$, and T, are known for a given system of charged particles. These may be regarded as the basic quantities characterizing a classical OCP. Let us note also that each of these parameters represents a quantity related to the discrete nature of the individual particles constituting the plasma: They represent, respectively, the mass, electric charge, average volume occupied, and average kinetic energy per particle. We may therefore regard them as the *discreteness parameters* of the plasma.

Having thus recognized the nature of these parameters, we can now introduce a theoretical limiting procedure by which the individuality of the particles is suppressed and the mediumlike properties of the plasma emphasized. Suppose a process in which we cut each particle into finer and finer pieces; in the limit, m, Ze, $1/n$, and T all approach zero. If in this limit we keep the fluidlike parameters, such as the mass density mn, charge density Zen, and kinetic energy density nT, finite, we must regard the discreteness parameters m, Ze, $1/n$, and T as infinitesimal quantities of the same order. Such a limiting procedure is called the *fluid limit*; it was introduced by Rostoker and Rosenbluth [1960]. In this limit the plasma becomes a continuous fluid.

It is important to note that the Debye length and the plasma frequency are kept invariant in this fluid limit. They therefore provide appropriate scales by which length and time may be measured in the resultant plasma fluid.

The notion of the discreteness parameters is especially useful in a theoretical treatment of kinetic equations for weakly coupled plasmas, which we shall consider in the next chapter. For practical purposes, however, it is often more convenient and mathematically transparent to deal with dimensionless parameters rather than those quantities with finite physical dimensions.

It is not difficult to prove that only a single family* of dimensionless parameters are derivable out of the four discreteness parameters (Problem 1.9). In this family, we seek to find a dimensionless parameter that is of the same order as the discreteness parameters; the answer is

$$\Delta \equiv \frac{1}{N_{\mathrm{D}}} = (3\Gamma)^{3/2} . \tag{1.54}$$

* Dimensionless parameters, A and B, belong to a single family if a relation $A = pB^q$ holds with non-vanishing real numbers, p and q.

This dimensionless parameter is likewise referred to as the discreteness parameter. As we saw in Eq. (1.30), it measures the ratio between the average potential and kinetic energies for a weakly coupled plasma. Being an extremely small number in these circumstances, it provides a useful expansion parameter in a theoretical treatment.

Utility of the discreteness parameters may be illustrated through a consideration of the frequency of Coulomb collisions, Eq. (1.28). Apart from its dependence on the Coulomb logarithm, the collision frequency is here found to be a quantity of the first order in the discreteness parameters. The Coulomb collisions stem from the discreteness of particles; in the fluid limit, the collision frequency vanishes.

These arguments provide a justification for a collisionless description of weakly coupled plasmas, where $\Gamma << 1$. In fact, the collective, mediumlike properties are the essential characters of such a plasma, since the fluid limit is a good approximation there. Later in Section 9.2, where anomalous transports are treated, we shall further consider the relation between elementary processes and discreteness.

PROBLEMS

1.1 Express the Fermi energy (1.2) in terms of the r_s parameter (1.6), and find

$$E_F = 50.11 r_s^{-2}\ (\text{eV}) = 5.82 \times 10^5 r_s^{-2}\ (\text{K}) .$$

1.2 The expression (1.2) for the Fermi energy is applicable for nonrelativistic degenerate electrons, $E_F < mc^2$, where m is the rest mass of an electron. Taking account of the relativistic relation between energy and momentum, show that the Fermi energy of degenerate electrons is generally given by

$$E_F = mc^2\left(\sqrt{1 + 1.96 \times 10^{-4} r_s^{-2}} - 1\right).$$

1.3 Complete the following table for the following plasmas at temperature T (K): arc discharge (AD), ionosphere (I), solar corona (SC), solar interior (SI), Jovian interior (JI), white-dwarf interior (WD), and outer crust of neutron star (NS), where n (cm^{-3}), Z, and A are the number density, effective charge number, and average mass number of the ions, respectively. Provide Γ and Y for the ions, and r_s and Θ for the electrons.

	$\log_{10}n$	$\log_{10}T$	Z	A	Γ	Y	r_s	Θ
AD	12	5	1	200				
I	6	3.2	1	2				
SC	5	6	1	1				
SI	25.9	7	1.2	1.3				
JI	24.3	3.9	1.2	1.3				
WD	30	8	6	12				
NS	28.7	7.7	26	56				

1.4 Work out Eq. (1.10) through Eq. (1.12).

1.5 In the frame of reference (Fig. 1.2) comoving with the center-of-mass velocity **V**, Eq. (1.9), no transfer of energy takes place (i.e., elastic collisions), so that

$$\frac{\mu|\mathbf{v}|^2}{2} = \frac{\mu|\mathbf{v} + \Delta\mathbf{p}/\mu|^2}{2},$$

where $\Delta\mathbf{p}$ is the momentum transferred. Carry out the Galilean transformation back to the laboratory frame, and show that the amount of energy exchanged at a collision is given by Eq. (1.13).

1.6 Solve the differential equation (1.19) with the boundary condition that $\Phi(r)$ vanishes at infinity, and find Eq. (1.20).

1.7 Derive Eq. (1.31).

1.8 Using the solution to Problem 1.6, or otherwise, derive Eq.(1.43).

1.9 Show that only a single family of dimensionless parameters can be constructed out of the combinations of quantities Ze, m, n, and T for the classical OCP, and that Γ is such a dimensionless parameter proportional to the strength of Coulomb coupling, e^2.

1.10 Show that only a single family of dimensionless parameters can be constructed out of the combinations of quantities e, m, n, and $\hbar$ for the degenerate electron gas, and that r_s is such a dimensionless parameter proportional to the strength of Coulomb coupling, e^2.

1.11 Verify that the collision frequency (1.28) is of the first order with respect to the discreteness parameters.

1.12 Consider a Wigner-Seitz model of a Coulomb solid, where a plasma particle is assumed to oscillate harmonically in the ion-sphere potential of E_{IS} in Eq. (1.33). Calculate the root-mean-square amplitude of the oscillation normalized by the ion-sphere radius, $\sqrt{\langle \Delta r^2 \rangle}/a$.

CHAPTER 2

Kinetic Equations

Plasma physics is concerned with a rich class of phenomena related to the dynamical processes in statistical mechanics. It is therefore essential to study the structures and properties of the kinetic equations which govern the dynamics of plasmas.

We begin this chapter with a formal exposition on the structures of the kinetic theory, with an aim at introducing Klimontovich's microscopic formalism and deriving the Bogoliubov-Born-Green-Kirkwood-Yvon (BBGKY) hierarchy equations for multiparticle distribution functions. The Vlasov equation for single-particle distributions is then derived from a formal solution to the BBGKY hierarchy; in the process, its validity and limitation will be discussed. This is particularly important in view of the fact that the Vlasov equations are used most frequently in the theoretical investigation of plasma phenomena. We shall next turn to a consideration of the collision term in the kinetic equation: This calls for a study of interparticle correlations, which are not included in a treatment based on the Vlasov equations. In these connections, such concepts as Bogoliubov's hierarchy of relaxation times and the Vlasov propagator will be introduced. Finally, the Balescu-Guernsey-Lenard (BGL) collision term will be derived and its properties investigated.

2.1 STRUCTURES OF KINETIC EQUATIONS

A theoretical foundation for the kinetic theory of classical plasmas is set forth in this section, starting with Klimontovich [1957,1967] formalism, followed by the Liouville equation [e.g., Bogoliubov 1962; Hansen & McDonald 1986], multiparticle distribution functions, and the BBGKY hierarchy equations [Bogoliubov 1946,1962; Born & Green 1946,1949; Kirkwood 1935,1946; Yvon 1935,1966].

A. The Klimontovich Equation

We consider a classical system containing N identical particles in a box of volume V; $n = N/V$ denotes the average number density. Each particle is characterized by the electric charge Ze and the mass m; we assume a smeared-out background of opposite charges so that the average space-charge field in the plasma may be canceled. A classical OCP model is thus adopted.

In the six-dimensional space consisting of the position $\mathbf{r}$ and the velocity $\mathbf{v}$, each particle has its own trajectory; for the ith particle, we write

$$\mathbf{X}_i(t) \equiv [\mathbf{r}_i(t), \mathbf{v}_i(t)] \,. \tag{2.1}$$

Since we are dealing with point particles, the microscopic density of the particles in the phase space may be expressed by the summation of six-dimensional delta functions as

$$N(\mathbf{X};t) \equiv \sum_i \delta[\mathbf{X} - \mathbf{X}_i(t)] \,, \tag{2.2}$$

where $\mathbf{X} \equiv (\mathbf{r},\mathbf{v})$. $N(\mathbf{X};t)$ may be called the *Klimontovich distribution function*.

The distribution function satisfies the continuity equation in the phase space:

$$\frac{dN}{dt} = \frac{\partial N}{\partial t} + \dot{\mathbf{X}} \cdot \frac{\partial N}{\partial \mathbf{X}} = 0 \,. \tag{2.3}$$

Writing the phase-space coordinates explicitly, we have

$$\frac{\partial N}{\partial t} + \mathbf{v} \cdot \frac{\partial N}{\partial \mathbf{r}} + \dot{\mathbf{v}} \cdot \frac{\partial N}{\partial \mathbf{v}} = 0 \tag{2.4}$$

where $\dot{\mathbf{v}}$ is the acceleration at the point $(\mathbf{r},\mathbf{v})$.

Equation (2.3) is simply a shorthand notation of Hamilton's equations of motion for the N individual particles. Its derivation proceeds as follows: Consider (2.1) at first as a *dynamical variable* depending on $6N$ coordinates $\{\mathbf{X}_j(t)\}$; $\mathbf{X}$ is a parameter. By virtue of the properties of the delta functions, its time derivative is calculated as

$$\dot{N} = \left(\sum_j \dot{\mathbf{X}}_j \cdot \frac{\partial}{\partial \mathbf{X}_j}\right) N = \sum_j \dot{\mathbf{X}}_j \cdot \frac{\partial}{\partial \mathbf{X}_j} \delta(\mathbf{X} - \mathbf{X}_j)$$

$$= -\sum_j \dot{\mathbf{X}}_j \cdot \frac{\partial}{\partial \mathbf{X}} \delta(\mathbf{X} - \mathbf{X}_j) = -\dot{\mathbf{X}} \cdot \frac{\partial}{\partial \mathbf{X}} N, \tag{2.5}$$

where $\dot{\mathbf{X}} = (\dot{\mathbf{r}}, \dot{\mathbf{v}})$. We now return to the original interpretation of (2.2) as a *distribution* in the $\mathbf{X}$ space. Since the time derivative (2.5) has originated from $\mathbf{X}_j(t)$, it represents a partial derivative with respect to time, and hence (2.3).

For the plasma problems, we take $\dot{\mathbf{v}}$ as the acceleration arising from Lorentz force on a particle at $\mathbf{X}$ due to microscopic electromagnetic fields, $\mathbf{E}(\mathbf{r},t)$ and $\mathbf{B}(\mathbf{r},t)$:

$$\dot{\mathbf{v}} = \frac{Ze}{m}\left[\mathbf{E}(\mathbf{r},t) + \frac{\mathbf{v}}{c} \times \mathbf{B}(\mathbf{r},t)\right]. \tag{2.6}$$

These fields consist of two separate contributions: those applied from external sources, and those produced by the microscopic fine-grained distribution (2.2) of the charged particles, i.e.,

$$\mathbf{E}(\mathbf{r},t) = \mathbf{E}_{\text{ext}}(\mathbf{r},t) + \mathbf{e}(\mathbf{r},t),$$

$$\mathbf{B}(\mathbf{r},t) = \mathbf{B}_{\text{ext}}(\mathbf{r},t) + \mathbf{b}(\mathbf{r},t). \tag{2.7}$$

The microscopic fine-grained fields, $\mathbf{e}(\mathbf{r},t)$ and $\mathbf{b}(\mathbf{r},t)$, are to be determined from a solution to the Maxwell equations:

$$\nabla \times \mathbf{e} + \frac{1}{c}\frac{\partial \mathbf{b}}{\partial t} = 0, \tag{2.8}$$

$$\nabla \times \mathbf{b} - \frac{1}{c}\frac{\partial \mathbf{e}}{\partial t} = 4\pi \frac{Ze}{c} \int d\mathbf{v}\mathbf{v}N(\mathbf{X};t), \tag{2.9}$$

$$\nabla \cdot \mathbf{e} = 4\pi Ze\left[\int d\mathbf{v}N(\mathbf{X};t) - n\right], \tag{2.10}$$

$$\nabla \cdot \mathbf{b} = 0. \tag{2.11}$$

For a given $N(\mathbf{X};t)$ one can generally solve this set of equations. The solution, when substituted in (2.6), would then amount to taking account of the electromagnetic interactions between the particles. For a nonrelativistic plasma, one may adopt an electrostatic approximation, within which the microscopic fields are written as

$$\mathbf{e}(\mathbf{r},t) = -Ze\frac{\partial}{\partial \mathbf{r}}\int d\mathbf{X}'\frac{N(\mathbf{X}';t)}{|\mathbf{r}-\mathbf{r}'|}, \quad \mathbf{b}(\mathbf{r},t) = \mathbf{0}. \tag{2.12}$$

An expression for the acceleration (2.6) is obtained in terms of $N(\mathbf{X},t)$ when Eq. (2.12) is substituted into (2.7). Equation (2.4) may thus be rewritten as

$$\left[\frac{\partial}{\partial t} + \mathcal{L}(\mathbf{X}) - \int d\mathbf{X}'\mathcal{V}(\mathbf{X},\mathbf{X}')N(\mathbf{X}')\right]N(\mathbf{X}) = 0. \tag{2.13}$$

Here $\mathcal{L}(\mathbf{X})$ is a single-particle operator defined by

$$\mathcal{L}(\mathbf{X}) = \mathbf{v}\cdot\frac{\partial}{\partial \mathbf{r}} + \frac{Ze}{m}\left[\mathbf{E}_{\text{ext}}(\mathbf{r},t) + \frac{\mathbf{v}}{c}\times\mathbf{B}_{\text{ext}}(\mathbf{r},t)\right]\cdot\frac{\partial}{\partial \mathbf{v}} \tag{2.14}$$

and

$$\mathcal{V}(\mathbf{X},\mathbf{X}') = \frac{(Ze)^2}{m}\left(\frac{\partial}{\partial \mathbf{r}}\frac{1}{|\mathbf{r}-\mathbf{r}'|}\right)\cdot\frac{\partial}{\partial \mathbf{v}} \tag{2.15}$$

is a two-particle operator arising from the Coulomb interaction. Equation (2.13) is called the *Klimontovich equation*; this equation describes space-time evolution of the microscopic distribution function.

B. The Liouville Distribution

The fine-grained distribution function, although precise in describing microscopic configurations of the many-particle system, would not by itself correspond to the coarse-grained quantities which we observe in the macroscopic world. To establish a connection between them, we need to introduce an averaging process based on the Liouville distribution over the $6N$-dimensional phase space (the Γ space).

The microscopic state of the system is expressed in the Γ space by a point

$$\{\mathbf{X}_i\} \equiv (\mathbf{X}_1, \mathbf{X}_2, \ldots, \mathbf{X}_N),$$

which we shall call a *system point*. Following a formal procedure of the ensemble theory in statistical mechanics, we may imagine $\mathcal{N}$ replicas which are macroscopically identical to the system under consideration, i.e., the systems with the same "macroscopic realizations". The choice of the number $\mathcal{N}$ is arbitrary, so that we may let it approach infinity whenever convenient.

Although identical from the macroscopic point of view, these replicas are generally different in their microscopic configurations; the system points are scattered over the Γ space. We may then define the Liouville distribution function $D(\{\mathbf{X}_i\};t)$ in the Γ space according to

$$D(\{\mathbf{X}_i\};t)d\{\mathbf{X}_i\} \equiv \lim_{\mathcal{N}\to\infty} \frac{\mathcal{N}_S}{\mathcal{N}},$$

where $\mathcal{N}_S$ refers to the number of those system points contained in an infinitesimal volume $d\{\mathbf{X}_i\}$ in the Γ space around $\{\mathbf{X}_i\}$. By definition, it satisfies the normalization condition

$$\int d\{\mathbf{X}_i\}D(\{\mathbf{X}_i\};t) = 1 .$$

The $\mathcal{N}$ system points distributed in the Γ space apparently do not interact each other; they behave like an ideal gas. The distribution function $D(\{\mathbf{X}_i\};t)$ thus satisfies a grand continuity equation of Liouville type:

$$\frac{\partial D}{\partial t} + \{\dot{\mathbf{X}}_i\} \cdot \frac{\partial D}{\partial \{\mathbf{X}_i\}} = 0 . \tag{2.16}$$

Along a trajectory in the phase space, the probability density is conserved (Liouville theorem).

With the aid of the Liouville distribution, we may now introduce a statistical average of a fine-grained quantity $A(\mathbf{X},\mathbf{X}',\ldots;\{\mathbf{X}_i\})$, defined at a set of points $(\mathbf{X},\mathbf{X}'\ldots)$ in the six-dimensional phase space, in the following way.

$$\langle A(\mathbf{X},\mathbf{X}',\ldots;t)\rangle = \int d\{\mathbf{X}_i\}D(\{\mathbf{X}_i\};t)A(\mathbf{X},\mathbf{X}',\ldots;\{\mathbf{X}_i\}) . \tag{2.17}$$

In view of the conservation property, this average may equivalently be transformed into an average over the initial distribution, so that

$$\langle A(\mathbf{X},\mathbf{X}',\ldots;t)\rangle$$

$$= \int d\{\mathbf{X}_i(0)\}D(\{\mathbf{X}_i(0)\};0)A(\mathbf{X},\mathbf{X}',\ldots;\{\mathbf{X}_i(\{\mathbf{X}_i(0)\};t)\}) \tag{2.18}$$

where $\{\mathbf{X}_i(\{\mathbf{X}_i(0)\};t)\}$ represents the coordinates of the system points in the Γ space at t under the condition that it was located at $\{\mathbf{X}_i(0)\}$ when $t=0$.

C. Multiparticle Distributions

The microscopic distribution functions, consisting of a summation of N six-dimensional delta functions, would not by themselves correspond to the coarse-grained quantities observed in the macroscopic world. The necessary correspondence can be established through a statistical average of a quantity depending on such microscopic distributions.

A *single-particle distribution function* is thus defined as*

$$F_1(\mathbf{X}) = \langle N(\mathbf{X};t)\rangle\,, \tag{2.19}$$

where $\mathbf{X} \equiv (\mathbf{r},\mathbf{v})$ is a six-dimensional coordinate representing the position and velocity of a particle; $\langle \cdots \rangle$ denotes a statistical average over the Liouville distribution.

The *s-particle distribution functions*, $F_s(1,\ldots,s)$, may be defined analogously: Thus, the two- and three-particle distribution functions are introduced according to

$$F_2(1,2) = \langle N(1;t)N(2;t)\rangle - \delta(1-2)F_1(1)\,, \tag{2.20}$$

$$\begin{aligned} F_3(1,2,3) = {} & \langle N(1;t)N(2;t)N(3;t)\rangle - \delta(1-2)\delta(1-3)F_1(1) \\ & - \delta(1-2)F_2(2,3) - \delta(2-3)F_2(3,1) - \delta(3-1)F_2(1,2)\,. \end{aligned} \tag{2.21}$$

To avoid complex notation, we adopt a shorthand notation and use the numerals 1, 2, 3, etc. in place of $\mathbf{X}$, $\mathbf{X}'$, $\mathbf{X}''$, etc. By definition, the distribution functions are symmetric with respect to interchange between the particles and satisfy conservation laws:

$$\int d\mathbf{X}_1 F_1(1) = N\,, \tag{2.22a}$$

$$\int d\mathbf{X}_2 F_2(1,2) = (N-1)F_1(1)\,, \tag{2.22b}$$

$$\int d\mathbf{X}_3 F_3(1,2,3) = (N-2)F_2(1,2)\,, \tag{2.22c}$$

and so on.

---

* As in Eqs. (2.19)–(2.21), the time variables are suppressed in the distribution functions for the remainder of this chapter for simplicity; each distribution function contains only a single time-variable.

The single-particle distribution describes the kinetic properties of the plasma under consideration. Higher-order distributions then characterize features of particle interaction. The interaction energy per unit volume of the OCP, for example, is calculated in terms of the two-particle distribution as

$$U_{\rm int}=\frac{(Ze)^2}{2V}\int_V d\mathbf{r}_1\int_V d\mathbf{r}_2\frac{1}{r_{12}}\left[\int d\mathbf{v}_1\int d\mathbf{v}_2F_2(1,2)-\frac{N(N-1)}{V^2}\right], \qquad (2.23)$$

where $r_{ij}=|\mathbf{r}_i-\mathbf{r}_j|$; the subtraction of $N(N-1)/V^2$ in the square brackets stems from the background charges in the OCP.

D. The BBGKY Hierarchy

The many-particle distribution functions $F_s(1,2,\ldots,s)$ have been defined and calculated in terms of the statistical average of products of the Klimontovich functions over the Liouville distribution; examples have been given in Eqs. (2.19)–(2.21). We now study the equations governing the evolution of those distribution functions.

The Klimontovich equation (2.13) may thus be written as

$$\left[\frac{\partial}{\partial t}+\mathcal{L}(1)-\int d\mathbf{X}_2\mathcal{V}(1,2)N(2;t)\right]N(1;t)=0\,. \qquad (2.24)$$

We now carry out the Liouville average of this equation. As Eq. (2.18) illustrates, the averaging process commutes with the differential operators involved in (2.24). Hence

$$\left[\frac{\partial}{\partial t}+\mathcal{L}(1)\right]F_1(1)=\int d\mathbf{X}_2\mathcal{V}(1,2)F_2(1,2)\,. \qquad (2.25)$$

In this derivation, one uses a relation,

$$\int d\mathbf{X}_1\delta(1-2)\mathcal{V}(1,2)F_s(1,\ldots,s)=0\,, \qquad (2.26)$$

which can be proven from symmetry considerations.

We may likewise start from an equation

$$\left[\frac{\partial}{\partial t}+\mathcal{L}(1)+\mathcal{L}(2)\right]N(1;t)N(2;t)$$

$$=\int d\mathbf{X}_3[\mathcal{V}(1,3)+\mathcal{V}(2,3)]N(1;t)N(2;t)N(3;t) \qquad (2.27)$$

derivable from a combination of the Klimontovich equations. Upon averaging this equation with respect to the Liouville distribution and making use of (2.25) and (2.26), we obtain

$$\left\{\frac{\partial}{\partial t} + \mathcal{L}(1) + \mathcal{L}(2) - [\mathcal{V}(1,2) + \mathcal{V}(2,1)]\right\} F_2(1,2)$$

$$= \int d\mathbf{X}_3[\mathcal{V}(1,3) + \mathcal{V}(2,3)]F_3(1,2,3)\,. \tag{2.28}$$

One can analogously extend the formalism to consider a product of an arbitrary number of the Klimontovich functions and carry out the statistical average of the resulting equation. We thus obtain the *BBGKY hierarchy equations* for the multiparticle distributions:

$$\left[\frac{\partial}{\partial t} + \sum_{i}^{s} \mathcal{L}(i) - \sum_{i\neq j}^{s} \mathcal{V}(i,j)\right] F_s(1,\ldots,s)$$

$$= \sum_{i}^{s} \int d\mathbf{X}_{s+1}\, \mathcal{V}(i,s+1)F_{s+1}(1,\ldots,s,s+1)\,. \tag{2.29}$$

Here $\mathcal{L}(i)$ is the single-particle operator (2.14) acting upon the ith particle under the influence of externally applied fields, $\mathbf{E}_{\text{ext}}(\mathbf{r},t)$ and $\mathbf{B}_{\text{ext}}(\mathbf{r},t)$; $\mathcal{V}(i,j)$ is the two-particle operator (2.15) representing the Coulomb interaction. These coupled set of equations provide a basis for the kinetic theory of plasmas.

In thermodynamic equilibrium, dependence of the distributions on space and velocity variables can be separated. For a homogeneous system one writes in terms of the Maxwellian velocity distribution (1.1) as

$$F_{1(1)} = nf_{\mathrm{M}}(\mathbf{v}_1)\,, \tag{2.30a}$$

$$F_2(1,2) = \frac{N(N-1)}{V^2} f_{\mathrm{M}}(\mathbf{v}_1)f_{\mathrm{M}}(\mathbf{v}_2)g(r_{12})\,, \tag{2.30b}$$

$$F_3(1,2,3) = \frac{N(N-1)(N-2)}{V^3} f_{\mathrm{M}}(\mathbf{v}_1)f_{\mathrm{M}}(\mathbf{v}_2)f_{\mathrm{M}}(\mathbf{v}_3)g_3(r_{12},r_{23},r_{31})\,. \tag{2.30c}$$

The spatial part, $g(r_{12})$, of the two-particle distribution is usually referred to as the *radial distribution function* [e.g., Hansen & McDonald 1986]. The density of Coulomb-interaction energy, Eq. (2.23), is thus calculated as

$$U_{\text{int}} = \frac{(Zen)^2}{2} \int d(\mathbf{r}_1 - \mathbf{r}_2) \frac{1}{r_{12}} [g(r_{12}) - 1] , \tag{2.31}$$

where $N >> 1$ has been assumed.

Substituting Eqs. (2.30) in the second ($s = 2$) equation of the BBGKY hierarchy (2.29), one obtains the *Born-Green equation* [Born & Green 1949] for an OCP:

$$\frac{\partial g(r_{12})}{\partial \mathbf{r}_1} = -\frac{(Ze)^2}{T} \left[\frac{\partial}{\partial \mathbf{r}_1} \frac{1}{r_{12}} \right] g(r_{12})$$

$$- \frac{(Ze)^2 n}{T} \int d\mathbf{r}_3 \left[\frac{\partial}{\partial \mathbf{r}_1} \frac{1}{r_{13}} \right] g_3(r_{12}, r_{23}, r_{31}) . \tag{2.32}$$

For a weakly coupled ($\Gamma << 1$) OCP, an approximate solution to this equation, applicable for both long and short ranges, has been obtained [O'Neil & Rostoker 1965; Totsuji & Ichimaru 1973] as

$$g(r) = \exp \left[-\frac{(Ze)^2}{Tr} \exp(-k_{\mathrm{D}} r) \right] . \tag{2.33}$$

2.2 THE VLASOV EQUATION

The BBGKY hierarchy (2.29) has a structure such that it does not close in itself: The equation for single-particle distribution depends on the two-particle distribution, the equation for two-particle distribution in turn requires knowledge of three-particle distribution, and so forth. We must therefore find a method of truncating such an infinite series of equations.

Such a truncation may be achieved if the fluid limit of Sec. 1.4C is approached. Since the discreteness parameter Δ defined by Eq. (1.54) takes on an extremely small value for an ordinary gaseous plasma, the BBGKY hierarchy may be solved through a power-series expansion with respect to the discreteness parameter. Since the external fields, $\mathbf{E}_{\text{ext}}(\mathbf{r},t)$ and $\mathbf{B}_{\text{ext}}(\mathbf{r},t)$, remain invariant in the fluid limit, these quantities may be regarded as zeroth order in the discreteness parameters. The operator $\mathcal{L}(i)$ may likewise be of the zeroth order, *if we treat only those phenomena where length, time, and thus velocity are measured in the scales of* λ_{D}, ω_p^{-1}, *and* $\sqrt{T/m}$. By the same token, the two-particle operator $\mathcal{V}(i,j)$ is viewed as on the first order in the discreteness.

Thus, we expand

$$F_s = F_s^{(0)} + F_s^{(1)} + F_s^{(2)} + \cdots . \tag{2.34}$$

In light of the relations (2.4), to the lowest order in the discreteness parameters, the BBGKY hierarchy is simplified as

$$\left[\frac{\partial}{\partial t}+\sum_{i=1}^{s}\mathcal{L}(i)\right]F_s^{(0)}(1,\ldots,s)$$

$$=\sum_{i=1}^{s}\int d\mathbf{X}_{s+1}\mathcal{V}(i,s+1)F_{s+1}^{(0)}(1,\ldots,s,s+1)\,. \qquad (2.35)$$

This hierarchy may be truncated through an *ansatz*

$$F_s^{(0)}(1,\ldots,s)=\prod_{i=1}^{s}[nf(i)]\,, \qquad (2.36)$$

a simple product of s single-particle distribution functions. It is not difficult then to find that Eq. (2.36) indeed satisfies Eq. (2.35) if the single-particle distribution function is determined as a solution to the equation

$$\left[\frac{\partial}{\partial t}+\mathcal{L}(1)-n\int d\mathbf{X}_2\mathcal{V}(1,2)f(2)\right]f(1)=0\,. \qquad (2.37)$$

This is called the *Vlasov* equation [Vlasov 1938, 1967]. As may be clear from the foregoing derivation, the Vlasov equation offers a precise description of the plasma in the fluid limit as long as the validity of the factorization in the form of (2.36) is assured in the same limit.

It is interesting to note a formal similarity between the Vlasov equation (2.37) and the Klimontovich equation (2.13) in Sec. 2.1A. Yet it is quite important also to note the fundamental difference in the physical contents between the two equations: While the Klimontovich equation deals with the microscopic distribution function (2.2), containing all the fine structures arising from the individuality of the particles, the Vlasov equation is concerned with a coarse-grained distribution function obtained from a statistical average of the microscopic distribution function. The fluctuations due to discreteness of the particles have not been retained in the Vlasov equation.

The Vlasov equation may be rewritten in a little more conventional form when Eqs. (2.14) and (2.15) are substituted into Eq. (2.37). We may then extend the equation to those general cases of multicomponent plasmas. Denoting by $f_\sigma(\mathbf{r},\mathbf{v};t)$ the single-particle distribution function for the particles of σ species normalized to unity, we have

$$\left\{\frac{\partial}{\partial t}+\mathbf{v}\cdot\frac{\partial}{\partial \mathbf{r}}+\frac{Z_\sigma e}{m_\sigma}\left[\mathbf{E}(\mathbf{r},t)+\frac{\mathbf{v}}{c}\times\mathbf{B}(\mathbf{r},t)\right]\cdot\frac{\partial}{\partial \mathbf{v}}\right\}f_\sigma(\mathbf{r},\mathbf{v},t)=0\,, \qquad (2.38)$$

where

$$\mathbf{E}(\mathbf{r},t)=\mathbf{E}_{\text{ext}}(\mathbf{r},t)-\sum_\sigma Z_\sigma e n_\sigma \frac{\partial}{\partial \mathbf{r}}\int d\mathbf{r}'\int d\mathbf{v}'\frac{f_\sigma(\mathbf{r}',\mathbf{v}',t)}{|\mathbf{r}-\mathbf{r}'|}\,, \qquad (2.39)$$

$$\mathbf{B}(\mathbf{r},t)=\mathbf{B}_{\text{ext}}(\mathbf{r},t)\,.$$

Equation (2.38) has been given a number of different names in addition to the Vlasov equation. It has been called the *time-dependent Hartree equation*, because the last term of (2.39) in fact represents the electrostatic Hartree field produced by the average distribution of charged particles; in Eq. (2.38), particle interactions have been taken into account only through such an average self-consistent field. We may also note in this connection that Eq. (2.36) corresponds to the *Hartree factorization*.

The electromagnetic fields, as Eqs. (2.39) stand, enable us to take into account only the electrostatic interactions between the particles. We can easily extend the equations, however, so that the electromagnetic interactions may also be included in the Vlasov formalism. The modification required is to use in Eqs. (2.39) those electromagnetic fields determined self-consistently from the charge and current distributions arising from $f_\sigma(\mathbf{r},\mathbf{v};t)$, as in Eqs. (2.7). We shall consider such problems in later chapters.

Equation (2.38) may also be called the *collisionless Boltzmann equation*, since the terms describing the interparticle collisions are apparently missing on its right-hand side. Collisional effects arise as a consequence of the field fluctuations away from the average Hartree field; in the fluid limit, such fluctuations, being on the order of Δ, are to be neglected, as we have observed in Sec. 1.4C through an examination of discreteness in the collision frequency, Eq. (1.28).

Because of such a failure to include the collisional effects, the Vlasov equation cannot describe an approach of the thermodynamic equilibrium. Indeed, in the absence of the external fields ($\mathbf{E}_{\text{ext}}=\mathbf{0}$, $\mathbf{B}_{\text{ext}}=\mathbf{0}$), one can prove that any velocity distribution function, uniform in space and stationary in time, can satisfy Eq. (2.38); the average Hartree fields always vanish in such a homogeneous situation because of the overall neutrality of the system. To obtain a kinetic theory of the plasma capable of describing an approach to equilibrium, one must depart from the fluid limit and include the effects arising from the discreteness of the particles. We shall study some of those aspects in the following sections.

2.3 EVOLUTION OF CORRELATIONS

Effects of collisions in a plasma depend on interparticle correlations, which were not taken into account in the Hartree factorization of Eq. (2.36). In this section, we investigate how such correlations evolve in weakly coupled plasmas, where an ordering with respect to the infinitesimal discreteness parameter is a valid concept.

A. Pair Correlation Function and Collision Term

In conjunction with the s-particle distribution functions $F_s(1, \ldots, s)$ considered in Section 2.1A, it is useful to introduce the s-particle correlation functions, $C_s(1, \ldots, s)$, along the lines analogous to cluster expansions in the theory of liquids [e.g., Mayer & Mayer 1940; Hansen & McDonald 1986]. These functions are successively defined and calculated according to

$$F_1(1) = nC_1(1)\,, \tag{2.40a}$$

$$F_2(1,2) = n^2[C_1(1)C_1(2) + C_2(1,2)]\,, \tag{2.40b}$$

$$F_3(1,2,3) = n^3[C_1(1)C_1(2)C_1(3) + C_1(1)C_2(2,3) + C_1(2)C_2(3,1) + C_1(3)C_2(1,2) + C_3(1,2,3)]\,, \tag{2.40c}$$

and so on. Physically, $C_s(1, \ldots, s)$ represent the totally correlated parts of $F_s(1, \ldots, s)$. The correlation functions are also symmetric with respect to interchange of two particles.

The factorization (2.36) was assumed to be valid in the fluid limit. It follows that $C_2(1,2)/C_1(1)C_1(2)$ should vanish in the same limit. Let us then assume that this ratio is on the order of Δ, Eq. (1.54), that is,

$$\frac{C_2(1,2)}{C_1(1)C_1(2)} \approx \Delta\,. \tag{2.41a}$$

We may also assume a systematic trend, so that

$$\frac{C_3(1,2,3)}{C_1(1)C_1(2)C_1(3)} \approx \Delta^2\,. \tag{2.41b}$$

Although these are the assumptions commonly adopted for a kinetic theory of weakly coupled plasmas, we must be aware that in some instances those assumptions are simply invalid. A most typical example is the short-range correlations: As Eq. (2.33) illustrates, the pair correlation function $C_2(1,2)$

assumes a value on the same order of magnitude as the product $C_1(1)C_1(2)$ of the single-particle distribution functions at short distances. Short-range correlations cannot be correctly analyzed through a theory based on the assumptions (2.41) even for those plasmas with $\Delta << 1$. In a strongly coupled plasma ($\Gamma \geq 1$), the correlation effects naturally play a major part (cf. Chapter 7).

With these reservations in mind, let us assume (2.41) and study the consequences resulting from them. We substitute Eqs. (2.40) into the first two equations of the BBGKY hierarchy (2.29) and retain the terms up to the first order in the discreteness parameters. The first equation thus reads

$$\left[\frac{\partial}{\partial t}+\mathcal{L}(1)-n\int dX_2\mathcal{V}(1,2)C_1(2)\right]C_1(1)=n\int d\mathbf{X}_2\mathcal{V}(1,2)C_2(1,2)\,. \quad (2.42)$$

The left-hand side of this equation is identical to that of the Vlasov equation, (2.37). Equation (2.42) contains, on its right-hand side, an additional term depending on the pair correlation between particles. This new term is of the first order in the discreteness parameters; it is a *collision term* describing the collisional effects between individual particles. The second equation of the BBGKY hierarchy contains C_1 and C_2; the terms involving C_3 as well as some others drop out because they are of the second order in the discreteness. The equation is then simplified with the aid of Eq. (2.42), and becomes

$$\left\{\frac{\partial}{\partial t}+\mathcal{L}(1)+\mathcal{L}(2)-n\int d\mathbf{X}_3[\mathcal{V}(1,3)+\mathcal{V}(2,3)]C_1(3)\right\}C_2(1,2)$$

$$-n\int d\mathbf{X}_3\mathcal{V}(1,3)C_1(1)C_2(2,3)-n\int d\mathbf{X}_3\mathcal{V}(2,3)C_1(2)C_2(3,1)$$

$$=[\mathcal{V}(1,2)+\mathcal{V}(2,1)]C_1(1)C_1(2)\,. \quad (2.43)$$

The left-hand side of this equation is homogeneous with respect to C_2. The right-hand side may be regarded as a driving term, which acts to create the particle correlations through Coulomb interactions. Equations (2.42) and (2.43) thereby form a set of coupled equations between the single-particle and pair correlation functions.

A formal solution to Eq. (2.43) is not difficult to obtain: Recovering the time arguments, we write

$$C_2(1,2;t)=\int d1'\int d2'\int_{-\infty}^{t}dt'\,\{[\mathcal{V}(1',2')+\mathcal{V}(2',1')]C_1(1';t')C_1(2';t')$$

$$\times U(1,1';t-t')U(2,2';t-t')\}\,. \quad (2.44)$$

By so doing, we have expressed the correlation between 1 and 2 (in the phase space) at time t as a superposition of the correlation effects created between 1′ and 2′ at t' which propagate to 1 and 2 during the time interval $t-t'$. It is thus natural to regard the function U as a particle propagator in the pair correlation function. The particular choice of integration domain $(-\infty,t)$ for t' reflects the notion of *causality*: The correlation at t arises as a consequence of the effects accumulated during the time interval $(-\infty,t)$, but *not after that.* A further discussion on this subject will be given in conjunction with the response-function formalism in Chap. 3.

Substitution of Eq. (2.44) in Eq. (2.43) should reveal that (2.44) is a solution to (2.43) *if* the propagator U satisfies the equation

$$\left[\frac{\partial}{\partial t}+\mathcal{L}(1)-n\int d\mathbf{X}_2\mathcal{V}(1,2)C_1(2;t)\right]U(1,1';t-t')$$
$$-n\int d\mathbf{X}_2\mathcal{V}(1,2)C_1(1;t)U(2,1';t-t')=0 \tag{2.45}$$

with the initial condition

$$U(1,1';0)=\delta(1-1')\,. \tag{2.46}$$

Let us note that Eq. (2.45) is a linearized version of the Vlasov equation. In fact, if we make a replacement, $f\to f+\delta f$, in Eq. (2.37) and linearize it with respect to δf, then we obtain (2.45) in which δf occupies the places of U; the propagator U obeys the *linearized Vlasov equation.*

Although we have succeeded in finding a formal solution for evolution of the pair correlation function, this is perhaps almost as far as we can proceed at the moment. We are still left with a coupled problem in that a solution to the linearized Vlasov equation (2.45) for U requires a knowledge for the time-dependent behavior of C_1 and that the time evolution of C_1 in turn depends on U through C_2 via Eq. (2.42). To resolve such a coupling, we need to introduce additional physical arguments, due to Bogoliubov, on the characteristic times in the dynamical processes.

B. Bogoliubov's Hierarchy of Characteristic Times

When treating dynamic problems in statistical physics, one often encounters a situation where a well-defined hierarchy exists among the magnitudes of various characteristic times, as noted by Bogoliubov [1962].

We begin with the relaxation time τ_0 associated with a hydrodynamic quantity; it may be estimated as a macroscopic distance L divided by the sound velocity c_s, the characteristic velocity in hydrodynamics:

$$\tau_0 \approx \frac{L}{c_s} . \qquad (2.47)$$

The characteristic time τ_1 for the single-particle distribution function to relax to its local-equilibrium values may be calculated as the mean free path l divided by the mean velocity of the particles, $\langle v \rangle \approx \sqrt{T/m} \approx c_s$:

$$\tau_1 \approx \frac{l}{c_s} . \qquad (2.48)$$

For slowly varying disturbances $L >> l$, one finds

$$\tau_0 >> \tau_1 . \qquad (2.49)$$

We may similarly consider a characteristic time τ_2 associated with the pair correlation function: It may be estimated as an average time for a particle to travel over a correlation distance. For the plasma under consideration, the correlation distance may be given by the Debye length; hence

$$\tau_2 \approx \frac{\lambda_D}{c_s} . \qquad (2.50)$$

Comparison between τ_1 and τ_2 may be facilitated if we realize that $1/\tau_1 \approx \nu_m$, the frequency of Coulomb collisions given by Eq. (1.28), and that $1/\tau_2 \approx \omega_p$, the plasma frequency of Eq. (1.47). We thus find

$$\tau_1 >> \tau_2 \qquad (2.51)$$

as long as the discreteness parameter (1.54) remains infinitesimally small.

The inequality (2.51) implies that the pair correlation functions relax much faster than the single-particle distribution functions. The slower processes associated with the evolution of the latter functions are sometimes referred to as the *kinetic processes*. In such a kinetic stage of development, the pair correlation functions have already completed their own relaxation; these functions must therefore be expressible as functionals of the single-particle distribution functions. In fact, one of the most important conclusions from these hierarchy considerations is that the pair and all higher-order correlation functions can be written as functionals of single-particle distribution functions in the kinetic processes.

C. The Vlasov Propagator

The conclusion reached in the previous section implies that the single-particle distribution function may be regarded as independent of time as far as the evolution of the pair correlation function goes. Accordingly the time dependence of C_1 may be neglected in Eq. (2.45). We then have a manageable problem in which coupling of time evolutions between C_1 and C_2 has been disentangled.

For simplicity, we consider a uniform plasma in the absence of external fields ($\mathbf{E}_{\text{ext}} = \mathbf{0}, \mathbf{B}_{\text{ext}} = \mathbf{0}$). Then we substitute

$$\mathcal{L}(1) = \mathbf{v}_1 \cdot \frac{\partial}{\partial \mathbf{r}_1}, \tag{2.52}$$

$$C_1(1;t) = f(\mathbf{v}_1), \tag{2.53}$$

in Eq. (2.45); $f(\mathbf{v})$ is the velocity distribution function with normalization

$$\int d\mathbf{v} f(\mathbf{v}) = 1 . \tag{2.54}$$

Taking account of the macroscopic neutrality of the system, we have

$$\left(\frac{\partial}{\partial t} + \mathbf{v}_1 \cdot \frac{\partial}{\partial \mathbf{r}_1}\right) U(1,1';t) - n \int d\mathbf{X}_2 \mathcal{V}(1,2) f(\mathbf{v}_1) U(2,1';t) = 0 . \tag{2.55}$$

With the initial condition (2.46), this equation may be solved through the Fourier-Laplace transformations.

For a translationally invariant system, the propagator $U(1,1';t)$ depends only on $\mathbf{r}_1 - \mathbf{r}'_1$ in space; we carry out a spatial Fourier transformation of the propagator with respect to $\mathbf{r}_1 - \mathbf{r}'_1$ as in the footnote of p. 19:

$$U_{\mathbf{k}}(\mathbf{v}_1, \mathbf{v}'_1;t) = \int_V d(\mathbf{r}_1 - \mathbf{r}'_1) U(1,1';t) \exp\left[-i\mathbf{k} \cdot (\mathbf{r}_1 - \mathbf{r}'_1)\right] . \tag{2.56}$$

We next introduce a one-sided Fourier transformation of $U_{\mathbf{k}}(\mathbf{v}_1, \mathbf{v}'_1;t)$ with respect to time via

$$\mathcal{U}_{\mathbf{k}}(\mathbf{v}, \mathbf{v}'_1;\omega) = \int_0^\infty dt U_{\mathbf{k}}(\mathbf{v}_1, \mathbf{v}'_1;t) \exp(i\omega t) . \tag{2.57a}$$

It follows that the function $\mathcal{U}_{\mathbf{k}}(\mathbf{v}_1,\mathbf{v}'_1;\omega)$ is analytic in the upper half of the complex ω plane; a positive imaginary part of ω may be sufficient to guarantee convergence of the integration (2.57a). $\mathcal{U}_{\mathbf{k}}(\mathbf{v}_1,\mathbf{v}'_1;\omega)$ is then continued analytically into the lower half of the ω plane; there it generally has singularities such as poles. Inverse transformation of (2.57a) is written as

$$U_{\mathbf{k}}(\mathbf{v}_1,\mathbf{v}'_1;t)=\frac{1}{2\pi}\int_{C+} d\omega\,\mathcal{U}_{\mathbf{k}}(\mathbf{v}_1,\mathbf{v}'_1;\omega)\exp(-i\omega t)\,, \tag{2.57b}$$

where the contour $C+$ extends from $-\infty$ to $+\infty$ along a path in the upper half of the ω plane in such a way that all the singularities lie below it. For $t>0$, we can close the contour with an infinite semicircle in the lower half-plane; Cauchy's theorem can be used for the inverse transformation. For $t<0$, we close the contour with an infinite semicircle in the upper half-plane; we then have

$$U_{\mathbf{k}}(\mathbf{v}_1,\mathbf{v}'_1;t)=0 \quad (t<0)\,. \tag{2.58}$$

As we see from the foregoing argument, the one-sided Fourier transformation is equivalent to the Laplace transformation; this transformation makes it possible to take explicit account of the initial conditions in a solution to a differential equation.

Let us thus carry out the Fourier-Laplace transformations to Eq. (2.55) with the initial condition (2.46). Since the operator $\mathcal{V}(1,2)$ may be expanded as

$$\mathcal{V}(1,2)=\frac{\omega_p^2}{n}\sum_{\mathbf{k}}\frac{i\mathbf{k}}{k^2}\cdot\frac{\partial}{\partial\mathbf{v}_1}\exp[i\mathbf{k}\cdot(\mathbf{r}_1-\mathbf{r}_2)]\,, \tag{2.59}$$

we have

$$\mathcal{U}_{\mathbf{k}}(\mathbf{v}_1,\mathbf{v}'_1;\omega)=\frac{i}{\omega-\mathbf{k}\cdot\mathbf{v}_1}\delta(\mathbf{v}_1-\mathbf{v}'_1)$$

$$-\frac{\omega_p^2}{k^2}\frac{1}{\omega-\mathbf{k}\cdot\mathbf{v}_1}\mathbf{k}\cdot\frac{\partial}{\partial\mathbf{v}_1}f(\mathbf{v}_1)\int d\mathbf{v}_2\,\mathcal{U}_{\mathbf{k}}(\mathbf{v}_2,\mathbf{v}'_1;\omega)\,. \tag{2.60}$$

Upon integrating both sides with respect to $\mathbf{v}_1$, we find

$$\int d\mathbf{v}_1\,\mathcal{U}_{\mathbf{k}}(\mathbf{v}_1,\mathbf{v}'_1;\omega)=\frac{i}{(\omega-\mathbf{k}\cdot\mathbf{v}'_1)\varepsilon(\mathbf{k},\omega)} \tag{2.61}$$

where

$$\varepsilon(\mathbf{k},\omega) = 1 + \frac{\omega_p^2}{k^2}\int d\mathbf{v}\,\frac{1}{\omega - \mathbf{k}\cdot\mathbf{v}}\,\mathbf{k}\cdot\frac{\partial}{\partial \mathbf{v}}f(\mathbf{v})\,. \tag{2.62}$$

This is the *Vlasov dielectric function*; it plays the essential part in describing the collective behavior of the plasma. We shall study the detailed properties of this function and the physical consequences in the subsequent chapters.

Substituting Eq. (2.61) in Eq. (2.60), we finally obtain a solution,

$$\mathcal{U}_{\mathbf{k}}(\mathbf{v}_1,\mathbf{v}'_1;\omega) = \frac{i}{\omega - \mathbf{k}\cdot\mathbf{v}_1}\,\delta(\mathbf{v}_1 - \mathbf{v}'_1)$$

$$- i\,\frac{\omega_p^2}{k^2}\,\frac{1}{(\omega - \mathbf{k}\cdot\mathbf{v}_1)\,(\omega - \mathbf{k}\cdot\mathbf{v}'_1)\varepsilon(\mathbf{k},\omega)}\,\mathbf{k}\cdot\frac{\partial}{\partial \mathbf{v}_1}f(\mathbf{v}_1)\,, \tag{2.63}$$

which we shall call the *Vlasov propagator*. Clearly, the first term on the right-hand side represents a free-particle propagator; the second term takes account of the effects of Coulomb interaction.

2.4 THE BGL COLLISION TERM

Having thus solved the linearized Vlasov equation for the propagator, we are now in a position to evaluate the pair correlation function and thereby to calculate the explicit form for the collision term on the right-hand side of Eq. (2.42).

A. Derivation

To begin, we carry out the spatial Fourier transformation for the pair correlation function (2.44) in the same way as Eq. (2.56), with the aid of Eqs. (2.52) and (2.59):

$$C_{2\mathbf{k}}(\mathbf{v}_1,\mathbf{v}_2) = i\,\frac{\omega_p^2}{n}\int d\mathbf{v}'_1\int d\mathbf{v}'_2\int_0^\infty dt\,\frac{\mathbf{k}}{k^2}\cdot\left[\left(\frac{\partial}{\partial \mathbf{v}'_1} - \frac{\partial}{\partial \mathbf{v}'_2}\right)f(\mathbf{v}'_1)f(\mathbf{v}'_2)\right]$$

$$\times\, U_{\mathbf{k}}(\mathbf{v}_1,\mathbf{v}'_1;t)U_{-\mathbf{k}}(\mathbf{v}_2,\mathbf{v}'_2;t)\,. \tag{2.64}$$

In terms of this function, the collision term may be expressed as

$$\left.\frac{\partial f(\mathbf{v})}{\partial t}\right]_c = -i\omega_p^2\sum_{\mathbf{k}}\frac{\mathbf{k}}{k^2}\cdot\frac{\partial}{\partial \mathbf{v}}\int d\mathbf{v}'C_{2\mathbf{k}}(\mathbf{v},\mathbf{v}')\,. \tag{2.65}$$

Substituting Eq. (2.57b) in Eq. (2.64) and performing the time and a frequency integrations, we obtain

$$C_{2\mathbf{k}}(\mathbf{v}_1,\mathbf{v}_2) = i\frac{\omega_p^2}{2\pi n}\int d\mathbf{v}'_1 \int d\mathbf{v}'_2 \int_{-\infty}^{\infty} d\omega \frac{\mathbf{k}}{k^2}\cdot\left[\left(\frac{\partial}{\partial \mathbf{v}'_1}-\frac{\partial}{\partial \mathbf{v}'_2}\right) f(\mathbf{v}'_1)f(\mathbf{v}'_2)\right]$$

$$\times\, \mathcal{U}_{\mathbf{k}}(\mathbf{v}_1,\mathbf{v}'_1;\omega)\,\mathcal{U}_{-\mathbf{k}}(\mathbf{v}_2,\mathbf{v}'_2;-\omega)\,. \qquad (2.66)$$

In the calculations we have noted that $\mathcal{U}_{\mathbf{k}}(\mathbf{v},\mathbf{v}';\omega)$ vanishes in the limit, $|\omega| \to \infty$. The contour of ω integration in Eq. (2.66) extends from $-\infty$ to $+\infty$; it sees all the singularities of $\mathcal{U}_{\mathbf{k}}(\mathbf{v}_1,\mathbf{v}'_1;\omega)$ from above. The function $\mathcal{U}_{-\mathbf{k}}(\mathbf{v}_2,\mathbf{v}'_2;-\omega)$, which is complex conjugate to $\mathcal{U}_{\mathbf{k}}(\mathbf{v}_2,\mathbf{v}'_2;\omega)$, however, obeys different boundary conditions: It is an analytic function in the lower half of the complex ω plane; it is then continued analytically into the upper half-plane, where it may have singularities. The integration contour therefore sees all the singularities of $\mathcal{U}_{-\mathbf{k}}(\mathbf{v}_2,\mathbf{v}'_2;-\omega)$ from below.

An explicit expression for the collision term may be obtained from Eq. (2.65) with the aid of Eqs. (2.63) and (2.66). The calculations are lengthy and require an extensive use of the analytic properties mentioned above; these are left to the reader as an exercise (Problem 2.6). The final result is the *BGL collision term,*

$$\left.\frac{\partial f(\mathbf{v})}{\partial t}\right]_c = \frac{\pi\omega_p^4}{n}\sum_{\mathbf{k}}\frac{\mathbf{k}}{k^2}\cdot\frac{\partial}{\partial \mathbf{v}}\int d\mathbf{v}' \frac{\mathbf{k}}{k^2}\cdot\left[\left(\frac{\partial}{\partial \mathbf{v}}-\frac{\partial}{\partial \mathbf{v}'}\right) f(\mathbf{v})f(\mathbf{v}')\right]$$

$$\times\frac{\delta(\mathbf{k}\cdot\mathbf{v}-\mathbf{k}\cdot\mathbf{v}')}{|\varepsilon(\mathbf{k},\mathbf{k}\cdot\mathbf{v})|^2} \qquad (2.67)$$

derived independently by Balescu [1960], Guernsey [1960], and Lenard [1960]. The collision term of this form with $\varepsilon(\mathbf{k},\mathbf{k}\cdot\mathbf{v}) = 1$ was obtained first by Landau [1936, 1937].

For a multicomponent plasma, the BGL collision term can be written as

$$\left.\frac{\partial f_\sigma(\mathbf{v})}{\partial t}\right]_c = \frac{\pi\omega_\sigma^2}{n_\sigma}\sum_{\sigma'}\sum_{\mathbf{k}}\omega_{\sigma'}^2\frac{\mathbf{k}}{k^2}\cdot\frac{\partial}{\partial \mathbf{v}}\int d\mathbf{v}'$$

$$\times\frac{\mathbf{k}}{k^2}\cdot\left[\left(\frac{m_{\sigma'}}{m_\sigma}\frac{\partial}{\partial \mathbf{v}}-\frac{\partial}{\partial \mathbf{v}'}\right) f_\sigma(\mathbf{v})f_{\sigma'}(\mathbf{v}')\right]\frac{\delta(\mathbf{k}\cdot\mathbf{v}-\mathbf{k}\cdot\mathbf{v}')}{|\varepsilon(\mathbf{k},\mathbf{k}\cdot\mathbf{v})|^2} \qquad (2.68)$$

where

$$\varepsilon(\mathbf{k},\omega) = 1 + \sum_\sigma \frac{\omega_\sigma^2}{k^2} \int d\mathbf{v} \frac{1}{\omega - \mathbf{k}\cdot\mathbf{v}} \mathbf{k} \cdot \frac{\partial}{\partial \mathbf{v}} f_\sigma(\mathbf{v}) \,. \tag{2.69}$$

is the Vlasov dielectric function of the multicomponent plasma, and

$$\omega_\sigma^2 = \frac{4\pi n_\sigma (Z_\sigma e)^2}{m_\sigma} \tag{2.70}$$

is the square of the plasma frequency of σ species.

There are a number of different ways to derive the BGL collision term; the methods used by Balescu, Guernsey, and Lenard were not the same as the one described here. Later in Chapter 7, we shall return to further considerations of collision terms in the fluctuation-theoretic approach. An analysis on the basis of the Fokker-Planck equation (e.g., Thompson & Hubbard 1960; Hubbard, J. 1961a) will be presented in Chapter 8.

B. Properties

It is instructive to investigate the k dependence of the terms involved in Eq. (2.67): Apart from the contributions arising from $|\varepsilon(\mathbf{k},\mathbf{k}\cdot\mathbf{v})|^2$, the terms under the k summation are proportional to k^{-3}. Transforming the summation into integration with the phase space volume $k^2 dk$, we then find that Eq. (2.67) would diverge logarithmically at both the lower and the upper ends of the k integration if $\varepsilon(\mathbf{k},\mathbf{k}\cdot\mathbf{v}) = 1$ was assumed as in the Landau collision term.

The physical origin of divergence in the small k domain is the same as that discussed in Section 1.3A, related to cummulative scattering with small momentum transfers. The divergence in this domain has been cured in the BGL collision term, since $\varepsilon(\mathbf{k},\mathbf{k}\cdot\mathbf{v})$ is in fact proportional to k^{-2} in the small k regime; physically, this amounts to taking into account the long-range screening in the plasma.

Divergence in the large k regime, however, still remains in the BGL collision term; $\varepsilon(\mathbf{k},\mathbf{k}\cdot\mathbf{v})$ approaches unity in the limit of large k. Cure for such a divergence may be obtained from two physically distinct sources: the classical orbital-deflection effect characterized by the *Landau length*,

$$b_{\mathrm{L}} = \frac{(Ze)^2}{T}, \tag{2.71}$$

and the quantum-mechanical diffraction effect due to the thermal de Broglie wavelengths of Eq. (1.8).

Classically, the origin of the short-range divergence may be traced to the assumptions (2.41). Since momentum transfer between particles is large at short distances, the major effects come from discreteness of the particles; expansion in powers of the discreteness parameter does not lead to a valid description of the phenomena in this regime. We recall on the one hand that the BGL theory gives a convergent result for collisions with small momentum transfers. On the other hand, as we saw in Section 1.3A, a theory based on binary collisions provides a correct description for short-range collisions with large momentum transfers. The ranges of validity for these two theories, in fact, greatly overlap when $\Delta \ll 1$. Here, a possibility exists that these two approaches may be unified and a fully convergent theory may be constructed with a precise determination for the argument in the Coulomb logarithm. Such a possibility was first recognized by J. Hubbard [1961b]. Subsequently, it has been developed into a unified theory by Kihara and Aono [1963, 1971].

Wyld and Pines [1962] derived a quantum-mechanical collision term through a calculation of the transition probability via a Coulomb interaction screened by a dielectric function. Passing over to the classical limit ($\hbar \to 0$), they have thus shown that the BGL term may be recovered if the wave vector k is kept finite (and hence the momentum transfer $\hbar k$ vanishes), and that the ordinary Boltzmann collision term may be obtained if the momentum transfer is kept finite. Quantum-mechanical transport equations [e.g., Ichimaru & Tanaka 1985; Ichimaru, Iyetomi & Tanaka 1987] with inclusion of the diffraction effect will be considered later in Chapter 7.

The BGL collision term (2.67) exhibits a number of remarkable properties [Lenard 1960]:

(a) If $f(\mathbf{v})$ is positive for $t = 0$, it is positive for $t > 0$.
(b) The particle density is independent of time.
(c) The mean velocity is independent of time.
(d) The mean kinetic energy is independent of time.
(e) Any Maxwellian distribution is a stationary solution.
(f) As $t \to \infty$, any solution approaches a Maxwellian distribution. Therefore, these are the only stationary solutions.

To prove those properties, we find it convenient to rewrite Eq. (2.67) so that

$$\left.\frac{\partial f(\mathbf{v})}{\partial t}\right]_c = -\frac{\partial}{\partial \mathbf{v}} \cdot \mathbf{J}(\mathbf{v}) \tag{2.72}$$

with

$$\mathbf{J}(\mathbf{v}) = \int d\mathbf{v}'\mathsf{Q}(\mathbf{v},\mathbf{v}') \cdot \left[\left(\frac{\partial}{\partial \mathbf{v}} - \frac{\partial}{\partial \mathbf{v}'}\right) f(\mathbf{v}) f(\mathbf{v}')\right], \tag{2.73}$$

$$\mathsf{Q}(\mathbf{v},\mathbf{v}') = -\frac{\pi\omega_p^4}{n} \sum_{\mathbf{k}} \frac{\mathbf{kk}}{k^4} \frac{\delta(\mathbf{k}\cdot\mathbf{v} - \mathbf{k}\cdot\mathbf{v}')}{|\varepsilon(\mathbf{k},\mathbf{k}\cdot\mathbf{v})|^2}. \tag{2.74}$$

A proof of (a) proceeds as follows. If $f(\mathbf{v})$ is positive everywhere initially and becomes negative at some later time, there must be an instant at which its minimum value first becomes negative. At such a point we have these four conditions: (i) $f(\mathbf{v}) = 0$; (ii) $\partial f(\mathbf{v})/\partial \mathbf{v} = \mathbf{0}$; (iii) $\partial^2 f(\mathbf{v})/\partial \mathbf{v}\partial \mathbf{v}$ is a nonnegative definite tensor; and (iv) $\partial f(\mathbf{v})/\partial t < 0$. Taking account of the first two conditions, we have

$$\frac{\partial f(\mathbf{v})}{\partial t} = -\int d\mathbf{v}' f(\mathbf{v}')\mathsf{Q}(\mathbf{v},\mathbf{v}') : \frac{\partial^2 f(\mathbf{v})}{\partial \mathbf{v}\partial \mathbf{v}}.$$

Since $\mathsf{Q}(\mathbf{v},\mathbf{v}')$ is a negative definite tensor, we thus find that conditions (iii) and (iv) are incompatible with the kinetic equation.

Properties (b)–(d) are related to the conservation properties. Property (b) means

$$\frac{d}{dt}\int d\mathbf{v} f(\mathbf{v}) = 0,$$

which may be shown through a direct integration of (2.72). Similarly, properties (c) and (d) can be proved through calculations of $(d/dt)\int d\mathbf{v}\mathbf{v}f(\mathbf{v})$ and $(d/dt)\int d\mathbf{v}v^2 f(\mathbf{v})$, with the aid of partial integrations. In the process we note the symmetry $\mathsf{Q}(\mathbf{v},\mathbf{v}') = \mathsf{Q}(\mathbf{v}',\mathbf{v})$ and an identity $(\mathbf{v} - \mathbf{v}') \cdot \mathsf{Q}(\mathbf{v},\mathbf{v}') = \mathbf{0}$.

Property (e) can be proved by a direct substitution. Property (f) is related to the H theorem of a standard kinetic theory [e.g., ter Haar 1960]. Proof of the properties (e) and (f) is left to the reader as an exercise (Problem 2.7).

It is worth noticing that the BGL collision term conserves the mean kinetic energy only and does not concern itself with the interaction energy. This stems from the fact that the theory has taken into account only the lowest-order effects in the discreteness parameters. As we noted in Eqs. (1.30) and (1.54), the average interaction energy is smaller by a factor of the discreteness parameter than the average kinetic energy. To construct a theory which would conserve the sum of both kinetic and interaction energies through the collision term, we must consider the higher-order correlation effects which are not included in the BGL theory; we shall do this later starting with Chapter 7.

PROBLEMS

2.1 Show that the internal energy per unit volume of an OCP, Eq. (2.23), can also be calculated as

$$U_{\text{int}} = \sum_{\mathbf{k}}{}' \frac{2\pi (Ze)^2}{k^2} \left[\langle \rho_{\mathbf{k}}(t) \rho_{-\mathbf{k}}(t) \rangle - n \right],$$

where $\rho_{\mathbf{k}}(t)$ is the Fourier component of the density fluctuations given by Eq. (1.40).

2.2 Derive Eq. (2.32).

2.3 Substitute Eq. (2.33) in Eq. (2.31), and show

$$\frac{U_{\text{int}}}{nT} = -\frac{\sqrt{3}}{2}\Gamma^{3/2} - 3\Gamma^3 \left[\frac{3}{8}\ln(3\Gamma) + \frac{\gamma}{2} + \frac{1}{8}\ln\frac{4}{9} - \frac{3}{8} \right]$$

($\gamma = 0.57721 \cdots$) in the expansion for $\Gamma << 1$. The formula (2.33) therefore leads to an interaction energy deviating slightly from the expansion Eq. (1.30) in the terms proportional to Γ^3.

2.4 When a single-particle distribution $f_\sigma(\mathbf{v})$ in a constant uniform magnetic field is a function of only $v_{\parallel}$ and $v_{\perp}$, the components of the velocity parallel and perpendicular to the magnetic field, show that it is a stationary solution to the Vlasov equation in the absence of an external perturbation.

2.5 Derive Eq. (2.60).

2.6 Carry out the calculations leading from Eq. (2.65) to Eq. (2.67).

2.7 Prove the properties (e) and (f) in Section 2.4B.

CHAPTER 3

Plasmas as Dielectric Media

Plasma has been defined as a statistical ensemble of mobile charged particles; these charged particles move randomly, interact with each other via electromagnetic forces, and respond to electromagnetic disturbances applied from external sources. A plasma is inherently capable of sustaining a rich class of electromagnetic phenomena.

A proper description of such an electromagnetic phenomenon may be obtained if one knows how a plasma will respond macroscopically to a given electromagnetic disturbance; with such knowledge, a complete scheme for describing the electromagnetic properties may be constructed. The functions characterizing those responses therefore play a central part in plasma electrodynamics; all the properties of plasma as a macroscopic medium are contained in the response functions.

The microscopic features of particle interactions in plasma are not lost in such a macroscopic description either. They are incorporated in the calculations of the response functions through the ways in which the particles, mutually interacting, adjust themselves to external disturbances. This approach may thus resemble that of a "black box"; those properties of plasma that depend on the nature of microscopic interactions between particles are revealed to the external observer through the course in which the plasma responds and adjusts itself to the external disturbance.

Since our interest lies in electromagnetic phenomena, it may be appropriate to select an electric field as an external disturbance to the plasma; typically then the response appears in the form of an induced current density in the plasma. The relationship between these two vectorial quantities is generally expressed by a conductivity tensor. If we further take into account the displacement current, the force-current relationship is more compactly expressed in terms of the dielectric tensor.

The main objective of this chapter is to lay a theoretical foundation for the linear-response formalism and the electrodynamic description of plasmas as continuous media: We shall first introduce the linear response functions and outline the fluctuation-dissipation theorem. Combining them, we establish a rigorous description for the fluctuations and correlations in plasmas under thermodynamic equilibria. A dielectric formulation for plasma is then obtained in terms of the density response functions. The dielectric tensor is defined subsequently; this will set the macroscopic electrodynamic equations with a solution in the form of the dispersion relations. The dispersion relations characterize the nature of longitudinal and transverse excitations in a dielectric medium. The dielectric tensors for Vlasov plasmas are explicitly calculated, both in the absence and in the presence of an external magnetic field. Finally the dielectric formulation for an electron gas at finite temperatures will be set forth within the framework of the random-phase approximation (RPA); strong coupling effects beyond the RPA are also described.

3.1 LINEAR RESPONSE FORMALISM

The linear response functions are introduced in terms of the functional derivatives of physically observable quantities. Rigorous descriptions of the fluctuations and correlations are then presented with the fluctuation-dissipation theorem. Finally the specific cases with the density-density response functions and the compressibility sum rules are treated in conjunction with the thermodynamic sum rules for the relaxation functions.

A. Linear Response Functions

Let $O[\varphi_{\text{ext}}(t)]$ be the expectation value of an observable in the plasma expressed as a functional of an external field, $\varphi_{\text{ext}}(t)$. Consider further an application of infinitesimal disturbance, $\delta\varphi_{\text{ext}}(t)$, superimposed on $\varphi_{\text{ext}}(t)$. The resultant increment, $\delta O(t)$, of the observable is then given in terms of its functional derivative (cf., Appendix A) as

$$\delta O(t) = \int_{-\infty}^{t} dt' \frac{\delta O[\varphi_{\text{ext}}(t)]}{\delta\varphi_{\text{ext}}(t')} \delta\varphi_{\text{ext}}(t') \,. \tag{3.1a}$$

The functional derivative, $\delta O[\varphi_{ext}(t)]/\delta\varphi_{ext}(t')$, thus represents the *linear response function* in the presence of the external field, $\varphi_{ext}(t)$. Combinations between the observables and the external fields can generally lead to a variety of the linear response functions.

Let us further assume that the plasma under consideration is invariant under translations in space and time, that is, homogeneous in space and stationary in time when the external field is absent, i.e., $\varphi_{ext}(t) = 0$. Moving over to the Fourier representation for the spatial variations (cf., the footnote of p. 19), we rewrite Eq. (3.1a) as

$$\delta O(\mathbf{k},t) = \int_{-\infty}^{t} dt' \left[\frac{\delta O}{\delta\varphi_{ext}}(\mathbf{k},t-t')\right]_0 \delta\varphi_{ext}(\mathbf{k},t') . \tag{3.1b}$$

The functional derivative

$$R(\mathbf{k},t-t') = \left[\frac{\delta O}{\delta\varphi_{ext}}(\mathbf{k},t-t')\right]_0 \tag{3.2}$$

evaluated at $\varphi_{ext}(t) = 0$ is a function of $t-t'$ in time, because the plasma is stationary.

The domain $(-\infty,t)$ of the t' integration in Eqs. (3.1) sets the *retarded* boundary conditions for the response function. It reflects the notion of causality. Starting with an unperturbed state at $t' = -\infty$, the increment $\delta O(\mathbf{k},t)$ is gradually induced in the system owing to the application of the perturbing field; in these circumstances, the effect $\delta O(\mathbf{k},t)$ cannot precede the cause $\delta\varphi_{ext}(\mathbf{k},t')$.

In a stationary system, the Fourier transformation of Eq. (3.1b) with respect to time gives

$$\delta O(\mathbf{k},\omega) = \mathcal{R}(\mathbf{k},\omega)\delta\varphi_{ext}(\mathbf{k},\omega) . \tag{3.3}$$

Here both $\delta O(\mathbf{k},t)$ and $\delta\varphi_{ext}(\mathbf{k},t')$ are transformed as usual according to

$$\mathcal{A}(\mathbf{k},\omega) = \int_{-\infty}^{\infty} dt A(\mathbf{k},t) \exp(i\omega t) . \tag{3.4}$$

For a real physical quantity, it has a symmetry

$$\mathcal{A}^*(\mathbf{k},\omega) = \mathcal{A}(-\mathbf{k},-\omega) \tag{3.5}$$

with a real value of the frequency ω, where the asterisk means the complex conjugate.

The response function $\mathcal{R}(\mathbf{k},\omega)$ in Eq. (3.3) is transformed, however, with the one-sided Fourier transformation (cf., Eq. 2.57a)

$$\mathcal{R}(\mathbf{k},\omega) = \int_0^\infty dt R(\mathbf{k},t) \exp(i\omega t) , \tag{3.6}$$

reflecting the principle of causality imposed in Eqs. (3.1). It follows that $\mathcal{R}(\mathbf{k},\omega)$ is analytic in the upper half of the complex plane, and is continued analytically into the lower half of the ω plane; there, $\mathcal{R}(\mathbf{k},\omega)$ generally has singularities. It has a symmetry relation

$$\mathcal{R}^*(\mathbf{k},\omega) = \mathcal{R}(-\mathbf{k},-\omega^*) . \tag{3.7}$$

Inverse transformation of $\mathcal{R}(\mathbf{k},\omega)$ is given by

$$R(\mathbf{k},t) = \frac{1}{2\pi}\int_{C_+} d\omega\, \mathcal{R}(\mathbf{k},\omega) \exp(-i\omega t) , \tag{3.8}$$

where C_+ is the contour of integration specified in conjunction with Eq. (2.57b). For $t < 0$ in particular, one finds $R(\mathbf{k},t) = 0$, another manifestation of causality.

Due to the analytic properties on the complex ω plane, the retarded response function obeys the dispersion relation of the Kramers-Kronig type

$$\int_{C_+} d\omega' \frac{\mathcal{R}(\mathbf{k},\omega') - \mathcal{R}(\mathbf{k},\infty)}{\omega' - \omega} = 0 , \tag{3.9}$$

if $\mathcal{R}(\mathbf{k},\omega) - \mathcal{R}(\mathbf{k},\infty)$ vanishes faster than $1/\omega$ in the limit $|\omega| \to \infty$. In Section 3.2B, the consequences of Eq. (3.9) will be elucidated for the dielectric response function.

B. Fluctuation-Dissipation Theorem

In the theory of the many-particle system in thermodynamic equilibrium, the fluctuation-dissipation theorem provides a rigorous connection between the spectral functions of fluctuations and the imaginary parts of the relevant linear response functions [Callen & Welton 1951; Kubo 1957]. The theorem relates the canonically (or grand canonically) averaged commutator and anticommutator of any pair of Hermitian operators, such as the number densities evaluated at two different points in space and time. The average of such a commutator is related to a response function, while the average of an anticommutator gives a correlation function, which turns into a spectral function of fluctuations after

Fourier transformations. It may therefore be said that the theorem possesses a form unique in physics, relating the properties of the system in equilibrium (i.e., fluctuations) with the parameters which characterize the irreversible processes, that is, the imaginary parts of the response functions. Here we summarize contents of the fluctuation-dissipation theorem.

Consider an external disturbance field

$$a(\mathbf{r},t) = a \exp\left[i(\mathbf{k}\cdot\mathbf{r} - \omega t) + \eta t\right] + \mathrm{cc} \tag{3.10}$$

applied to a system in thermodynamic equilibrium which would be uniform without the disturbance (cc stands for the complex conjugate and η is a positive infinitesimal). The disturbance couples with the system via an extra Hamiltonian

$$H_{\mathrm{ext}}(t) = -\int_V d\mathbf{r} A(\mathbf{r}) a(\mathbf{r},t) \tag{3.11a}$$

where A represents a physical quantity observable in the system. The total Hamiltonian written as

$$H_{\mathrm{tot}} = H + H_{\mathrm{ext}}(t) \tag{3.11b}$$

then drives the system out of equilibrium according to the Heisenberg equation of motion [e.g., Pines 1963; Fetter & Walecka 1971].

A physical quantity B, which can be the same as A, of the system is perturbed and thereby deviates from its average value by $\delta B(\mathbf{r},t)$. Within the framework of the linear response formalism in Section 3.1A, we may express the deviation as

$$\delta B(\mathbf{r},t) = B \exp\left[i(\mathbf{k}\cdot\mathbf{r} - \omega t) + \eta t\right] + \mathrm{cc}\,. \tag{3.12}$$

In this case

$$K_{BA}(\mathbf{k},\omega) = \frac{B}{a} \tag{3.13a}$$

gives a linear response function in its general form. Explicit calculations with the Hamiltonian (3.11) yield

$$K_{BA}(\mathbf{k},\omega) = \frac{i}{\hbar}\int d\mathbf{r}\int_0^\infty dt \langle B(\mathbf{r}'+\mathbf{r},t'+t)A(\mathbf{r}',t')$$

$$- A(\mathbf{r}',t')B(\mathbf{r}'+\mathbf{r},t'+t)\rangle \exp\left[-i(\mathbf{k}\cdot\mathbf{r} - \omega t)\right] \tag{3.13b}$$

(see Appendix B for its derivation). Here $A(\mathbf{r},t)$ and $B(\mathbf{r},t)$ are the Heisenberg operators evolving with the unperturbed Hamiltonian H, and $\langle \cdots \rangle$ refers to the expectation value in the unperturbed equilibrium state. The linear response functions therefore depend only on the system properties without perturbations.

The physical quantities A and B fluctuate in space and time even in a system under thermodynamic equilibrium. The correlation function between them is defined as

$$C_{BA}(\mathbf{r},t) = \frac{1}{2}\langle B(\mathbf{r}'+\mathbf{r},t'+t)A(\mathbf{r}',t') + A(\mathbf{r}',t')B(\mathbf{r}'+\mathbf{r},t'+t)\rangle \,. \tag{3.14}$$

Fourier transformation of such a correlation function yields a spectral function

$$S_{BA}(\mathbf{k},\omega) = \frac{1}{2\pi}\int d\mathbf{r}\int_{-\infty}^{\infty} dt C_{BA}(\mathbf{r},t)\exp\left[-i(\mathbf{k}\cdot\mathbf{r}-\omega t)\right] \,. \tag{3.15}$$

It is then connected with the linear response function (3.13) via the *fluctuation-dissipation theorem*

$$S_{BA}(\mathbf{k},\omega) = -\frac{i\hbar}{4\pi}\coth\left(\frac{\hbar\omega}{2T}\right)[K_{BA}(\mathbf{k},\omega) - K_{AB}(-\mathbf{k},-\omega)] \tag{3.16}$$

(see Appendix C for its derivation). The general response functions $K_{BA}(\mathbf{k},\omega)$ possess symmetry properties [e.g., Forster 1975] analogous to Eq. (3.7) so that the square brackets in Eq. (3.16) take on pure imaginary values when $A = B$.

In the limit of $T \to 0$, Eq. (3.16) becomes

$$S_{BA}(\mathbf{k},\omega) = -\frac{i\hbar}{4\pi}\frac{\omega}{|\omega|}[K_{BA}(\mathbf{k},\omega) - K_{AB}(-\mathbf{k},-\omega)] \,. \tag{3.17}$$

In the classical limit $\hbar \to 0$, on the other hand, Eq. (3.16) takes the form,

$$S_{BA}(\mathbf{k},\omega) = -\frac{iT}{2\pi\omega}[K_{BA}(\mathbf{k},\omega) - K_{AB}(-\mathbf{k},-\omega)] \,. \tag{3.18}$$

It is instructive to consider a special case of the response problems where the external disturbance in the spatial Fourier component takes the form

$$a(\mathbf{k},t) = \begin{cases} a(\mathbf{k})\exp(\eta t) & (t \le 0) \\ 0 & (t > 0) \end{cases} \,. \tag{3.19}$$

The Fourier component $\delta B(\mathbf{k},t=0)$ of the induced fluctuation is then calculated in terms of the *relaxation functions*

$$K_{BA}''(\mathbf{k},\omega)=\frac{1}{2\hbar}\int d\mathbf{r}\int_{-\infty}^{\infty} dt\langle B(\mathbf{r}'+\mathbf{r},t'+t)A(\mathbf{r}',t')$$

$$-A(\mathbf{r}',t')B(\mathbf{r}'+\mathbf{r},t'+t)\rangle \exp\left[-i(\mathbf{k}\cdot\mathbf{r}-\omega t)\right] \quad (3.20)$$

as

$$\frac{\delta B(\mathbf{k},t=0)}{a(\mathbf{k})}=\frac{1}{\pi}\int_{-\infty}^{\infty} d\omega\frac{K_{BA}''(\mathbf{k},\omega)}{\omega}$$

$$=\langle B(\mathbf{k},t=0)A(-\mathbf{k},t=0)-A(-\mathbf{k},t=0)B(\mathbf{k},t=0)\rangle/2T \quad (3.21)$$

(Problems 3.3 and 3.4).

The relaxation functions (3.20) lead to a spectral representation of the causal response functions

$$K_{BA}^{c}(\mathbf{k},z)=\frac{1}{\pi}\int_{-\infty}^{\infty} d\omega\frac{K_{BA}''(\mathbf{k},\omega)}{\omega-z} \quad (\mathrm{Im}\ z\neq 0)\,, \quad (3.22)$$

from which the usual retarded response functions are obtained as

$$K_{BA}(\mathbf{k},\omega)=\lim_{\eta\to 0} K_{BA}^{c}(\mathbf{k},z)\,|_{z=\omega+i\eta}\,. \quad (3.23)$$

Thermodynamic sum rules for the relaxation functions are finally obtained from the long-wavelength limit of Eq. (3.21) as

$$\frac{\partial B}{\partial a}\equiv\lim_{\mathbf{k}\to 0}\frac{\delta B(\mathbf{k},t=0)}{a(\mathbf{k})}=\lim_{\mathbf{k}\to 0}\frac{1}{\pi}\int_{-\infty}^{\infty} d\omega\frac{K_{BA}''(\mathbf{k},\omega)}{\omega}\,. \quad (3.24)$$

C. Density-Density Response Functions

The dielectric formulation for a multicomponent plasma [Ichimaru et al. 1985] is constructed from the foregoing general theory in terms of the density-density response functions, which may be obtained by setting $A=n_\tau$ and $B=n_\sigma$, the number densities of particles in the τ and σ species. Since the derivative of the grand-canonical thermodynamic potential is given by

$$d\Omega=-SdT-PdV-\sum_{\sigma}N_\sigma d(\mu_\sigma+\varphi_\sigma) \quad (3.25)$$

with S, P, and N_σ designating the entropy, the total pressure, and the number of particles in the σ species [e.g., Landau & Lifshitz 1969], we may generally take $a = \mu_\sigma + \varphi_\sigma$, the electrochemical potential, in Eq. (3.11a).

The density-density response functions are denoted by

$$\chi_{\sigma\tau}(\mathbf{k},\omega) = -K_{\sigma\tau}(\mathbf{k},\omega)\,. \tag{3.26a}$$

The spectral functions of the density fluctuations, or the *dynamic structure factors*, are then calculated with the fluctuation-dissipation theorem Eq. (3.16) as

$$S_{\sigma\tau}(\mathbf{k},\omega) = \frac{i\hbar}{4\pi}\coth\left(\frac{\hbar\omega}{2T}\right)[\chi_{\sigma\tau}(\mathbf{k},\omega) - \chi_{\tau\sigma}(-\mathbf{k},-\omega)]\,. \tag{3.26b}$$

Integration of the dynamic structure factors over the entire frequency domain

$$S_{\sigma\tau}(\mathbf{k}) = \frac{1}{\sqrt{n_\sigma n_\tau}}\int_{-\infty}^{\infty} d\omega\, S_{\sigma\tau}(\mathbf{k},\omega) \tag{3.27}$$

yields the *static structure factors* corresponding to the spectra of density fluctuations in the wave-vector space.

Integration over the frequencies in Eq. (3.27) may be carried out by deformation of the contours as follows: We first introduce the causal density response functions $\chi^{c}_{\sigma\tau}(\mathbf{k},\omega)$ as in Eq. (3.22). Equation (3.26b) is then integrated as

$$S_{\sigma\tau}(\mathbf{k}) = -\frac{\hbar}{4\pi i\sqrt{n_\sigma n_\tau}}\mathcal{P}\int_{-\infty}^{\infty} d\omega\coth\left(\frac{\hbar\omega}{2T}\right)[\chi^{c}_{\sigma\tau}(\mathbf{k},\omega+i\eta) - \chi^{c}_{\sigma\tau}(\mathbf{k},\omega-i\eta)]$$

$$= -\frac{\hbar}{4\pi i\sqrt{n_\sigma n_\tau}}\int_{C_1} dz\coth\left(\frac{\hbar z}{2T}\right)\chi^{c}_{\sigma\tau}(\mathbf{k},z)\,.$$

Here $\mathcal{P}$ means the principal part of integration, and C_1 refers to the contour of integration depicted in Fig. 3.1. Since $\chi^{c}_{\sigma\tau}(\mathbf{k},z)$ is separately analytic in the upper and lower halves of the complex z-plane [Pines & Nozières 1966], we can deform the contour from C_1 to C_2 to obtain

$$S_{\sigma\tau}(\mathbf{k}) = -\frac{T}{\sqrt{n_\sigma n_\tau}}\sum_{l=-\infty}^{\infty}\chi^{c}_{\sigma\tau}(\mathbf{k},z_l)\,, \tag{3.28a}$$

where

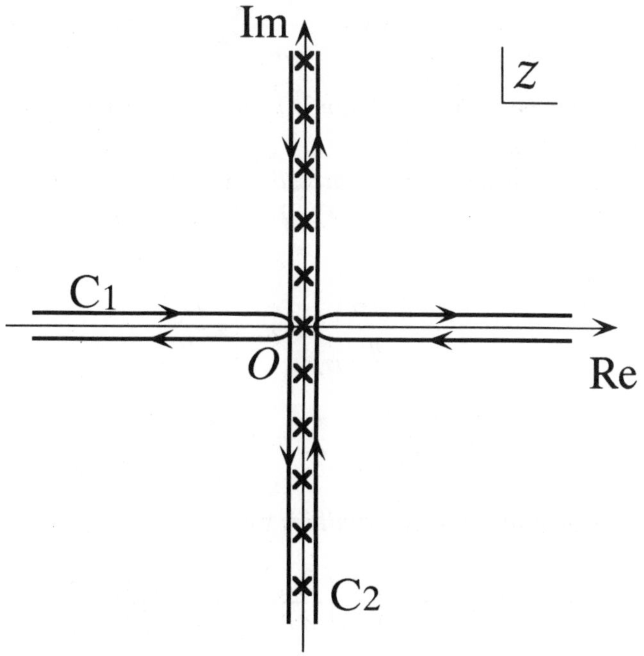

Figure 3.1 Contours of integrations, C_1 and C_2, for Eq. (3.28a). Crosses refer to the poles of $\coth(\hbar z/2T)$.

$$z_l = \frac{2\pi i T}{\hbar} l \quad (l = 0, \pm 1, \pm 2, \ldots) \tag{3.29}$$

are the poles of $\coth(\hbar z/2T)$ on the imaginary z-axis. In the classical limit of Eq. (3.18), only the $l = 0$ term contributes in Eq. (3.28a), so that

$$S_{\sigma\tau}(\mathbf{k}) = -\frac{T}{\sqrt{n_\sigma n_\tau}} \chi_{\sigma\tau}(\mathbf{k}, 0) . \tag{3.28b}$$

The static structure factor is related to the radial distribution function in Eq. (2.30b) via the Fourier transformation as

$$g_{\sigma\tau}(r) = 1 + h_{\sigma\tau}(r)$$

$$= 1 + \frac{1}{\sqrt{n_\sigma n_\tau}} \sum_{\mathbf{k}} [S_{\sigma\tau}(k) - \delta_{\sigma\tau}] \exp(i\mathbf{k}\cdot\mathbf{r}) . \quad (3.30)$$

Here $h_{\sigma\tau}(r)$ is the *pair-correlation function* associated with $g_{\sigma\tau}(r)$, and $\delta_{\sigma\tau}$ represents Kronecker's delta.

Finally we consider the *compressibility sum rules* derived from the thermodynamic sum rules of Eq. (3.24). To this end we note thermodynamic relations

$$\left(\frac{\partial P}{\partial \mu_\sigma}\right)_{T,V} = n_\sigma \left(\equiv \frac{N_\sigma}{V}\right), \quad (3.31a)$$

$$n_\sigma \left(\frac{\partial P}{\partial n_\sigma}\right)_{T,V} = \sum_\tau n_\sigma n_\tau \left(\frac{\partial^2 (F/V)}{\partial n_\sigma \partial n_\tau}\right)_{T,V}, \quad (3.31b)$$

and define *isothermal compressibilities* by

$$\left(\kappa_T\right)_{\sigma\tau} = \frac{1}{2}\left[\frac{1}{n_\sigma}\left(\frac{\partial n_\sigma}{\partial P}\right)_{T,V} + \frac{1}{n_\tau}\left(\frac{\partial n_\tau}{\partial P}\right)_{T,V}\right]. \quad (3.32)$$

Here F/V is the Helmholtz free energy per unit volume, which is assumed to be a function of T and n_σ. The compressibility sum rules

$$\lim_{\mathbf{k}\to 0} \chi_{\sigma\tau}(\mathbf{k},0) = -n_\sigma n_\tau (\kappa_T)_{\sigma\tau} \quad (3.33a)$$

are then derived from Eq. (3.24). The compressibility sum rules provide an important consistency check for the thermodynamic treatments in dense plasmas.

In an OCP, the density perturbation produced by a change in the chemical potential creates an electrostatic space-charge potential, since the neutralizing charge-background remains unperturbed. Hence, accounting for the additional contribution of the space-charge field to the electrochemical potential, the compressibility sum rule now takes the form

$$\lim_{\mathbf{k}\to 0} \frac{\chi(\mathbf{k},0)}{1 + v(k)\chi(\mathbf{k},0)} = -n^2 \kappa_T . \quad (3.33b)$$

Here $v(k) = 4\pi (Ze)^2/k^2$ and $\chi(\mathbf{k},\omega)$ is the density-density response function for the OCP. In a multicomponent plasma, the induced space-charge field should analogously be accounted for in the application of the compressibility sum rule (3.33a).

3.2 DIELECTRIC FORMULATION

The dielectric response functions are a class of the linear response functions that emerge with the electrostatic approximation to the plasma, that is, when Coulomb interaction is assumed between the particles. These describe the dielectric properties associated with density fluctuations in nonrelativistic plasmas.

A. Dielectric Response Function

Consider the case where an external electrostatic potential field of infinitesimal strength,

$$\delta\Phi_{\rm ext}(\mathbf{r},t) = \Phi_{\rm ext}(\mathbf{k},\omega)\exp\left[i(\mathbf{k}\cdot\mathbf{r}-\omega t)\right] + \mathrm{cc}\,, \tag{3.34}$$

is applied to a uniform plasma. It will disturb the plasma and thereby create density fluctuations. We are here concerned only with a linear response, so that the space-charge field induced in the plasma may be written in terms of the Fourier component as $Ze\rho_{\rm ind}(\mathbf{k},\omega)$. Within the electrostatic approximation (cf., Eq. 2.39), the potential field associated with such a space-charge distribution is given by

$$\Phi_{\rm ind}(\mathbf{k},\omega) = \frac{4\pi Ze}{k^2}\rho_{\rm ind}(\mathbf{k},\omega)\,. \tag{3.35}$$

The *dielectric response function*, $\varepsilon(\mathbf{k},\omega)$, of the plasma is then defined via the linear-response relation

$$\Phi_{\rm ind}(\mathbf{k},\omega) = \left[\frac{1}{\varepsilon(\mathbf{k},\omega)} - 1\right]\Phi_{\rm ext}(\mathbf{k},\omega)\,. \tag{3.36}$$

The total potential field, $\Phi(\mathrm{k},\omega) = \Phi_{\rm ext}(\mathbf{k},\omega) + \Phi_{\rm ind}(\mathbf{k},\omega)$, thus changes its strength from the external value, $\Phi_{\rm ext}(\mathbf{k},\omega)$, by the dielectric screening factor $1/\varepsilon(\mathbf{k},\omega)$.

B. Analytic Properties

The dielectric response function satisfies a number of general relationship stemming from the analytic properties as a retarded response function. In this subsection, we consider the Kramers-Kronig relations and the *f*-sum rule as applied to the dielectric response function.

In the limit of high frequencies $|\omega| \to \infty$, we note that the dielectric response behaves asymptotically as

$$\frac{1}{\varepsilon(\mathbf{k},\omega)} \to 1 + \left(\frac{\omega_p}{\omega}\right)^2 \tag{3.37}$$

[e.g., Landau & Lifshitz 1960]. This can be readily proved through Eq. (1.46), for example, where we may take $|\omega|$ in the long-wavelength regime to be larger than both the single-particle excitation frequency $|\mathbf{k} \cdot \mathbf{v}_j|$ in the first terms on the right-hand side and the frequencies of multiparticle configurations represented by the second terms; the latter decay faster than k^2 for long wavelengths.

In light of the asymptotic behavior (3.37), the Kramers-Kronig relation (3.9) for the response function $1/\varepsilon(\mathbf{k},\omega)$ reads

$$\int_{-\infty}^{\infty} d\omega' \frac{1}{\omega' - \omega + i\eta}\left[\frac{1}{\varepsilon(\mathbf{k},\omega')} - 1\right] = 0\,. \tag{3.38}$$

Here a positive infinitesimal η is written explicitly in the denominator $\omega' - \omega + i\eta$ to account for the contour of integration C_+. The contribution from the infinite semicircle vanishes because of (3.37).

With the aid of the Dirac formula,

$$\lim_{\eta \to +0} \frac{1}{x \pm i\eta} = \mathcal{P}\frac{1}{x} \mp i\pi\delta(x)\,, \tag{3.39}$$

where $\mathcal{P}$ stands for the *principal part*, we may now split Eq. (3.38) into the real and imaginary parts, to obtain

$$\mathrm{Re}\left[\frac{1}{\varepsilon(\mathbf{k},\omega)}\right] - 1 = \frac{1}{\pi}\int_{-\infty}^{\infty} d\omega' \frac{\mathcal{P}}{\omega' - \omega}\,\mathrm{Im}\left[\frac{1}{\varepsilon(\mathbf{k},\omega')}\right], \tag{3.40a}$$

$$\mathrm{Im}\left[\frac{1}{\varepsilon(\mathbf{k},\omega)}\right] = \frac{1}{\pi}\int_{-\infty}^{\infty} d\omega' \frac{\mathcal{P}}{\omega' - \omega}\,\mathrm{Re}\left[1 - \frac{1}{\varepsilon(\mathbf{k},\omega')}\right]. \tag{3.40b}$$

These are the *Kramers-Kronig relations* for $1/\varepsilon(\mathbf{k},\omega)$.

Equation (3.40a) may be transformed as

$$\frac{1}{2}\,\mathrm{Re}\left[\frac{1}{\varepsilon(\mathbf{k},\omega)} + \frac{1}{\varepsilon(\mathbf{k},-\omega)}\right] - 1 = \frac{1}{\pi}\int_{-\infty}^{\infty} d\omega' \frac{\mathcal{P}}{(\omega')^2 - \omega^2}\,\omega'\,\mathrm{Im}\left[\frac{1}{\varepsilon(\mathbf{k},\omega')}\right]. \tag{3.41}$$

Comparing the asymptotic limit of Eq. (3.41) at $|\omega| \to \infty$ with (3.37), we then find

$$\int_{-\infty}^{\infty} d\omega\omega \operatorname{Im}\left[\frac{1}{\varepsilon(\mathbf{k},\omega)}\right] = -\pi\omega_p^2 . \tag{3.42}$$

This is the *f-sum rule* for the dielectric response function.

C. Dynamic Screening and Stopping Power

Consider a test particle with an electric charge Z_0e, mass M_0, and velocity $\mathbf{v}_0$ injected in the plasma. The associated space-charge density is

$$Z_0e\rho_{\text{ext}}(\mathbf{r},t) = Z_0e\delta(\mathbf{r}-\mathbf{v}_0t) , \tag{3.43a}$$

so that the Fourier components are

$$Z_0e\rho_{\text{ext}}(\mathbf{k},\omega) = 2\pi Z_0e\delta(\omega - \mathbf{k}\cdot\mathbf{v}_0) . \tag{3.43b}$$

The induced potential field is calculated from the inverse Fourier transformations of Eq. (3.36) as

$$\Phi_{\text{ind}}(\mathbf{r},t) = 4\pi Z_0e \sum_{\mathbf{k}}{}' \frac{1}{k^2}\left[\frac{1}{\varepsilon(\mathbf{k},\mathbf{k}\cdot\mathbf{v}_0)} - 1\right] \exp\left[i\mathbf{k}(\mathbf{r}-\mathbf{v}_0t)\right] . \tag{3.44}$$

This equation describes the velocity-dependent dynamic screening effect of the plasma.

If the test particle is at rest ($\mathbf{v}_0 = \mathbf{0}$), then the induced potential takes a form independent of time,

$$\Phi_{\text{ind}}(\mathbf{r}) = 4\pi Z_0e \sum_{\mathbf{k}}{}' \frac{1}{k^2}\left[\frac{1}{\varepsilon(\mathbf{k},0)} - 1\right] \exp\left(i\mathbf{k}\cdot\mathbf{r}\right) . \tag{3.45}$$

We recall that a problem in this category was treated in Section 1.3B. The function $\varepsilon(\mathbf{k},0)$ will be referred to as the *static screening function.*

When $\mathbf{v}_0 \neq \mathbf{0}$, the position of the test charge is shifted off the center of the induced potential (3.44). This may be seen clearly through an examination of Eq. (3.44) under the inversion, $\mathbf{r}-\mathbf{v}_0t \to -(\mathbf{r}-\mathbf{v}_0t)$; the presence of $\operatorname{Im}[1/\varepsilon(\mathbf{k},\mathbf{k}\cdot\mathbf{v}_0)]$, which is antisymmetric with respect to $\mathbf{k}$ (cf. Eq. 3.7), makes Eq. (3.44) asymmetric under such an inversion. As a result, the test charge is decelerated by the plasma [Fermi 1924; Bethe 1930; Bloch 1933;

Bohr 1948]. Let us therefore formulate such a stopping power of the plasma against a moving test charge.

The electric field due to the induced space charge is obtained from Eq. (3.44) as

$$\mathbf{E}_{\text{ind}}(\mathbf{r},t) = -4\pi Z_0 e \sum_{\mathbf{k}}{}' \frac{i\mathbf{k}}{k^2}\left[\frac{1}{\varepsilon(\mathbf{k},\mathbf{k}\cdot\mathbf{v}_0)} - 1\right] \exp\left[i\mathbf{k}\cdot(\mathbf{r}-\mathbf{v}_0 t)\right],$$

whence the force acting on the test charge is

$$\mathbf{F}(\mathbf{v}_0) = Z_0 e \mathbf{E}_{\text{ind}}(\mathbf{v}_0 t, t) = 4\pi (Z_0 e)^2 \sum_{\mathbf{k}}{}' \frac{\mathbf{k}}{k^2} \operatorname{Im}\left[\frac{1}{\varepsilon(\mathbf{k},\mathbf{k}\cdot\mathbf{v}_0)}\right]. \tag{3.46}$$

The *stopping power* is defined as the loss rate of the kinetic energy, $W = \frac{1}{2} M_0 v_0^2$, per unit flight length along its trajectory, i.e.,

$$-\frac{dW}{dx} = -\hat{\mathbf{v}}_0 \cdot \mathbf{F}(\mathbf{v}_0) = -\frac{(Z_0 e)^2}{\pi v_0^2} \int \frac{dk}{k} \int_{-kv_0}^{kv_0} d\omega\, \omega \operatorname{Im}\left[\frac{1}{\varepsilon(\mathbf{k},\omega)}\right], \tag{3.47}$$

where we have transformed the k summation into integration, and changed an integration variable from $\mathbf{k}\cdot\mathbf{v}_0$ to ω; $\hat{\mathbf{v}}_0$ is the unit vector in the direction of $\mathbf{v}_0$.

In the form of Eq. (3.47), we find an important application of the f-sum rule (3.42). When the velocity of the test charge is so large that

$$v_0 >> \sqrt{\frac{T}{m}}, \tag{3.48}$$

we may replace the upper and lower limits of the ω integration in Eq. (3.47) by $+\infty$ and $-\infty$, respectively. With the aid of (3.42), then, Eq. (3.47) becomes

$$-\frac{dW}{dx} = \left(\frac{Z_0 e \omega_p}{v_0}\right)^2 \int_{k_{\min}}^{k_{\max}} \frac{dk}{k} = \left(\frac{Z_0 e \omega_p}{v_0}\right)^2 \ln\left(\frac{k_{\max}}{k_{\min}}\right). \tag{3.49}$$

We thus have a sum rule due to Bethe [1930] on the stopping power, meaning that the total energy loss of a fast particle depends only on the total number of scatterers.

Equation (3.49) offers another example of the Coulomb logarithm, which depends on the ratio between the upper and lower bounds of the k integration. In light of the arguments in Section 1.4B and especially of Fig. 1.4, the lower bound is given by

$$k_{\text{min}} = \frac{\omega_p}{v_0} . \tag{3.50}$$

The collective effects are the major contribution to the stopping power in the long-wavelength regime (1.51).

Classically, k_{max} may be chosen in the vicinity of the inverse of a distance of closest approach between the test particle and a plasma particle [Bohr 1948]

$$k_{\text{max}} = \frac{\alpha \mu v_0^2}{| Z_0 Z e^2 |} , \tag{3.51}$$

where μ is the reduced mass, $\alpha = 2 \exp(-\gamma)$, and $\gamma = 0.57721 \cdots$ (Euler's constant).

When the velocity of the test charge becomes extremely large, so that

$$v_0 > \frac{| Z_0 Z e^2 |}{\hbar} , \tag{3.52}$$

then the de Broglie wavelength begins to exceed the classical distance of closest approach. In these circumstances, it has been found [Lindhard 1954] that

$$k_{\text{max}} = \frac{2 \mu v_0}{\hbar} . \tag{3.53}$$

D. Collective Modes

The response relation (3.36) carries an important consequence with regard to the nature of density-fluctuation excitations in plasmas. The relation implies that no space-charge field would be induced when $\Phi_{\text{ext}}(\mathbf{k},\omega) = 0$. However, if the situation is such that the frequency and the wave vector coincide with those at one of the poles of $1/\varepsilon(\mathbf{k},\omega) - 1$, or equivalently, if $\mathbf{k}$ and ω satisfy the equation

$$\varepsilon(\mathbf{k},\omega) = 0 , \tag{3.54}$$

then such a space-charge field could be spontaneously excited in the plasma without application of an external disturbance. Equation (3.54) therefore provides the dispersion relation for density-fluctuation excitations associated with the *collective modes* in the plasma.

Since the dielectric response function $\varepsilon(\mathbf{k},\omega)$ is a complex function with real and imaginary parts

$$\varepsilon(\mathbf{k},\omega) = \varepsilon_1(\mathbf{k},\omega) + i\varepsilon_2(\mathbf{k},\omega)\,, \tag{3.55}$$

Eq. (3.54) determines a complex frequency

$$\omega = \omega_k - i\gamma_k\,. \tag{3.56}$$

The collective mode is a long-lived, well-defined excitation only when the decay rate γ_k (> 0) is much smaller than the real part ω_k of the frequency, that is, when

$$\left|\frac{\gamma_k}{\omega_k}\right| \ll 1\,. \tag{3.57}$$

In these circumstances, the dielectric function may have a Taylor expansion near the collective mode (3.56), so that

$$\varepsilon(\mathbf{k},\omega) \approx \left|\frac{\partial\varepsilon_1}{\partial\omega}\right|_{\omega=\omega_k} (\omega - \omega_k + i\gamma_k)\,. \tag{3.58}$$

Hence one finds useful expressions applicable near a collective mode $\omega = \omega_k$:

$$\gamma_k = \left|\frac{\partial\varepsilon_1}{\partial\omega}\right|^{-1}_{\omega=\omega_k} \varepsilon_2(\mathbf{k},\omega_k)\,, \tag{3.59}$$

$$\mathrm{Im}\left[\frac{1}{\varepsilon(\mathbf{k},\omega)}\right] = -\pi\left|\frac{\partial\varepsilon_1}{\partial\omega}\right|^{-1}_{\omega=\omega_k} \delta(\omega - \omega_k)\,. \tag{3.60}$$

The formula (3.39) has been used in the derivation of (3.60).

E. Density Fluctuations

For a plasma in thermodynamic equilibrium, the fluctuation-dissipation theorem and the density response formalism, summarized in Section 3.1, can be applied to an establishment of the relation between the dynamic structure factor $S(\mathbf{k},\omega)$ and the dielectric response function $\varepsilon(\mathbf{k},\omega)$, defined in Eq. (3.36).

The external potential field, Eq. (3.34), is coupled with the density fluctuations in a plasma according to

$$H_{\mathrm{ext}}(t) = -Ze\int_V d\mathbf{r}\rho(\mathbf{r})\Phi_{\mathrm{ext}}(\mathbf{r},t)\,. \tag{3.61}$$

Upon comparison between Eqs. (3.11a), (3.35), (3.36), and (3.61), the density-density response function (3.26a) is found to be

$$\chi(\mathbf{k},\omega) = -\frac{1}{v(k)}\left[1-\frac{1}{\varepsilon(\mathbf{k},\omega)}\right], \tag{3.62}$$

where $v(k) = 4\pi(Ze)^2/k^2$. In light of the general symmetry relation (3.7), one has

$$\frac{1}{\varepsilon(\mathbf{k},\omega)} - \frac{1}{\varepsilon(-\mathbf{k},-\omega)} = 2i\,\mathrm{Im}\left[\frac{1}{\varepsilon(\mathbf{k},\omega)}\right] \tag{3.63}$$

for a real value of ω. It thus follows from Eq. (3.26b) that the dynamic structure factor is given by

$$S(\mathbf{k},\omega) = -\frac{\hbar k^2}{8\pi^2(Ze)^2}\coth\left(\frac{\hbar\omega}{2T}\right)\mathrm{Im}\left[\frac{1}{\varepsilon(\mathbf{k},\omega)}\right]. \tag{3.64}$$

This expression based on the fluctuation-dissipation theorem is rigorous as long as the plasma is in thermodynamic equilibrium.

Since $\hbar\omega << T$ for a classical plasma, Eq. (3.64) becomes

$$S(\mathbf{k},\omega) = -\frac{T}{\pi v(k)\omega}\mathrm{Im}\left[\frac{1}{\varepsilon(\mathbf{k},\omega)}\right]. \tag{3.65}$$

The static structure factor is then obtained through integration of Eq. (3.65) over the frequencies. The integration is performed exactly through the Kramers-Kronig relation (3.40a) as

$$S(\mathbf{k}) = \frac{T}{nv(k)}\left[1-\frac{1}{\varepsilon(\mathbf{k},0)}\right], \tag{3.66}$$

a result identical to Eq. (3.28b).

The radial distribution function Eq. (3.30) is calculated as

$$g(r) = 1 + \frac{1}{n}\sum_{\mathbf{k}}\left\{\frac{k^2}{k_\mathrm{D}^2}\left[1-\frac{1}{\varepsilon(\mathbf{k},0)}\right]-1\right\}\exp(i\mathbf{k}\cdot\mathbf{r}). \tag{3.67}$$

A useful expression for the interaction energy (2.23) per unit volume of the classical OCP is thus obtained as [e.g., Ichimaru 1982]

$$U_{\text{int}} = \frac{n}{2} \sum_{\mathbf{k}}{}' v(k)[S(\mathbf{k}) - 1]$$

$$= \frac{T}{2} \sum_{\mathbf{k}}{}' \left[1 - \frac{1}{\varepsilon(\mathbf{k},0)} - \frac{nv(k)}{T} \right]. \tag{3.68a}$$

The compressibility sum rule (3.33b) takes the form

$$\lim_{\mathbf{k}\to 0} \frac{1}{v(k)} [\varepsilon(\mathbf{k},0) - 1] = n \left(\frac{\partial n}{\partial P} \right)_{T,V}. \tag{3.68b}$$

Finally, the f-sum rule (3.42) reads

$$\int_{-\infty}^{\infty} d\omega\, \omega^2 S(\mathbf{k},\omega) = \frac{nTk^2}{m} \tag{3.69}$$

in terms of the spectral function of fluctuations in a classical plasma.

We shall return to detailed discussions further on the density-fluctuation excitations and their physical consequences in Chapter 7.

F. Polarization Potential Approach

In Section 1.4B we introduced the notion of the RPA, whereby the nonlinear terms of the fluctuations were ignored. The Vlasov formalism adopts such an RPA in that the correlation effects are neglected through the Hartree factorization (2.36). The Vlasov dielectric function (2.62) may thus be looked upon as a dielectric response in the RPA for a classical plasma. For a quantum plasma, the Lindhard [1954] dielectric functions, to be treated in Section 3.5, are the RPA response functions.

The RPA dielectric functions exhibit severe difficulties when they are applied to the analyses of strong correlations. An illustrative case is offered by the Debye-Hückel radial-distribution function

$$g_{\text{DH}}(r) = 1 - \frac{(Ze)^2}{Tr} \exp\left(-\frac{r}{\lambda_{\text{D}}} \right), \tag{3.70}$$

which may be derived either from Eqs. (1.18) and (1.20) or by an application of Eq. (2.62) to Eq. (3.67) with the Maxwellian distribution (1.1) (Problem 3.8). In the short-range regime such that $r < (Ze)^2/T$, Eq. (3.70) would predict $g_{\text{DH}}(r) < 0$. This behavior is physically unacceptable, because a probability distribution should not be negative.

The origin of this difficulty may be traced to the fact that the RPA dielectric functions do not take account of interparticle correlations in the procedure of calculations. The correlation effects are particularly pronounced at short distances even in a weakly coupled plasma. (Recall the footnote on page 13.) As a result, Vlasov or Lindhard expression is conspicuously inaccurate in the prediction of short-range correlations.

A theoretical method of going beyond the RPA description and thereby accounting for the strong interparticle correlations rigorously in the framework of the dielectric formulation is provided by a *polarization potential approach* [Pines 1966; Ichimaru 1982], which we consider in this section.

In Section 3.2A, the dielectric response function, $\varepsilon(\mathbf{k},\omega)$, has been defined via Eq. (3.36) in relation to a linear response against the external test-charge field (3.34). Let $\Phi_{\rm pol}(\mathbf{k},\omega)$ describe an effective interaction potential produced by the charge-density fluctuations $\rho_{\rm ind}(\mathbf{k},\omega)$; it may be expressed as

$$\Phi_{\rm pol}(\mathbf{k},\omega) = \frac{4\pi Ze}{k^2}[1 - G(\mathbf{k},\omega)]\rho_{\rm ind}(\mathbf{k},\omega) . \tag{3.71}$$

Such a polarization potential generally differs from the bare Coulomb potential, $4\pi Ze/k^2\rho_{\rm ind}(\mathbf{k},\omega)$, because the exchange and Coulombic correlations between the particles act to modify the potentials of the induced space charges. The difference is measured by the *dynamic local-field correction* $G(\mathbf{k},\omega)$ in Eq. (3.71).

We here observe that the potential of the external charge is in effect renormalized in the plasma as

$$\Phi_{\rm ren}(\mathbf{k},\omega) = \Phi_{\rm ext}(\mathbf{k},\omega) + \Phi_{\rm pol}(\mathbf{k},\omega) . \tag{3.72}$$

The screened density-response function, $\chi_0(\mathbf{k},\omega)$, of the plasma is thus defined and introduced in accord with

$$\rho_{\rm ind}(\mathbf{k},\omega) = Ze\chi_0(\mathbf{k},\omega)\Phi_{\rm ren}(\mathbf{k},\omega) . \tag{3.73}$$

Substitution of Eq. (3.73) in Eqs. (3.35) and (3.36) yields

$$\varepsilon(\mathbf{k},\omega) = 1 - \frac{v(k)\chi_0(\mathbf{k},\omega)}{1 + v(k)G(\mathbf{k},\omega)\chi_0(\mathbf{k},\omega)} \tag{3.74}$$

where $v(k) = 4\pi(Ze)^2/k^2$. The density-density response function (3.62) is likewise expressed as

$$\chi(\mathbf{k},\omega) = \frac{\chi_0(\mathbf{k},\omega)}{1 - v(k)[1 - G(\mathbf{k},\omega)]\chi_0(\mathbf{k},\omega)} . \tag{3.75}$$

Equations (3.74) and (3.75) express the dielectric response function and the density-density response function in terms of the two functions, $\chi_0(\mathbf{k},\omega)$ and $G(\mathbf{k},\omega)$. Such a freedom of choice in the expressions can be exploited to derive a physically meaningful representation with a clear separation of the higher-order correlation effects in the response functions. To achieve this end, we may set the local field correction so that the polarization potential (3.71) absorbs *all* the correlation effects arising from the induced charge-density field. The correlation effects in the response of the plasma are now shifted into the renormalized potential (3.72); hence the screened response in Eq. (3.73) is given by a *free-particle polarizability* of the uncorrelated particles. The RPA response is obtained by setting $G(\mathbf{k},\omega) = 0$ in Eq. (3.74). The local field corrections then represent the higher-order effects in the correlations beyond the RPA. The free-particle polarizabilities will be evaluated in Section 3.4 for various classical plasmas and in Section 3.5 for quantum plasmas.

It is particularly instructive to apply the dielectric formulation of Eq. (3.74) in the evaluation for the static structure factor Eq. (3.66) in the classical OCP. Since

$$\chi_0(\mathbf{k},0) = -\frac{n}{T} \tag{3.76}$$

for a classical plasma in thermodynamic equilibrium (cf. Eq. 2.62 with Eq. 1.1, or Eq. 4.1 below), we have

$$S(k) = \frac{1}{1 + \frac{n}{T} v(k)[1 - G(k)]} , \tag{3.77}$$

where

$$G(k) \equiv G(\mathbf{k},0) \tag{3.78}$$

is the static (i.e., $\omega = 0$) evaluation of the dynamic local-field correction. As far as the classical systems in thermodynamic equilibrium are concerned, only the static local-field corrections enter the calculations of the thermodynamic functions via Eq. (3.68a).

In light of Eqs. (1.52) and (3.76), the compressibility sum rule (3.68b) now takes the form

$$G(k) \to -\left[\frac{\partial P_{\text{ex}}}{\partial (nT)}\right]_{T,V} \frac{k^2}{k_{\text{D}}^2}, \tag{3.79}$$

where

$$P_{\text{ex}} = P - nT, \tag{3.80}$$

the excess pressure beyond the ideal-gas contribution. More generally, we shall show later in Section 11.3 that the static local-field corrections correspond to the direct correlation functions obtained as the second density-functional derivatives of the interaction parts in the free energy and that Eq. (3.77) turns into the Ornstein-Zernike relations in the theory of classical liquids [e.g., Hansen & McDonald 1986].

Situations differ manifestly in the quantum systems, however. For an evaluation of a thermodynamic function in a quantum system, it is necessary to sum the contributions of the density-density response functions at the infinite number of poles Eq. (3.28) on the complex-ω plane associated with coth $(\hbar\omega/2T)$ as in Eq. (3.27a); all the dynamic information is essential.

The local field corrections thus play a central part in describing strong-coupling effects for highly correlated plasmas. The dielectric formulation with the local field corrections can provide a precise description of the exchange and Coulomb-induced correlations in quantum and classical plasmas [e.g., Ichimaru, Iyetomi & Tanaka 1987] when it is coupled with the fluctuation-dissipation theorem of Section 3.1. We shall consider these aspects in detail starting with Chapter 7.

3.3 THE DIELECTRIC TENSOR

The dielectric tensor contains essentially all the information about the linear electromagnetic properties of a plasma. It is calculated as a linear response function of a plasma to an electromagnetic disturbance [e.g., Rukhadze & Silin 1961, 1962; Stix 1962; Akhiezer et al. 1967; Ginzburg 1970]. In this section its definition and the resulting dispersion relations will be studied.

A. Definition

Let us generalize the test-charge problem treated in the preceding section and now consider a situation in which charge- and current-densities, $Ze\rho_{\text{ext}}(\mathbf{r},t)$ and $\mathbf{J}_{\text{ext}}(\mathbf{r},t)$, are applied to a uniform and stationary plasma. They satisfy the continuity equation

$$Ze\frac{\partial \rho_{\text{ext}}}{\partial t}+\nabla\cdot\mathbf{J}_{\text{ext}}=0\,. \tag{3.81}$$

The Maxwell equations are

$$\nabla\times\mathbf{E}+\frac{1}{c}\frac{\partial\mathbf{B}}{\partial t}=\mathbf{0}\,, \tag{3.82a}$$

$$\nabla\times\mathbf{B}-\frac{1}{c}\frac{\partial\mathbf{E}}{\partial t}=\frac{4\pi}{c}(\mathbf{J}+\mathbf{J}_{\text{ext}})\,, \tag{3.82b}$$

$$\nabla\cdot\mathbf{E}=4\pi Ze(\rho+\rho_{\text{ext}})\,, \tag{3.82c}$$

$$\nabla\cdot\mathbf{B}=0\,. \tag{3.82d}$$

The induced charge- and current-densities, $Ze\rho(\mathbf{r},t)$ and $\mathbf{J}(\mathbf{r},t)$, satisfy a continuity equation similar to Eq. (3.81).

In the spirit of the linear response formalism for a uniform and stationary plasma, quantities depending on the space-time coordinates are Fourier-transformed and expressed as functions of $\mathbf{k}$ and ω. We define the *conductivity tensor*, $\sigma(\mathbf{k},\omega)$, as the quantity connecting two vectors, $\mathbf{J}(\mathbf{k},\omega)$ and $\mathbf{E}(\mathbf{k},\omega)$, via

$$\mathbf{J}(\mathbf{k},\omega)=\sigma(\mathbf{k},\omega)\cdot\mathbf{E}(\mathbf{k},\omega)\,. \tag{3.83}$$

The *dielectric tensor* is then defined and introduced by

$$\boldsymbol{\varepsilon}(\mathbf{k},\omega)=\mathsf{I}-\frac{4\pi}{i\omega}\sigma(\mathbf{k},\omega)\,, \tag{3.84}$$

where I is the unit tensor represented as

$$\mathsf{I}=\begin{pmatrix}1&0&0\\0&1&0\\0&0&1\end{pmatrix}. \tag{3.85}$$

In terms of the dielectric tensor and the Fourier components of the field variables, Eqs. (3.82) are expressed as

$$\mathbf{k}\times\mathbf{E}-\frac{\omega}{c}\mathbf{B}=\mathbf{0}\,, \tag{3.86a}$$

$$\mathbf{k}\times\mathbf{B}+\frac{\omega}{c}\boldsymbol{\varepsilon}\cdot\mathbf{E}=-\frac{4\pi i}{c}\mathbf{J}_{\text{ext}}\,, \tag{3.86b}$$

$$\mathbf{k}\cdot\boldsymbol{\varepsilon}\cdot\mathbf{E}=-4\pi iZe\rho_{\text{ext}}\,, \tag{3.86c}$$

$$\mathbf{k}\cdot\mathbf{B}=0\,. \tag{3.86d}$$

B. Dispersion Relations

We proceed to find a formal solution to the basic set of electromagnetic equations (3.86). Substitution of (3.86a) in (3.86b) yields

$$\left[\boldsymbol{\varepsilon}-\left(\frac{kc}{\omega}\right)^2\mathsf{I}_{\mathrm{T}}\right]\cdot\mathbf{E}=\frac{4\pi}{i\omega}\mathbf{J}_{\text{ext}}\,. \tag{3.87}$$

The *dispersion tensor* is defined by

$$\Delta(\mathbf{k},\omega)\equiv\boldsymbol{\varepsilon}(\mathbf{k},\omega)-\left(\frac{kc}{\omega}\right)^2\mathsf{I}_{\mathrm{T}}\,, \tag{3.88}$$

where

$$\mathsf{I}_{\mathrm{T}}\equiv\mathsf{I}-\frac{\mathbf{kk}}{k^2} \tag{3.89}$$

is the transeverse projection tensor with respect to the direction of the wave vector $\mathbf{k}$; we then have

$$\Delta\cdot\mathbf{E}=\frac{4\pi}{i\omega}\mathbf{J}_{\text{ext}}\,. \tag{3.90}$$

When no external disturbances are applied to the plasma, that is, when $\mathbf{J}_{\text{ext}}(\mathbf{k},\omega)=\mathbf{0}$, Eq. (3.90) becomes

$$\Delta\cdot\mathbf{E}=\mathbf{0}\,. \tag{3.91}$$

Equation (3.91) possesses nontrivial solutions only if the determinant constructed from the elements of the tensor $\Delta(\mathbf{k},\omega)$ vanishes:

$$\det|\,\Delta(\mathbf{k},\omega)\,|=0\,. \tag{3.92}$$

This equation thus determines the frequency-wave vector dispersion relations for propagation of the electromagnetic waves in the plasma.

C. Isotropic Media

The dispersion relations derived above are generally applicable to the linear electromagnetic phenomena in a homogeneous and stationary plasma. If in addition the medium is assumed isotropic, that is, invariant under rotation of the coordinate axes, then the dispersion relations are simplified substantially.

We note that the dielectric tensor $\boldsymbol{\varepsilon}(\mathbf{k},\omega)$ in these circumstances contains only one element of directional dependence through that of the wave vector $\mathbf{k}$; the tensor character of $\boldsymbol{\varepsilon}(\mathbf{k},\omega)$ must be determined from a general consideration so as to reflect such a dependence. Tensors of rank two, which can be constructed out of the vector $\mathbf{k}$ and which are mutually independent, are the longitudinal projection tensor

$$\mathbf{I}_{\rm L} \equiv \frac{\mathbf{kk}}{k^2} \tag{3.93}$$

and the transverse projection tensor $\mathbf{I}_{\rm L}$ defined by Eq. (3.89). The dielectric tensor must therefore be expressed as a linear combination of these two tensors:

$$\boldsymbol{\varepsilon}(\mathbf{k},\omega) = \varepsilon_{\rm L}(k,\omega)\mathbf{I}_{\rm L} + \varepsilon_{\rm T}(k,\omega)\mathbf{I}_{\rm T} \,. \tag{3.94}$$

The coefficients, $\varepsilon_{\rm L}(k,\omega)$ and $\varepsilon_{\rm T}(k,\omega)$, introduced here depend only on the magnitude of $\mathbf{k}$ and not on its direction. Inversely, when the dielectric tensor is given for an isotropic medium, the functions, $\varepsilon_{\rm L}(\mathbf{k},\omega)$ and $\varepsilon_{\rm T}(\mathbf{k},\omega)$, are calculated according to

$$\varepsilon_{\rm L}(k,\omega) = \frac{\mathbf{k}\cdot\boldsymbol{\varepsilon}(\mathbf{k},\omega)\cdot\mathbf{k}}{k^2} \,, \tag{3.95}$$

$$\varepsilon_{\rm T}(k,\omega) = \frac{1}{2}\,[{\rm Tr}\;\boldsymbol{\varepsilon}(\mathbf{k},\omega) - \varepsilon_{\rm L}(k,\omega)] \tag{3.96}$$

where Tr means the summation over the diagonal elements.

The dispersion tensor (3.88) now reads

$$\Delta(\mathbf{k},\omega) = \varepsilon_{\rm L}(k,\omega)\mathbf{I}_{\rm L} + \left[\varepsilon_{\rm T}(k,\omega) - \left(\frac{kc}{\omega}\right)^2\right]\mathbf{I}_{\rm T} \,. \tag{3.97}$$

Thus, it is also diagonalized into purely longitudinal and transverse parts. The dispersion relation (3.92) splits itself into two equations:

$$\varepsilon_{\rm L}(k,\omega)=0\,, \tag{3.98}$$

$$\varepsilon_{\rm T}(k,\omega)=\left(\frac{kc}{\omega}\right)^2 . \tag{3.99}$$

The physical meaning of these equations may become clear if Eq. (3.97) and an identity

$$\mathbf{E}\equiv \mathbf{I}_{\rm L}\cdot\mathbf{E}+\mathbf{I}_{\rm T}\cdot\mathbf{E}$$

are substituted in Eq. (3.91). We thus find $\mathbf{I}_{\rm L}\cdot\mathbf{E}\neq\mathbf{0}$ when Eq. (3.98) is satisfied; Eq. (3.98) therefore represents the dispersion relation for the longitudinal mode. Similarly, $\mathbf{I}_{\rm T}\cdot\mathbf{E}\neq\mathbf{0}$ when Eq. (3.99) is satisfied; hence this equation gives the dispersion relation for the transverse mode in the plasma.

3.4 THE VLASOV PLASMAS

We now calculate the dielectric tensor for a plasma in a kinetic theory based on the Vlasov equation of Section 2.2. The results are applicable to classical plasmas in weak correlations.

A. The Vlasov-Maxwell Equations

The single-particle distribution function $f_\sigma(\mathbf{r},\mathbf{v};t)$ is defined for particles of the σ species as in the Vlasov equation (2.38); its normalization has been specified by Eq. (2.54), so that the local number density, for example, is given by

$$n_\sigma(\mathbf{r},t)=n_\sigma\int d\mathbf{v}f_\sigma(\mathbf{r},\mathbf{v};t) \tag{3.100}$$

where $n_\sigma\equiv N_\sigma/V$ is the average number density of the particles.

The electromagnetic fields, $\mathbf{E}(\mathbf{r},t)$ and $\mathbf{B}(\mathbf{r},t)$, in Eq. (2.38) are the sums of those applied from outside sources and those induced through internal distribution of the particles. Those fields satisfy the set of Maxwell equations (3.82).

The sets of Vlasov-Maxwell equations, (2.38) and (3.82), therefore provide a complete kinetic description of a plasma in the fluid limit. In this section we use these equations to calculate the dielectric tensor for a plasma with and without an external magnetic field [e.g., Sitenko & Stepanov 1956; Bernstein 1958].

B. Integration along Unperturbed Trajectories

The plasma under consideration is assumed to be in a homogeneous and stationary state at $t=-\infty$; it is characterized by the velocity distribution functions $f_\sigma(\mathbf{v})$. A uniform and constant magnetic field, $\mathbf{B}=B\hat{\mathbf{z}}$, is applied to the plasma from an external source. Here and hereafter, $\hat{\mathbf{z}}$ and analogous notation $\hat{\mathbf{x}}$ and $\hat{\mathbf{y}}$ denote the unit vectors in the directions along the respective Cartesian axes.

The electric field disturbance is turned on adiabatically as

$$\delta\mathbf{E}\exp[i(\mathbf{k}\cdot\mathbf{r}-\omega t)+\eta t]+\mathrm{cc}\,. \tag{3.101}$$

The positive infinitesimal η here serves to assure the adiabatic turn-on of the disturbance and thereby to guarantee the retarded response of the system; we let $\eta\to+0$ eventually. Associated with the electric field (3.101), there is a magnetic field disturbance determined from (3.86a) as

$$\begin{aligned}&\delta\mathbf{B}\exp[i(\mathbf{k}\cdot\mathbf{r}-\omega t)+\eta t]+\mathrm{cc}\\&=\frac{c}{\omega}\mathbf{k}\times\delta\mathbf{E}\exp[i(\mathbf{k}\cdot\mathbf{r}-\omega t)+\eta t]+\mathrm{cc}\,.\end{aligned} \tag{3.102}$$

In response to these disturbances, the distribution functions depart from their stationary values. We recall that any distribution functions uniform in space can be a solution to the Vlasov equation (2.38) in the absense of external perturbations. Within the confines of the linear response theory, the resultant distributions may thus be written as

$$f_\sigma(\mathbf{v})+\left\{\delta f_\sigma(\mathbf{v})\exp[i(\mathbf{k}\cdot\mathbf{r}-\omega t)+\eta t]+\mathrm{cc}\right\}. \tag{3.103}$$

Substituting Eqs. (3.101) to (3.103) in Eq. (2.38) and retaining only those terms linear in $\delta\mathbf{E}$ and $\delta f_\sigma(\mathbf{v})$, we find

$$\begin{aligned}&\left[\frac{\partial}{\partial t}+\mathbf{v}\cdot\frac{\partial}{\partial\mathbf{r}}+\Omega_\sigma(\mathbf{v}\times\hat{\mathbf{z}})\cdot\frac{\partial}{\partial\mathbf{v}}\right]\delta f_\sigma(\mathbf{v})\exp[i(\mathbf{k}\cdot\mathbf{r}-\omega t)+\eta t]\\&=-\frac{Z_\sigma e}{m_\sigma}\frac{\partial f_\sigma}{\partial\mathbf{v}}\cdot\mathsf{T}(\mathbf{k},\omega;\mathbf{v})\cdot\delta\mathbf{E}\exp[i(\mathbf{k}\cdot\mathbf{r}-\omega t)+\eta t]\,,\end{aligned} \tag{3.104}$$

where

$$\Omega_\sigma=\frac{Z_\sigma eB}{m_\sigma c} \tag{3.105}$$

is the *cyclotron frequency* for a particle of the σ species, and

$$\mathbf{T}(\mathbf{k},\omega;\mathbf{v}) \equiv \left(1 - \frac{\mathbf{k}\cdot\mathbf{v}}{\omega}\right)\mathbf{I} + \frac{\mathbf{k}\mathbf{v}}{\omega}\,. \tag{3.106}$$

This tensor enables one to sum the plane-wave contributions of Eqs. (3.101) and (3.102) and thereby to express the electromagnetic effects in terms of $\delta\mathbf{E}$ alone.

The current density induced by the electromagnetic-field fluctuations (3.101) and (3.102) is expressed as

$$\delta\mathbf{J} = \sum_{\sigma} Z_\sigma e n_\sigma \int d\mathbf{v}\mathbf{v}\,\delta f_\sigma(\mathbf{v})\,. \tag{3.107}$$

In the absence of the external magnetic field, i.e., when $\mathbf{B} = \mathbf{0}$, the increment $\delta f_\sigma(\mathbf{v})$ of the distribution, necessary for the calculations of Eq. (3.107) and the conductivity tensor through Eq. (3.83), can be easily obtained from Eq. (3.104) as

$$\delta f_\sigma(\mathbf{v}) = i\frac{Z_\sigma e}{m_\sigma}\frac{1}{\mathbf{k}\cdot\mathbf{v} - \omega - i\eta}\frac{\partial f_\sigma(\mathbf{v})}{\partial\mathbf{v}}\cdot\mathbf{T}(\mathbf{k},\omega;\mathbf{v})\cdot\delta\mathbf{E}\,. \tag{3.108}$$

In the presence of the external magnetic field, the calculation of the increment $\delta f_\sigma(\mathbf{v})$ from Eq. (3.104) is somewhat complicated because of the spiraling orbitals of the particles in the magnetic field. We need to find a method to "unwind" the spirals in order to follow the time evolution of the response in the plasma.

In these connections, we find it instructive to investigate the nature of the differential operators appearing on the left-hand side of Eq. (3.104), so that an appropriate integral operator may be found to invert such differentiations. We thus consider a differential operator generally defined as

$$\frac{d}{dt} \equiv \frac{\partial}{\partial t} + \frac{d\mathbf{r}}{dt}\cdot\frac{\partial}{\partial\mathbf{r}} + \frac{d\mathbf{v}}{dt}\cdot\frac{\partial}{\partial\mathbf{v}} \tag{3.109}$$

where $\mathbf{r} = \mathbf{r}(t)$ and $\mathbf{v} = \mathbf{v}(t)$ describe a given trajectory of a particle in the phase space $(\mathbf{r},\mathbf{v})$. Clearly, Eq. (3.109) represents a differentiation with respect to time along the prescribed particle orbital in the phase space.

The $d\mathbf{r}/dt$ and $d\mathbf{v}/dt$ are the velocity and the acceleration of the particle; let us choose

$$\frac{d\mathbf{r}}{dt} = \mathbf{v}\,, \quad \frac{d\mathbf{v}}{dt} = \Omega_\sigma(\mathbf{v}\times\hat{\mathbf{z}})\,. \tag{3.110}$$

These equations determine an unperturbed free orbital of a charged particle in a uniform magnetic field. With the knowledge that the particle position and velocity at time t are

$$\mathbf{r} = x\hat{\mathbf{x}} + y\hat{\mathbf{y}} + z\hat{\mathbf{z}}\,, \quad \mathbf{v} = v_x\hat{\mathbf{x}} + v_y\hat{\mathbf{y}} + v_z\hat{\mathbf{z}}\,, \tag{3.111}$$

those at t' are obtained from a solution to Eq. (3.110) as

$$\mathbf{r}' = x'\hat{\mathbf{x}} + y'\hat{\mathbf{y}} + z'\hat{\mathbf{z}}\,, \quad \mathbf{v}' = v'_x\hat{\mathbf{x}} + v'_y\hat{\mathbf{y}} + v'_z\hat{\mathbf{z}}\,. \tag{3.112}$$

The solution may be compactly expressed with the aid of tensors (represented in the Cartesian coordinates)

$$\mathsf{H}_\sigma(t) = \begin{pmatrix} \sin\Omega_\sigma t & 1-\cos\Omega_\sigma t & 0 \\ -(1-\cos\Omega_\sigma t) & \sin\Omega_\sigma t & 0 \\ 0 & 0 & \Omega_\sigma t \end{pmatrix}, \tag{3.113}$$

$$\begin{aligned} \mathsf{B}_\sigma(t) &= \frac{1}{\Omega_\sigma}\frac{d\mathsf{H}_\sigma(t)}{dt} \\ &= \begin{pmatrix} \cos\Omega_\sigma t & \sin\Omega_\sigma t & 0 \\ -\sin\Omega_\sigma t & \cos\Omega_\sigma t & 0 \\ 0 & 0 & 1 \end{pmatrix} \end{aligned} \tag{3.114}$$

as

$$\mathbf{v}' = \mathsf{B}_\sigma(t'-t)\cdot\mathbf{v}\,, \tag{3.115}$$

$$\mathbf{r}' = \mathbf{r} + \frac{1}{\Omega_\sigma}\mathsf{H}_\sigma(t'-t)\cdot\mathbf{v}\,. \tag{3.116}$$

As Eqs. (3.109) and (3.110) imply, the differential operators on the left-hand side of Eq. (3.104) altogether describe a time differentiation along the unperturbed orbital (3.115) and (3.116). We may thus invert this differentiation by integrating Eq. (3.104) with respect to time along the unperturbed particle trajectory as

$$\begin{aligned} &\delta f_\sigma(\mathbf{v})\exp[i(\mathbf{k}\cdot\mathbf{r}-\omega t)+\eta t] \\ &= -\frac{Z_\sigma e}{m_\sigma}\int_{-\infty}^{t} dt' \left[\frac{\partial f_\sigma(\mathbf{v}')}{\partial \mathbf{v}'}\right]\cdot \mathsf{T}(\mathbf{k},\omega,\mathbf{v}')\cdot\delta\mathbf{E}\exp[i(\mathbf{k}\cdot\mathbf{r}'-\omega t')+\eta t'] \end{aligned}$$

or

$$\delta f_\sigma(\mathbf{v}) = -\frac{Z_\sigma e}{m_\sigma}\int_0^\infty d\tau \left[\frac{\partial f_\sigma(\mathbf{v}')}{\partial \mathbf{v}'}\right] \cdot \mathbf{T}(\mathbf{k},\omega,\mathbf{v}') \cdot \delta\mathbf{E} \exp\left[-i\varphi(\tau) - \eta\tau\right] \tag{3.117}$$

where

$$\varphi(\tau) = \mathbf{k}\cdot(\mathbf{r}-\mathbf{r}') - \omega\tau, \quad \tau = t - t' . \tag{3.118}$$

An explicit evaluation of the integral in Eq. (3.117) consequently leads to the calculations of the induced-current density (3.107) and thereby the dielectric tensor via Eqs. (3.83) and (3.84).

C. Dielectric Tensor

The unperturbed velocity distribution functions $f_\sigma(\mathbf{v})$ in Eq. (3.117) may be written as functions of the components, $v_{||}$ and $v_\perp$, of the velocity parallel and perpendicular to the external magnetic field. (See Problem 2.4.) We thus set

$$\mathbf{v} = v_\perp \cos\theta\,\hat{\mathbf{x}} + v_\perp \sin\theta\,\hat{\mathbf{y}} + v_{||}\,\hat{\mathbf{z}} \tag{3.119}$$

and then Eq. (3.115) becomes

$$\begin{aligned} v_x' &= v_\perp \cos(\Omega_\sigma\tau + \theta)\,, \\ v_y' &= v_\perp \sin(\Omega_\sigma\tau + \theta)\,, \\ v_z' &= v_{||}\,. \end{aligned} \tag{3.120}$$

The induced-current density (3.107) is now calculated as

$$\delta\mathbf{J} = -\sum_\sigma \frac{(Ze)^2 n_\sigma}{m_\sigma}\int_0^{2\pi} d\theta \int_0^\infty v_\perp dv_\perp \int_{-\infty}^\infty dv_{||}$$

$$\times \int_0^\infty d\tau\,\mathbf{v}\left[\frac{\partial f_\sigma(\mathbf{v}')}{\partial \mathbf{v}'}\right] \cdot \mathbf{T}(\mathbf{k},\omega;\mathbf{v}') \cdot \partial\mathbf{E} \exp\left[-i\varphi(\tau) - \eta t\right]$$

whence the dielectric tensor is obtained as

$$\boldsymbol{\varepsilon}(\mathbf{k},\omega) = \mathbf{I} + \sum_\sigma \frac{\omega_\sigma^2}{i\omega}\int_0^{2\pi} d\theta \int_0^\infty v\,dv_\perp \int_{-\infty}^\infty dv_{||}$$

$$\times \int_0^\infty d\tau\,\mathbf{v}\left[\frac{\partial f_\sigma(\mathbf{v}')}{\partial \mathbf{v}'}\right] \cdot \mathbf{T}(\mathbf{k},\omega;\mathbf{v}') \exp\left[-i\varphi(\tau) - \eta t\right] .$$

Since

$$\frac{d}{d\tau}\exp[-i\varphi(\tau)-\eta\tau]=-i(\mathbf{k}\cdot\mathbf{v}'-\omega)\exp[-i\varphi(\tau)-\eta\tau],$$

we may carry out a partial integration with respect to τ for the first half of the integral term, to obtain

$$\boldsymbol{\varepsilon}(\mathbf{k},\omega)=\left(1-\frac{\omega_p^2}{\omega^2}\right)\mathbf{I}+\sum_\sigma\frac{\omega_\sigma^2}{\omega^2}\int_0^{2\pi}d\theta\int_0^\infty v_\perp dv_\perp\int_{-\infty}^\infty dv_{\parallel}$$

$$\times\int_0^\infty d\tau\,\mathbf{v}\left[\frac{\partial}{\partial\tau}\frac{\partial f_\sigma(\mathbf{v}')}{\partial\mathbf{v}'}-i\mathbf{k}\cdot\frac{\partial f_\sigma(\mathbf{v}')}{\partial\mathbf{v}'}\mathbf{v}'\right]\exp[-i\varphi(\tau)-\eta\tau]\,. \tag{3.121}$$

Here the plasma frequency is defined in terms of the partial plasma frequencies (2.70) as

$$\omega_p^2=\sum_\sigma\omega_\sigma^2\,. \tag{3.122}$$

In carrying through these integrations, we specifically choose the wave vector $\mathbf{k}$ on the x–z plane (see Fig. 3.2) and write

$$\mathbf{k}=k_\perp\hat{\mathbf{x}}+k_{\parallel}\hat{\mathbf{z}}\,. \tag{3.123}$$

We also note that

$$\frac{\partial f_\sigma(\mathbf{v}')}{\partial\mathbf{v}'}=\frac{\partial f_\sigma}{\partial v_\perp}[\cos(\Omega_\sigma\tau+\theta)\hat{\mathbf{x}}+\sin(\Omega_\sigma\tau+\theta)\hat{\mathbf{y}}]+\frac{\partial f_\sigma}{\partial v_{\parallel}}\hat{\mathbf{z}}\,. \tag{3.124}$$

Since

$$\varphi(\tau)=z[\sin(\Omega_\sigma\tau+\theta)-\sin\theta]+k_{\parallel}v_{\parallel}\tau-\omega\tau, \tag{3.125}$$

$$z\equiv\frac{k_\perp v_\perp}{\Omega_\sigma}, \tag{3.126}$$

a decomposition

$$\exp[-i\varphi(\tau)-\eta\tau]=\sum_{n=-\infty}^{\infty}\sum_{n'=-\infty}^{\infty}J_n(z)J_{n'}(z)$$

$$\times\exp\{-i[n(\Omega_\sigma\tau+\theta)-n'\theta+k_{\parallel}v_{\parallel}\tau-\omega\tau]-\eta\tau\} \tag{3.127}$$

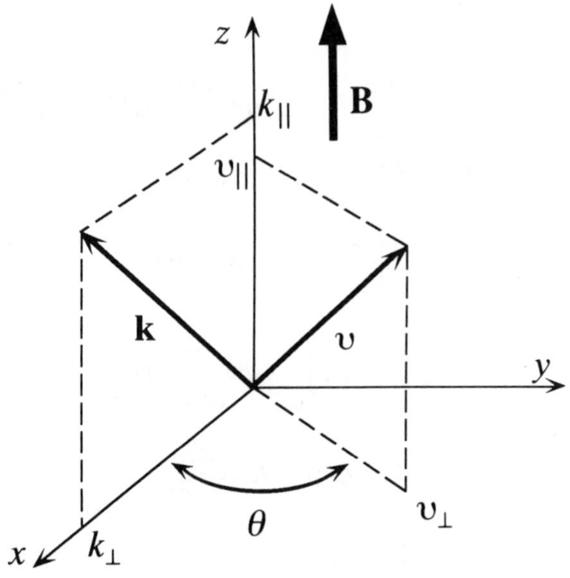

Figure 3.2 Configuration of various vectors.

is obtained.*

Expressions (3.119), (3.120), (3.124), and (3.127) are substituted in Eq. (3.121); integration with respect to θ then yields selection rules for the integers n and n' in the summations of Eq. (3.127). After a substantial amount of algebra, which we leave to the reader as an exercise (Problem 3.13), we find

$$\varepsilon(\mathbf{k},\omega)=\left(1-\frac{\omega_p^2}{\omega^2}\right)\mathbf{I}-\sum_\sigma\frac{\omega_\sigma^2}{\omega^2}\sum_{n=-\infty}^{\infty}\int d\mathbf{v}$$

$$\times\left(\frac{n\Omega_\sigma}{v_\perp}\frac{\partial f_\sigma}{\partial v_\perp}+k_{||}\frac{\partial f_\sigma}{\partial v_{||}}\right)\frac{\Pi_\sigma(v_\perp,v_{||};n)}{n\Omega_\sigma+k_\perp v_\perp-\omega-i\eta} \quad (3.128)$$

* Note that

$$\exp(-iz\sin\varphi)=\sum_{n=-\infty}^{\infty}J_n(z)\exp(-in\varphi)$$

where J_n is the Bessel function of the nth order.

where

$$\Pi_\sigma(v,v_{\|};n)=\begin{bmatrix} \dfrac{n^2\Omega_\sigma^2}{k_\perp^2}J_n^2 & iv_\perp\dfrac{n\Omega_\sigma}{k_\perp}J_nJ_n' & v_{\|}\dfrac{n\Omega_\sigma}{k_\perp}J_n^2 \\ -iv_\perp\dfrac{n\Omega_\sigma}{k_\perp}J_nJ_n' & v^2(J_n')^2 & -iv_{\|}v_\perp J_nJ_n' \\ v_{\|}\dfrac{n\Omega_\sigma}{k_\perp}J_n^2 & iv_{\|}v_\perp J_nJ_n' & v_{\|}^2J_n^2 \end{bmatrix}, \quad (3.129)$$

$$\int d\mathbf{v}\equiv 2\pi\int_0^\infty v_\perp dv_\perp\int_{-\infty}^\infty dv_{\|}, \quad (3.130)$$

$$J_n=J_n(z)\,,\quad J_n'=\frac{dJ_n(z)}{dz}, \quad (3.131)$$

and z has been defined by Eq. (3.126).

In the absence of the external magnetic field, it is straightforward to derive

$$\begin{aligned}\boldsymbol{\varepsilon}(\mathbf{k},\omega)&=\mathbf{I}-\sum_\sigma\frac{\omega_\sigma^2}{\omega}\int d\mathbf{v}\frac{\mathbf{v}}{\mathbf{k}\cdot\mathbf{v}-\omega-i\eta}\frac{\partial f_\sigma}{\partial\mathbf{v}}\cdot\mathbf{T}(\mathbf{k},\omega;\mathbf{v})\\ &=\left(1-\frac{\omega_p^2}{\omega^2}\right)\mathbf{I}-\sum_\sigma\frac{\omega_\sigma^2}{\omega^2}\int d\mathbf{v}\frac{\mathbf{v}\mathbf{v}}{\mathbf{k}\cdot\mathbf{v}-\omega-i\eta}\mathbf{k}\cdot\frac{\partial f_\sigma}{\partial\mathbf{v}}\end{aligned} \quad (3.132)$$

from Eq. (3.108).

We remark that the last terms in Eqs. (3.128) and (3.132) decreases as ω^{-3} in the limit of high frequencies. The dielectric tensors therefore behave asymptotically as

$$\boldsymbol{\varepsilon}(\mathbf{k},\omega)\rightarrow\left[1-\frac{\omega_p^2}{\omega^2}\right]\mathbf{I}\qquad(\omega\rightarrow\infty). \quad (3.133)$$

The asymptotic behavior (3.37) for the dielectric response function is a special case of (3.132), as we shall see in the next subsection.

D. Dielectric Response

The dielectric response function, defined through Eqs. (3.35) and (3.36), can be derived from the dielectric tensor defined via Eqs. (3.83) and (3.84).

To do so, we first note that the induced space-charge density is expressed as

$$\delta\rho_e = \sum_\sigma Z_\sigma e n_\sigma \int d\mathbf{v}\,\delta f_\sigma(\mathbf{v})\,. \tag{3.134}$$

This charge density is connected with the induced current (3.83) or (3.107) via the continuity equation of the type (3.81). In the electrostatic approximation, on which the dielectric response is based, the electric fields in Eq. (3.83) are given by (minus) the gradients of the electrostatic potentials, which are the sums between the external potentials of Eq. (3.34) and the induced potential fields of Eq. (3.36).

These relations are combined to yield a general expression,

$$\varepsilon(\mathbf{k},\omega) = \frac{\mathbf{k}\cdot\boldsymbol{\varepsilon}(\mathbf{k},\omega)\cdot\mathbf{k}}{k^2}\,, \tag{3.135}$$

a form identical with Eq. (3.95). In the absence of the external magnetic field, one finds from Eq. (3.132),

$$\varepsilon(\mathbf{k},\omega) = 1 - \sum_\sigma \frac{\omega_\sigma^2}{k^2}\int d\mathbf{v}\,\frac{1}{\mathbf{k}\cdot\mathbf{v}-\omega-i\eta}\,\mathbf{k}\cdot\frac{\partial f_\sigma(\mathbf{v})}{\partial\mathbf{v}}\,. \tag{3.136}$$

With a uniform external magnetic field, one likewise obtains from Eq. (3.128)

$$\varepsilon(\mathbf{k},\omega) = 1 - \sum_\sigma \frac{\omega_\sigma^2}{k^2}\sum_{n=-\infty}^{\infty}\int d\mathbf{v}$$

$$\times\left(\frac{n\Omega_\sigma}{v_\perp}\frac{\partial f_\sigma}{\partial v_\perp} + k_{||}\frac{\partial f_\sigma}{\partial v_{||}}\right)\frac{J_n^2(z)}{n\Omega_\sigma + k_{||}v_{||} - \omega - i\eta}\,. \tag{3.137}$$

These are *Vlasov dielectric response functions.* It can be proved that those dielectric responses satisfy the asymptotic properties of Eq. (3.37).

3.5 ELECTRON GAS AT FINITE TEMPERATURES

An electron gas is an OCP of electrons (electric charge $-e$ and mass m) embedded in a uniform neutralizing background of positive charges. We consider a homogeneous electron gas with average number density n $(= N/V)$ and temperature T. The system is characterized basically by two dimensionless plasma parameters,

$$\Gamma_e = \frac{e^2}{aT} \tag{3.138}$$

and the Fermi-degeneracy parameter Eq. (1.3), where a is the Wigner-Seitz radius defined by Eq. (1.4). Alternatively we may use the usual r_s parameter defined by Eq. (1.6), expressed in terms of those two parameters as

$$r_s = \frac{\Gamma_e \Theta}{2\lambda^2}, \tag{3.139}$$

where $\lambda \equiv (4/9\pi)^{1/3} = 0.521 \cdots$.

The conduction electrons in metals constitute a typical example of strongly coupled degenerate plasmas [e.g., Ichimaru 1982]. Since $\Theta << 1$, the Coulomb coupling parameter is given by r_s, which takes on values between 2 and 6 for such metallic electrons. It thus makes an interesting physical system in that correlation effects produced by the exchange and Coulomb interactions together with the quantum-mechanical interference effects between electrons affect the system properties in a fundamental way.

Those electrons with finite degrees of Fermi degeneracy and Coulomb coupling are likewise found in a variety of physical systems. Examples include the interior of main-sequence stars, the giant planets such as Jupiter, those plasmas in projected inertial-confinement fusion experiments, and liquid metals. In this section, we consider fundamentals of the dielectric formulation for such a quantum plasma at finite temperatures.

A. Fermi Distributions

The electron is a fermion with spin quantum number 1/2. The quantum states designated by the momentum $\mathbf{p} = \hbar\mathbf{k}$ and spin σ are occupied by the electrons, according to the *Fermi distribution* [e.g., Landau & Lifshitz 1969]

$$f_\sigma(\mathbf{p}) = \frac{1}{\exp[(\varepsilon_k - \mu_0)/T] + 1}. \tag{3.140}$$

Here

$$\varepsilon_k = \frac{(\hbar k)^2}{2m} \tag{3.141}$$

is the kinetic energy of an electron, and μ_0 is the chemical potential of the free-electron gas, which is determined from the normalization,

$$\int \frac{d\mathbf{p}}{(2\pi\hbar)^3} f_\sigma(\mathbf{p}) = n_\sigma, \tag{3.142}$$

where n_σ denotes the number density of the electrons with spin σ. For the electrons in a paramagnetic state, which we assume in this section, $n_\sigma = n/2$, a half of the total number density of the electrons.

In the treatment of a free-electron gas at finite temperatures, it is useful to define the *Fermi integrals*

$$I_\nu(\alpha) \equiv \int_0^\infty dt \frac{t^\nu}{\exp(t-\alpha)+1}. \tag{3.143}$$

The normalization condition (3.142) is then expressed as

$$I_{1/2}(\alpha) = \frac{2}{3}\Theta^{-3/2}, \tag{3.144}$$

where

$$\alpha = \frac{\mu_0}{T}. \tag{3.145}$$

The Fermi pressure P_F of the free-electron gas is likewise expressed as

$$\frac{P_F}{nT} = \Theta^{3/2} I_{3/2}(\alpha). \tag{3.146a}$$

The Helmholtz free energy F_0 of the ideal-gas part is then calculated as

$$\frac{F_0}{NT} = \alpha - \frac{P_F}{nT}, \tag{3.146b}$$

where N is the total number of the electrons.

Useful fitting formulas for the chemical potential and the Fermi pressure are:

$$\frac{\mu_0}{T} = -\frac{3}{2}\ln\Theta + \ln\frac{4}{3\sqrt{\pi}} + \frac{A\Theta^{-(b+1)} + B\Theta^{-(b+1)/2}}{1 + A\Theta^{-b}}, \tag{3.147}$$

with $A = 0.25954$, $B = 0.072$, and $b = 0.858$; and

$$\frac{P_F}{nT} = 1 + \frac{2}{5}\frac{X\Theta^{-(y+1)} + Y\Theta^{-(y+1)/2}}{1 + X\Theta^{-y}}, \tag{3.148}$$

with $X = 0.27232$, $Y = 0.145$, and $y = 1.044$. Maximum deviations of Eq. (3.147) from the exact values determined from Eq. (3.144) are about 0.19% at

$\Theta \approx 0.05$; those of Eq. (3.148) from the exact values determined from Eq. (3.146a) are about 0.26% at $\Theta \approx 5$.

In the classical limit, i.e., when $\Theta >> 1$, the Fermi integrals may be expanded as [e.g., Pathria 1972]

$$I_\nu(\alpha) = \Gamma(\nu+1) \sum_{s=1}^{\infty} (-1)^{s+1} \exp(s\alpha) s^{-(\nu+1)} , \tag{3.149}$$

where $\Gamma(z)$ is the gamma function. In this limit one thus has

$$\frac{\mu_0}{T} = -\frac{3}{2} \ln \Theta + \ln \frac{4}{3\sqrt{\pi}} , \tag{3.150}$$

$$\frac{P_{\mathrm{F}}}{nT} = 1 . \tag{3.151}$$

In the quantum limit of strong degeneracy, i.e., when $\Theta << 1$,

$$I_\nu(\alpha) = \frac{\alpha^{\nu+1}}{\nu+1} \left[1 + \sum_{s=1}^{\infty} 2(1 - 2^{1-2s}) \, \zeta(2s) \frac{(\nu+1)!}{(\nu+1-2n)!} \alpha^{-2s} \right] , \tag{3.152}$$

where $\zeta(z)$ is the zeta function. Hence

$$\mu_0 = E_{\mathrm{F}} , \tag{3.153}$$

$$P_{\mathrm{F}} = \frac{2}{5} n E_{\mathrm{F}} . \tag{3.154}$$

B. Density-Density Response

A dielectric response of such an electron gas may be formulated by the application of a weak external potential field, Eq. (3.34). The Hamiltonian of the system (with a unit volume) is given by [see e.g., Pines 1963; Fetter & Walecka 1971]

$$H_{\mathrm{tot}} = H + H_{\mathrm{ext}} , \tag{3.155}$$

$$H = \sum_{\mathbf{p}\sigma} \frac{p^2}{2m} c^+_{\mathbf{p}\sigma} c_{\mathbf{p}\sigma} + \frac{1}{2} \sum_{\mathbf{q}}{}' v(q) [\rho_{\mathbf{q}} \rho_{-\mathbf{q}} - n] , \tag{3.156}$$

$$H_{\mathrm{ext}} = e \Phi_{\mathrm{ext}}(\mathbf{k}, \omega) \rho_{-\mathbf{k}} \exp(-i\omega t + \eta t) + \mathrm{hc} . \tag{3.157}$$

Here $v(q) = 4\pi e^2/q^2$, hc stands for Hermitian conjugate, and $\sum_{\mathbf{q}}{}'$ implies the $\mathbf{q}$–summation with omission of the $q = 0$ term. The operators, $c^+_{\mathbf{p}\sigma}$ and $c_{\mathbf{p}\sigma}$, are the creation and annihilation operators of an electron with momentum $\mathbf{p}$ and spin σ, which satisfy the anticommutator relations for the fermions:

$$\{c^+_{\mathbf{p}\sigma}, c^+_{\mathbf{p}'\sigma'}\} = \{c_{\mathbf{p}\sigma}, c_{\mathbf{p}'\sigma'}\} = 0\,, \tag{3.158}$$

$$\{c^+_{\mathbf{p}\sigma}, c_{\mathbf{p}'\sigma'}\} = \delta_{\mathbf{pp}'}\delta_{\sigma\sigma'}\,. \tag{3.159}$$

In these equations, $\{A,B\} = AB + BA$; $\delta_{\mathbf{pp}'}$ and $\delta_{\sigma\sigma'}$ are Kronecker's deltas.

For the calculation of the density response, it is convenient to work with operators representing electron-hole pairs

$$\rho_{\mathbf{pk}\sigma} = c^+_{\mathbf{p}\sigma} c_{\mathbf{p}+\hbar k\sigma}\,. \tag{3.160}$$

Fourier components of the density fluctuations in Eqs. (3.156) and (3.157) are then given by

$$\rho_{\mathbf{k}} = \sum_{\mathbf{p}\sigma} \rho_{\mathbf{pk}\sigma}\,. \tag{3.161}$$

The occupation-number operator for the state $(\mathbf{p},\sigma)$ is expressed as

$$n_{\mathbf{p}\sigma} = \rho_{\mathbf{p}k=0\sigma}\,. \tag{3.162}$$

The Heisenberg equation of motion for $\rho_{\mathbf{pk}\sigma}$ is

$$i\hbar \frac{\partial}{\partial t}\rho_{\mathbf{pk}\sigma} = [\rho_{\mathbf{pk}\sigma}, H_{\mathrm{tot}}] \tag{3.163}$$

where [,] denotes a commutator. Explicit calculations with the aid of Eqs. (3.156) and (3.157) yield

$$i\hbar \frac{\partial}{\partial t}\rho_{\mathbf{pk}\sigma} = \hbar\omega_{\mathbf{pk}}\,\rho_{\mathbf{pk}\sigma} \tag{3.164a}$$

$$+\frac{1}{2} v(k)\left\{\rho_{\mathbf{k}}, n_{\mathbf{p}\sigma} - n_{\mathbf{p}+\hbar k\sigma}\right\} \tag{3.164b}$$

$$+\frac{1}{2}\sum_{\mathbf{q}(\neq k)}{}' v(q)\left\{\rho_{\mathbf{q}}, \rho_{\mathbf{pk}-\mathbf{q}\sigma} - \rho_{\mathbf{p}+\hbar\mathbf{q}\mathbf{k}-\mathbf{q}\sigma}\right\} \tag{3.164c}$$

$$+\,e\Phi_{\mathrm{ext}}(\mathbf{k},\omega)(n_{\mathbf{p}\sigma} - n_{\mathbf{p}+\hbar\mathbf{k}\sigma})\exp(-i\omega t + \eta t)\,, \tag{3.164d}$$

where

$$\omega_{\mathbf{pk}} = \frac{\mathbf{k}\cdot\mathbf{p}}{m} + \frac{\hbar k^2}{2m} \tag{3.165}$$

are the excitation frequencies for the electron-hole pairs.

We now set $\rho_{\mathbf{pk}\sigma}$ as a sum of the unperturbed part $\rho^{(0)}_{\mathbf{pk}\sigma}$ and the induced perturbation $\delta\rho_{\mathbf{pk}\sigma}\exp(-i\omega t + \eta t)$, and linearize Eq. (3.164) with respect to the perturbations. The induced density fluctuations and the density-density response function (3.26a) are then calculated as

$$\begin{aligned}\rho_{\text{ind}}(\mathbf{k},\omega) &= \left\langle \sum_{\mathbf{p}\sigma} \delta\rho_{\mathbf{pk}\sigma} \right\rangle \\ &= -e\chi(\mathbf{k},\omega)\Phi_{\text{ext}}(\mathbf{k},\omega)\,.\end{aligned} \tag{3.166}$$

Here $\langle \cdots \rangle$ means a statistical average over the unperturbed states.

Finally the dielectric response function is given by

$$\frac{1}{\varepsilon(\mathbf{k},\omega)} = 1 + v(k)\chi(\mathbf{k},\omega)\,, \tag{3.167}$$

according to Eq. (3.62).

C. The Hartree-Fock Approximation

The three terms (3.164a)–(3.164c) on the right-hand side govern the evolution of density-fluctuation excitations in the electron gas. The term (3.164a) describes free motions of the electron-hole pairs, and physically corresponds to the first term on the right-hand side of Eq. (1.46). The terms (3.164b) and (3.164c), on the other hand, stem from the Coulomb interaction. The former term represents a mean-field contribution equivalent to the second, plasma-oscillation term on the left-hand side of Eq. (1.46), while the latter describes the effect of nonlinear coupling between the density fluctuations as in the last term of Eq. (1.46).

In the *Hartree-Fock approximation*, the Coulomb-interaction terms (3.164b) and (3.164c) are ignored in the calculations of the dielectric response function. The Hartree-Fock density-density response function is thus calculated as

$$\chi^{\text{HF}}(\mathbf{k},\omega) = \chi_0(\mathbf{k},\omega)\,, \tag{3.168}$$

where

$$\chi_0(\mathbf{k},\omega) = \frac{2}{\hbar}\sum_{\mathbf{p}} \frac{1}{\omega - \omega_{\mathbf{pk}} + i\eta}[f_\sigma(\mathbf{p}) - f_\sigma(\mathbf{p} + \hbar\mathbf{k})] \tag{3.169}$$

is the *free-electron polarizability* [Khanna & Glyde 1976; Gouedard & Deutsch 1978] with the Fermi distribution $f_\sigma(\mathbf{p}) = \langle n_{\mathbf{p}\sigma}\rangle$ given by Eq. (3.140). The polarizability (3.169) has been evaluated with the retarded boundary conditions set forth in Section 3.1.

D. The Lindhard Dielectric Function

In the RPA, the nonlinear coupling term (3.164c) between the density fluctuations is neglected. The RPA dielectric function for the electron gas or the *Lindhard dielectric function* [Lindhard 1954] is thus calculated as

$$\varepsilon(\mathbf{k},\omega) = 1 - v(k)\chi_0(\mathbf{k},\omega)\,. \tag{3.170}$$

For the electron gas in the ground state (i.e., at $T = 0$), the static dielectric function is obtained from Eqs. (3.169) and (3.170) as

$$\varepsilon(\mathbf{k},0) = 1 + \frac{k_{\mathrm{TF}}^2}{k^2}\left\{\frac{1}{2} + \frac{k_{\mathrm{F}}}{2k}\left(1 - \frac{k^2}{4k_{\mathrm{F}}^2}\right)\ln\left|\frac{k + 2k_{\mathrm{F}}}{k - 2k_{\mathrm{F}}}\right|\right\}. \tag{3.171}$$

Here

$$k_{\mathrm{F}} = (3\pi^2 n)^{1/3} \tag{3.172}$$

is the *Fermi wave number* and

$$k_{\mathrm{TF}} = \sqrt{\frac{6\pi n e^2}{E_{\mathrm{F}}}} = 0.814\sqrt{r_s}\,k_{\mathrm{F}} \tag{3.173}$$

is the *Thomas-Fermi screening parameter.*

The stopping power of the metallic electrons can be analyzed in the way described in Section 3.2C. For the metallic electrons with strong Fermi degeneracy $\Theta << 1$, a condition analogous to (3.48) is given by

$$v_0 >> v_{\mathrm{F}}\,, \tag{3.174}$$

where $v_{\mathrm{F}} = (\hbar/m)(3\pi^2 n)^{1/3}$ is the Fermi velocity. With (3.174) the *f*-sum rule calculation of Eq. (3.49) remains valid. On the basis of Lindhard dielectric function Eq. (3.170), the stopping power has been calculated [Lindhard & Winther 1964] as

$$-\frac{dW}{dx}=\frac{4\pi Z_0^2 e^4 n}{mv_0^2}\left[\ln\frac{2mv_0^2}{\hbar\omega_p}-\frac{3}{5}\left(\frac{v_\mathrm{F}}{v_0}\right)^2-\frac{3}{14}\left(\frac{v_\mathrm{F}}{v_0}\right)^4-\cdots\right]. \tag{3.175}$$

The first term on the right-hand side is the Bethe sum-rule contribution. Corrections to the second terms due to finite degeneracies [Maynard & Deutsch 1982] and strong coupling effects between electrons [Tanaka & Ichimaru 1985] have been evaluated.

The collective mode of the electron gas is analyzed as in Section 3.2D. The resulting plasma oscillations in a degenerate electron gas are referred to as *plasmons*. The plasmon dispersion coefficient α defined in the long-wavelength expansion

$$\omega_k=\omega_p+\alpha\frac{\hbar}{m}k^2+\cdots \tag{3.176}$$

is then calculated from the solution to Eq. (3.54) with the RPA dielectric function Eq. (3.170) as

$$\alpha_\mathrm{RPA}=0.6\frac{E_\mathrm{F}}{\hbar\omega_p}=0.638r_s^{-1/2}. \tag{3.177}$$

The dispersion coefficient for metallic electrons has been measured by scattering of electron or X-ray beams [Raether 1980; Schülke et al. 1987].

E. Thermodynamic Functions

The thermodynamic functions of electron gas are formulated in terms of the dielectric response function Eq. (3.167). The internal energy U_int per unit volume is calculated as the statistical average of the second term on the right-hand side of Eq. (3.156); the internal energy per electron in units of thermal energy T is thus obtained as

$$u_\mathrm{ex}(\Gamma_e,\Theta)\equiv\frac{U_\mathrm{int}}{nT}=\frac{\Gamma_e}{\pi\lambda k_\mathrm{F}}\int_0^\infty dk[S(k)-1] \tag{3.178}$$

where $S(k)$ is the static structure factor of the electrons (cf., Eq. 3.28a). The excess free energy per electron in the same units is then calculated through the coupling-constant integrations [e.g., Fetter & Walecka 1971] as

$$f_{\text{ex}}(\Gamma_e,\Theta) \equiv \frac{F_{\text{ex}}}{NT} = \int_0^{\Gamma_e} d\Gamma \, \frac{u_{\text{ex}}(\Gamma,\Theta)}{\Gamma} \tag{3.179}$$

(Problem 3.21). The excess pressure is given [e.g., Ichimaru, Iyetomi & Tanaka 1987] by

$$p_{\text{ex}}(\Gamma_e,\Theta) \equiv \frac{P_{\text{ex}}}{nT} = \frac{1}{3}\, u_{\text{ex}}(\Gamma_e,\Theta) - \frac{2\Theta}{3}\left(\frac{\partial f_{\text{ex}}}{\partial \Theta}\right)_{\Gamma_e} . \tag{3.180}$$

In the Hartree-Fock approximation Eq. (3.168), the static structure factor is calculated as

$$S^{\text{HF}}(k) = \frac{3}{2}\,\Theta \sum_{l=-\infty}^{\infty} \Phi(x,l) \tag{3.181}$$

with $x = k/k_{\text{F}}$ where

$$\Phi(x,l) = -\frac{2E_{\text{F}}}{3n}\,\chi_0^f(k,z_l)$$

$$= \frac{1}{2x}\int_0^{\infty} dy \, \frac{y}{\exp\,[(y^2/\Theta) - \alpha] + 1} \ln \left| \frac{(2\pi l\Theta)^2 + (x^2 + 2xy)^2}{(2\pi l\Theta)^2 + (x^2 - 2xy)^2} \right| \tag{3.182}$$

is a dimensionless free-electron polarizability evaluated with the causal boundary conditions (3.22) [e.g., Pines & Nozières 1966].

The Hartree-Fock excess internal energy evaluated through Eq. (3.178) has been accurately parametrized by Perrot and Dharma-wardana [1984] as

$$u_{\text{ex}}^{\text{HF}}(\Gamma_e,\Theta) = -a_{\text{HF}}(\Theta)\Gamma_e \tag{3.183}$$

with

$$a_{\text{HF}}(\Theta) = \frac{(9\pi/4)^{1/3}}{\pi}\,\frac{a_0}{a_1}\tanh\left(\frac{1}{\Theta}\right), \tag{3.184a}$$

$$a_0 = 0.75 + 3.04363\Theta^2 - 0.09227\Theta^3 + 1.7035\Theta^4 , \tag{3.184b}$$

$$a_1 = 1 + 8.31051\Theta^2 + 5.1105\Theta^4 . \tag{3.184c}$$

In the limit of strong degeneracy ($\Theta \ll 1$), the Hartree-Fock excess free energy per electron is

$$\frac{F_{\rm ex}^{\rm HF}}{N} = -\frac{(9\pi/4)^{1/3}}{\pi}\frac{3}{2r_{\rm s}} = -\frac{0.916}{r_{\rm s}} \ ({\rm Ry}) \tag{3.185}$$

where

$$\mathrm{Ry} = \frac{me^4}{2\hbar^2} = 13.6\ (\mathrm{eV}) \tag{3.186}$$

is the *Rydberg energy* representing the binding energy of a hydrogen atom in the ground state. The first term 0.75 on the right-hand side of Eq. (3.184b) corresponds to Eq. (3.185).

In the RPA, the interaction energy [Montroll & Ward 1958; Englert & Brout 1960; Gupta & Rajagopal 1980; Dharma-wardana & Taylor 1981] can be evaluated analogously by substituting

$$\chi^{\rm RPA}(\mathbf{k},\omega) = \frac{\chi_0(\mathbf{k},\omega)}{1 - v(k)\chi_0(\mathbf{k},\omega)} \tag{3.187}$$

in Eqs. (3.28a) and (3.178). For a degenerate electron gas at high densities (i.e., $r_{\rm s} << 1$, meaning weak coupling), Gell-Mann and Brueckner [1957] thus obtained the first few terms of the ground-state energy in the $r_{\rm s}$ expansion [e.g., Pines 1963] as

$$\frac{F^{\rm GB}}{N} = \frac{2.21}{r_{\rm s}^2} - \frac{0.916}{r_{\rm s}} + 0.062 \ln r_{\rm s} - 0.096 \ln r_{\rm s} + \cdots \ ({\rm Ry})\,. \tag{3.188}$$

The first and second terms on the right-hand side derive from Eqs. (3.146b) and (3.185); the last term contains the second-order exchange contributions.

The RPA may not be valid for those electrons at metallic densities with $2 \le r_{\rm s} \le 6$, for example; the nonlinear term (3.164c), neglected in the RPA, needs to be retained. In the polarization potential approach of Section 3.2F, such an effect beyond the RPA is taken into account through the local field corrections $G(\mathbf{k},\omega)$.

For a strongly coupled electron liquid, the study of the local field corrections is therefore a subject of foremost importance; we shall do this later starting with Chapter 7. Here for the sake of deriving explicit expressions for the thermodynamic functions applicable to such an electron liquid at finite temperatures, we introduce a local field correction in the static approximation Eq. (3.78), due to Singwi, Tosi, Land, and Sjölander [1968], leaving its derivation in Problem 3.23 and later in Chapter 7. The STLS local-field correction is expressed as a functional of the static structure factor,

$$G^{\mathrm{STLS}}(k) = -\frac{1}{n}\sum_{\mathbf{q}}{}' \frac{\mathbf{k}\cdot\mathbf{q}}{q^2}\left[S(|\mathbf{k}-\mathbf{q}|)-1\right]. \tag{3.189}$$

In this approximation, the density-density response is given by

$$\chi^{\mathrm{STLS}}(\mathbf{k},\omega) = \frac{\chi_0(\mathbf{k},\omega)}{1 - v(k)[1-G^{\mathrm{STLS}}(k)]\chi_0(\mathbf{k},\omega)}. \tag{3.190}$$

The set of equations (3.28a), (3.189), and (3.190), assisted by Eq. (3.182), thus offers a complete, self-consistent scheme for the determination of $S(k)$ and $G(k)$, through which the thermodynamic properties of the electron liquids may be analyzed [Tanaka & Ichimaru 1986a].

The excess internal energies Eq. (3.178) calculated in the STLS scheme have been parametrized in analytic formulas as

$$u_{\mathrm{ex}}^{\mathrm{STLS}}(\Gamma_e,\Theta) = -\Gamma_e\frac{u_0}{u_1}, \tag{3.191}$$

$$u_0 = a(\Theta) + b(\Theta)\Gamma_e^{1/2} + c(\Theta)\Gamma_e, \tag{3.192a}$$

$$u_1 = 1 + d(\Theta)\Gamma_e^{1/2} + e(\Theta)\Gamma_e, \tag{3.192b}$$

with

$$a(\Theta) = a_{\mathrm{HF}}(\Theta), \tag{3.193}$$

$$b(\Theta) = \Theta^{1/2}\tanh(1/\Theta^{1/2})\frac{b_0}{b_1}, \tag{3.194}$$

$$b_0 = 0.341308 + 12.070873\Theta^2 + 1.148889\Theta^4,$$

$$b_1 = 1 + 10.495346\Theta^2 + 1.326623\Theta^4,$$

$$c(\Theta) = [0.872496 + 0.025248\exp(-1/\Theta)]e(\Theta), \tag{3.195}$$

$$d(\Theta) = \Theta^{1/2}\tanh(1/\Theta^{1/2})\frac{d_0}{d_1}, \tag{3.196}$$

$$d_0 = 0.614925 + 16.996055\Theta^2 + 1.489056\Theta^4,$$

$$d_1 = 1 + 10.10935\Theta^2 + 1.22184\Theta^4,$$

$$e(\Theta) = \Theta\tanh(1/\Theta)\frac{e_0}{e_1}, \tag{3.197}$$

$$e_0 = 0.539409 + 2.522206\Theta^2 + 0.178484\Theta^4 ,$$
$$e_1 = 1 + 2.555501\Theta^2 + 0.146319\Theta^4$$

[Ichimaru, Iyetomi & Tanaka 1987].

With the expression (3.191), the coupling-constant integration of Eq. (3.179) is performed to yield

$$f_{\rm ex}^{\rm STLS}(\Gamma_e,\Theta) = -\frac{c}{e}\Gamma_e - \frac{2}{e}\left(b - \frac{cd}{e}\right)\Gamma_e^{1/2} - \frac{1}{e}\left[\left(a - \frac{c}{e}\right) - \frac{d}{e}\left(b - \frac{cd}{e}\right)\right]$$
$$\times \ln \left| e\Gamma_e + d\Gamma_e^{1/2} + 1 \right| + \frac{2}{e(4e-d^2)^{1/2}}\left[d\left(a - \frac{c}{e}\right) + \left(2 - \frac{d^2}{e}\right)\left(b - \frac{cd}{e}\right)\right]$$
$$\times \left\{ \tan^{-1}\left[\frac{2e\Gamma_e^{1/2} + d}{(4e^2 - d^2)^{1/2}}\right] - \tan^{-1}\left[\frac{d}{(4e - d^2)^{1/2}}\right]\right\}. \tag{3.198}$$

The condition that $4e - d^2 > 0$ is satisfied for any Θ.

In the classical limit $\Theta >> 1$, the formulas (3.191) and (3.198) can accurately reproduce the OCP thermodynamic functions in Chapter 11. The functions $b(\Theta)$, $c(\Theta)$, $d(\Theta)$, and $e(\Theta)$ vanish at $\Theta = 0$ in such a way that Eqs. (3.191) and (3.198) become functions of r_s alone. Ceperley and Alder [1980] calculated the excess free energy of the degenerate electron liquid at various r_s values by the Green's function Monte Carlo (GFMC) method. In Table 3.1, the values of the correlation energy

$$E_c \equiv \frac{F_{\rm ex} - F_{\rm ex}^{\rm HF}}{N} \tag{3.199}$$

obtained for the ground state in the GFMC [Ceperley & Alder 1980], STLS [Singwi et al. 1968], and RPA schemes are compared.

Table 3.1 Correlation energies (mRy) of electron liquids at $T = 0$ in the GFMC, STLS, and RPA schemes.

r_s	GFMC	STLS	RPA
1	−119	−124	−157
2	−90.2	−92	−124
5	−56.3	−56	−85
10	−37.22	−36	

Spin-dependent correlations and thermodynamic functions for electron liquids at arbitrary degeneracy and spin polarization have been investigated through a solution to another self-consistent set of integral equations in the modified-convolution approximation, which we shall study in Chapter 7. Analytic expressions for the thermodynamic functions, analogous to Eqs. (3.191) and (3.198), have been derived; phase boundary curves, arising from divergence of the isothermal compressibility and of the spin susceptibility, are thereby obtained [Tanaka & Ichimaru 1989].

PROBLEMS

3.1 Prove Eq. (3.7).

3.2 Through consultation with an original article [e.g., Callen & Welton 1951; Kubo 1957] on the fluctuation-dissipation theorem or otherwise, show a derivation for Eq. (3.16).

3.3 Derive the first relation of Eq. (3.21) from an explicit calculation of the system's response to Eq. (3.19) with the aid of Eqs. (3.23) and (3.24).

3.4 In the presence of external disturbances a_k, let the total Hamiltonian of the system be

$$H_{\text{tot}} = H - \sum_k A_{-k} a_k \, .$$

Fluctuations induced by the disturbances are then calculated as

$$\delta B(k,t=0) = \langle B_k; \{H_{\text{tot}}\} \rangle - \langle B_k; H \rangle \, ,$$

where

$$\langle B_k; H \rangle = \frac{\text{Tr}\,[B_k \exp(-H/T)]}{\text{Tr}\exp(-H/T)}$$

is the canonical average with the Hamiltonian H. Show that the derivative $\partial\delta B(k,t=0)/\partial a_k$ gives the second expression in Eq. (3.21).

3.5 In the test-charge problem of Section 3.2C, assume the dielectric function given by

$$\varepsilon(\mathbf{k},\omega) = 1 - \left(\frac{\omega_p}{\omega}\right)^2 + i\eta\frac{\omega}{|\omega|}$$

(η is a positive infinitesimal) and find that the induced space-charge distribution is calculated as

$$Ze\rho_{\mathrm{ind}}(\mathbf{r},t) = \frac{Z_0 e}{\lambda_p}\,\delta(x)\delta(y)\sin[(z - v_0 t)/\lambda_p]\,\theta(v_0 t - z)\,.$$

Here $\theta(v_0 t - z)$ is the unit step function, $\lambda_p = v_0/\omega_p$, and the z-axis is chosen in the direction of $\mathbf{v}_0$.

3.6 In the case where a beam of energetic α particles is injected in an electron gas, calculate the energy of an α particle at which $k_{\max}$ of Eq. (3.51) equals $k_{\max}$ of Eq. (3.53). Examine the result in relation to the condition (3.48) for classical OCPs.

3.7 Prove Eq. (3.69).

3.8 Derive Eq. (3.70) in the two ways explained in the text.

3.9 Prove Eq. (3.79).

3.10 The total energy absorbed by the medium in the presence of an electric field disturbance $\mathbf{E}(\mathbf{r},t)$ is given by

$$E_{\mathrm{abs}} = \int d\mathbf{r} \int_{-\infty}^{\infty} \mathbf{E}(\mathbf{r},t)\cdot\mathbf{J}(\mathbf{r},t)\,.$$

Show that this can also be written in terms of the spectral components as

$$E_{\mathrm{abs}} = \frac{V}{2\pi}\sum_{\mathbf{k}}\int_{-\infty}^{\infty} d\omega\,\mathbf{E}^*(\mathbf{k},\omega)\cdot\frac{1}{2}[\sigma(\mathbf{k},\omega) + \sigma^+(\mathbf{k},\omega)]\cdot\mathbf{E}(\mathbf{k},\omega)$$

where $\sigma^+(\mathbf{k},\omega)$ is the Hermitian conjugate tensor of $\sigma(\mathbf{k},\omega)$, that is,

$$[\sigma^+(\mathbf{k},\omega)]_{ij} = [\sigma(\mathbf{k},\omega)]^*_{ji}\,.$$

[It follows from this calculation that the conductivity is anti-Hermitian ($\sigma = -\sigma^+$) when there is no absorption.]

3.11 For an isotropic medium, instead of $\varepsilon_{\mathrm{L}}(k,\omega)$ and $\varepsilon_{\mathrm{T}}(k,\omega)$ of Section 3.3C, it is customary to use another set of scalar functions, $\varepsilon(k,\omega)$ and $\mu(k,\omega)^{-1}$, to set relations

$$\mathbf{D}(k,\omega) = \varepsilon(k,\omega)\mathbf{E}(k,\omega)\,,$$

$$\mathbf{H}(k,\omega) = \mu(k,\omega)^{-1}\mathbf{B}(k,\omega)\,.$$

In fact, this may be the approach most frequently followed in electrodynamics. Set

$$\mathbf{D} = \left[\,\mathsf{I} - \frac{4\pi}{i\omega}(\mathsf{I} - \tau\,\mathsf{I}_{\mathrm{T}})\cdot\sigma\right]\cdot\mathbf{E}\,,$$

$$\mathbf{H} = \mathbf{B} + \tau\frac{4\pi}{ick^2}(\mathbf{k}\times\sigma)\cdot\mathbf{E}\,,$$

choose the value of τ properly, and show that the two sets of scalar functions have the relations

$$\varepsilon(k,\omega) = \varepsilon_{\mathrm{L}}(k,\omega)$$

$$\mu(k,\omega)^{-1} = 1 + \left(\frac{\omega}{ck}\right)^2\left[\varepsilon_{\mathrm{L}}(k,\omega) - \varepsilon_{\mathrm{T}}(k,\omega)\right].$$

3.12 In a fluid-dynamic approach [see e.g., Krall & Trivelpiece 1973], the behavior of various components in a plasma may be described by the continuity equations

$$\frac{\partial n_\sigma}{\partial t} + \frac{\partial}{\partial \mathbf{r}}\cdot(n_\sigma \mathbf{u}_\sigma) = 0$$

and the equations of motion

$$n_\sigma\left(\frac{\partial}{\partial t} + \mathbf{u}_\sigma\cdot\frac{\partial}{\partial \mathbf{r}}\right)\mathbf{u}_\sigma = -\gamma_\sigma\frac{T_\sigma}{m_\sigma}\frac{\partial n_\sigma}{\partial \mathbf{r}} + \frac{Z_\sigma e}{m_\sigma}n_\sigma\left(\mathbf{E} + \frac{\mathbf{u}_\sigma}{c}\times\mathbf{B}\right) - \frac{n_\sigma\mathbf{u}_\sigma}{\tau_\sigma}\,.$$

Here $\mathbf{u}_\sigma$ is the flow velocity, τ_σ the relaxation time, and γ_σ a constant of order unity. For the unperturbed state, we take a homogeneous plasma with a constant uniform magnetic field $\mathbf{B}$ in the z direction; we assume no net flows of the particles (i.e., $\mathbf{u}_\sigma = \mathbf{0}$) in this state. Calculate the dielectric tensor for this system and show that it is expressed as

$$\boldsymbol{\varepsilon}(\mathbf{k},\omega) = \mathbf{I} + \sum_\sigma \frac{\omega_\sigma^2}{\omega^2(1 - i\omega\tau_\sigma)} \frac{i\tau_\sigma\omega\,\mathbf{b}}{\omega(1 - i\omega\tau_\sigma) + i(\gamma_\sigma T_\sigma \tau_\sigma/m_\sigma)\mathbf{k}\cdot\mathbf{b}\cdot\mathbf{k}}$$

$$\cdot\left\{\left[\omega(1 - i\omega\tau_\sigma) + i\frac{\gamma_\sigma T_\sigma \tau_\sigma}{m_\sigma}\mathbf{k}\cdot\mathbf{b}\cdot\mathbf{k}\right]\mathbf{I} - i\frac{\gamma_\sigma T_\sigma \tau_\sigma}{m_\sigma}\mathbf{kk}\cdot\mathbf{b}\right\}$$

where

$$\mathbf{b} = \left[(1 - i\omega\tau_\sigma)^2 + (\Omega_\sigma \tau_\sigma)^2\right]^{-1}$$

$$\times\begin{bmatrix} (1 - i\omega\tau_\sigma)^2 & \Omega_\sigma\tau_\sigma(1 - i\omega\tau_\sigma) & 0 \\ -\Omega_\sigma\tau_\sigma(1 - i\omega\tau_\sigma) & (1 - i\omega\tau_\sigma)^2 & 0 \\ 0 & 0 & (1 - i\omega\tau_\sigma)^2 + (\Omega_\sigma\tau_\sigma)^2 \end{bmatrix}.$$

Examine the asymptotic property (3.133) for this dielectric tensor.

3.13 Carry out calculations leading from Eq. (3.121) to Eq. (3.128). The following formulas may be useful in these derivations:

$$\frac{2n}{z} J_n(z) = J_{n-1}(z) + J_{n+1}(z) ,$$

$$2J'_n(z) = J_{n-1}(z) - J_{n+1}(z) ,$$

$$\sum_{n=-\infty}^{\infty} J_n^2(z) = 1 ,$$

$$\sum_{n=-\infty}^{\infty} [nJ_n(z)]^2 = z^2/2 .$$

3.14 Derive Eq. (3.135).

3.15 Derive Eq. (3.137).

3.16 Show directly from Eqs. (3.136) and (3.137) that these functions have the asymptotic property (3.37).

3.17 When the velocity distribution functions $f_\sigma(\mathbf{v})$ depend only on v^2, show that the static dielectric function $\varepsilon(\mathbf{k},0)$ takes on the same values between Eqs. (3.136) and (3.137). In conjunction with the internal-energy formula Eq. (3.68a) and the excess free-energy formula Eq. (3.179), relate this result with the Bohr-Van Leeuwen theorem which states the absence of diamagnetism for a classical OCP in thermodynamic equilibrium.

3.18 Carry out computations of Eqs. (3.144) and (3.146a), and thereby examine the numerical accuracy for the formulas (3.147) and (3.148).

3.19 Derive Eq. (3.164).

3.20 When $T=0$, the Fermi distribution (3.140) becomes a unit step function. Calculate Eq. (3.169) with this distribution and show Eq. (3.171).

3.21 The Helmholtz free energy of electron gas per unit volume is given in terms of the logarithm of the partition function as

$$\frac{F}{V} = -nT \ln\left\{\mathrm{Tr}\exp\left(-\frac{H}{T}\right)\right\}$$

where H is the Hamiltonian (3.156). The Hamiltonian is the sum of the kinetic energy and the interaction energy (per unit volume), the latter of which is proportional to the Coulomb coupling parameter e^2 or Γ. Show a derivation of Eq. (3.179) by differentiation and subsequent integration of the equation above with respect to the coupling parameter.

3.22 Show that $\chi_0^e(k,z_l)$ calculated from Eq. (3.169) at $T=0$ can also be expressed as

$$4\pi v(k)\chi_0^e(k,z_l) = -\left(\frac{4}{9\pi}\right)^{1/3}\frac{r_s}{\pi^2 q}Q_q\left(\frac{s}{v_F}\right)$$

where $s=-iz_l/k$, $\mathbf{q}=\mathbf{k}/k_F$, and

$$Q_q(u) = \int d\mathbf{p}\int_{-\infty}^{\infty} dt \exp\left[ituq - |t|\left(\frac{1}{2}q^2 + \mathbf{q}\cdot\mathbf{p}\right)\right].$$

By this expression, Gell-Mann and Brueckner [1957] evaluated the RPA contributions to the correlation energy Eq. (3.188) in the ground state via Eq. (3.187).

3.23 Derive the STLS local-field correction Eq. (3.189) through the following procedure: Consider the BBGKY hierarchy Eq. (2.29) at $s = 1$, where one sets

$$\mathbf{E}_{\text{ext}}(\mathbf{r},t) = -i\mathbf{k}\Phi_{\text{ext}}(\mathbf{k},\omega)\exp[i(\mathbf{k}\cdot\mathbf{r} - \omega t)] + \text{cc},$$

$$\mathbf{B}_{\text{ext}}(\mathbf{r},t) = \mathbf{0}$$

in Eq. (2.14). In the presence of the external disturbance prescribed above, the single-particle distribution $F_1(1)$ is perturbed as in Eq. (3.103). Assume in the calculation of the density-density response that the two-particle distribution is expressed in accord with Eq. (2.30b) as

$$F_2(1,2) = F_1(1)F_1(2)g(r_{12}),$$

where $g(r_{12})$ remains *unperturbed* and is thus related with the unperturbed static structure factor $S(k)$ via Eq. (3.30). Obtain Eq. (3.189) through a comparison of the resultant density-density response with Eq. (3.190).

CHAPTER 4

Electromagnetic Properties of Vlasov Plasmas in Thermodynamic Equilibria

In Chapters 2 and 3, we have studied basic theoretical tools for the investigation of various plasma phenomena. In particular, we have learned that the dielectric tensor is capable of describing both the longitudinal electrostatic and the transverse electromagnetic properties of a plasma. In this chapter, we consider the dielectric properties of Vlasov plasmas in thermodynamic equilibria. We shall therefore consider the dielectric tensor evaluated with a Maxwellian velocity distribution; a function of a complex variable, the plasma dispersion function, will thereby be introduced.

In the first half of this chapter, the longitudinal dielectric properties will be studied. Propagation and the Landau damping of collective modes will be investigated both in OCPs and in two-component plasmas. In the latter plasmas, positive ions as well as electrons participate in the dynamic processes. It will be found that a new mode of wave propagation, the ion-acoustic wave, is possible in addition to the ordinary plasma oscillation in such a system. Finally we shall consider the cases in which a uniform magnetic field is applied to the plasma. Both the plasma oscillation and the ion-acoustic wave are then modified; an additional mode of propagation associated with the cyclotron motion of charged particles appears. The latter is called the Bernstein mode.

In the second half, we proceed to consider the propagation of electromagnetic waves in plasmas. We shall begin with the simple cases of an isotropic plasma without an external magnetic field. An important feature of wave propagation in such a plasma is a cutoff at the plasma frequency; an electromagnetic wave with a frequency less than the plasma frequency cannot propagate in the plasma. The cases of a plasma in an external magnetic field are subsequently treated. The cyclotron motion of charged particles around the magnetic lines of force brings about anisotropy to the medium properties; it introduces an additional frequency, the cyclotron frequency, to characterize the electromagnetic properties of the plasma. For a given electromagnetic phenomenon, we thus deal with a combined effect of at least three characteristic frequencies: the plasma frequency, the cyclotron frequency, and the frequency of the electromagnetic wave. Resonances at the cyclotron frequencies are remarkable features in such a magnetized plasma.

In the representation with circular polarizations, another remarkable feature emerges in that a magnetized plasma permits propagation of electromagnetic waves with frequencies much lower than the plasma frequency: helicon, Alfvén wave, and whistler are typical examples of such low-frequency modes of propagation. A phenomenon corresponding to Landau damping now appears in the form of Doppler-shifted cyclotron resonances. Finally we shall consider wave propagation perpendicular to the magnetic field. Two possible modes of propagation, ordinary and extraordinary modes, will be investigated.

4.1 DIELECTRIC RESPONSES IN MAXWELLIAN PLASMAS

The Maxwellian velocity distribution (1.1) is applicable for a classical plasma in thermodynamic equilibrium. We begin with a calculation of the dielectric response for such a plasma without a magnetic field. This will be followed by a treatment of the dielectric tensor in a magnetized plasma.

A. The Plasma Dispersion Function

In the absence of an external magnetic field, the x axis is arbitrarily chosen in the direction of the wave vector. For an OCP, Eq. (3.136) reads

$$\varepsilon(\mathbf{k},\omega) = 1 + \left(\frac{k_D}{k}\right)^2 \sqrt{\frac{m}{2\pi T}} \int_{-\infty}^{\infty} dv_x \frac{kv_x}{kv_x - \omega - i\eta} \exp\left(-\frac{mv_x^2}{2T}\right)$$

$$= 1 + \left(\frac{k_D}{k}\right)^2 W\left(\frac{\omega}{k\sqrt{T/m}}\right) \tag{4.1}$$

where k_D is the Debye wave number defined by Eq. (1.52). In the last step, *the W function* has been newly introduced according to

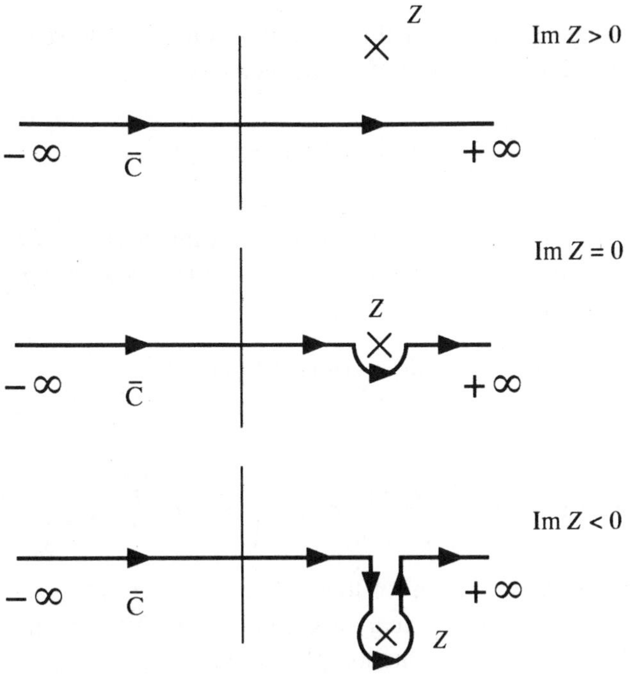

Figure 4.1 Integration contour $\overline{\mathrm{C}}$.

$$W(Z) = \frac{1}{\sqrt{2\pi}} \int_{-\infty}^{\infty} dx \frac{x}{x - Z - i\eta} \exp\left(-\frac{x^2}{2}\right).$$

This is a well-defined, analytic function in the upper half of the complex Z plane; it is then continued analytically into the lower half-plane. Such an analytical continuation may be realized when the contour of x integration is deformed so that the point Z always stays above it. We thus write

$$W(Z) = \frac{1}{\sqrt{2\pi}} \int_{\overline{\mathrm{C}}} dx \frac{x}{x - Z} \exp\left(-\frac{x^2}{2}\right) \tag{4.2}$$

where $\overline{\mathrm{C}}$ is the integration contour as specified in Fig. 4.1. The W function obeys the differential equation

$$\frac{dW}{dZ} = \left(\frac{1}{Z} - Z\right) W - \frac{1}{Z}$$

with the boundary condition $W(0) = 1$.

For an actual calculation of Eq. (4.2), it is convenient to start with the case where Im $Z > 0$. For real values of x, one then has

$$\frac{1}{x-Z} = i\int_0^\infty dt \exp\left[-i(x-Z)t\right] .$$

This identity is substituted in Eq. (4.2) and the x integration is first carried out. The remaining t integration may then be performed by splitting the integrand into two parts via

$$\exp(iZt) = \cos(Zt) + i\sin(Zt) .$$

We thus find

$$W(Z) = 1 - Z\exp\left(-\frac{Z^2}{2}\right)\int_0^Z dy \exp\left(\frac{y^2}{2}\right) + i\sqrt{\frac{\pi}{2}}Z\exp\left(-\frac{Z^2}{2}\right). \tag{4.3}$$

Finally, we use this expression for those Z in the lower half plane as well, and thereby complete the required analytical continuaton. The W function Eq. (4.3) is closely related to the error function of a complex argument.

In Fig. 4.2, the typical behavior of the W function is illustrated as a function of X, the real part of Z; the imaginary part Y is varied as a parameter. It is important to note that the curves do not cross the negative real axis when $Y > 0$; one or more intersections start to appear when $Y < 0$.

For analytical purposes, it is useful to find series expansions of $W(Z)$ for small and large arguments. When $|Z| < 1$, $W(Z)$ can be expressed in a convergent series

$$W(Z) = i\sqrt{\frac{\pi}{2}}Z\exp\left(-\frac{Z^2}{2}\right) + 1 - Z^2 + \frac{Z^4}{3} - \cdots + \frac{(-1)^{n+1}Z^{2n+2}}{(2n+1)!!} + \cdots \tag{4.4}$$

where

$$(2n+1)!! = (2n+1)(2n-1)\cdots 3\cdot 1 . \tag{4.5}$$

In the regime of large $|Z|$, an asymptotic series expansion,

$$W(Z) = i\sqrt{\frac{\pi}{2}}Z\exp\left(-\frac{Z^2}{2}\right) - \frac{1}{Z^2} - \frac{3}{Z^4} - \cdots - \frac{(2n-1)!!}{Z^{2n}} - \cdots, \tag{4.6}$$

is applicable.

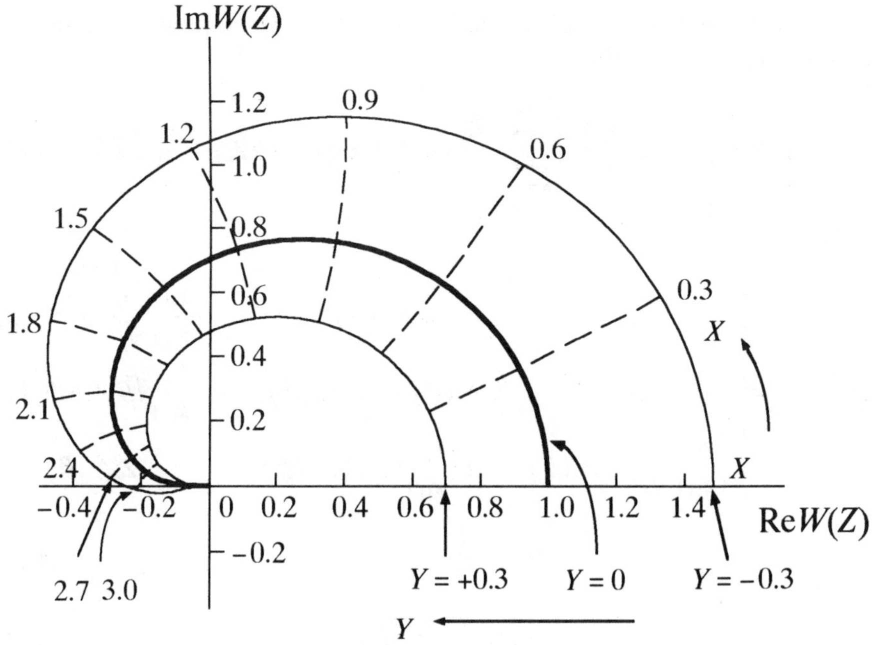

Figure 4.2 Behavior of the function $W(Z)$ on the complex plane (Z=X+iY).

B. Plasmas in Constant External Magnetic Field

We now proceed to consider the cases of a plasma with a constant uniform magnetic field. The system is no longer isotropic. For analyses of electromagnetic properties, all the elements of the dielectric tensor must be evaluated.

The explicit expression for the dielectric tensor may be obtained by substituting the Maxwellian (cf., Problem 2.4):

$$f_M(v_\perp, v_\parallel) = \left(\frac{m}{2\pi T}\right)^{3/2} \exp\left[-\frac{m(v_\perp^2 + v_\parallel^2)}{2T}\right] \tag{4.7}$$

in Eq. (3.128). The calculation is facilitated by recalling a relation between the Bessel functions:

$$2a^2 \int_0^\infty dx x J_n(px) J_n(qx) \exp(-a^2x^2) = \frac{1}{2a^2} I_n\left(\frac{pq}{2a^2}\right) \exp\left(-\frac{p^2+q^2}{4a^2}\right) \tag{4.8}$$

where I_n is the modified Bessel function of the nth order.

The expression for the dielectric tensor of the Vlasov plasmas in thermodynamic equilibrium is thus

$$\varepsilon(\mathbf{k},\omega) = \mathbf{I} - \left(\frac{\omega_p}{\omega}\right)^2 \left\{ \sum_{n=-\infty}^{\infty} \frac{Z_0}{Z_n} \Pi(\beta, Z_n; n)[1 - W(Z_n)] - Z_0^2 \hat{\mathbf{z}}\hat{\mathbf{z}} \right\} \tag{4.9}$$

where

$$\Pi(\beta, Z_n; n) = \begin{bmatrix} \frac{n^2}{\beta}\Lambda_n(\beta) & in\Lambda_n'(\beta) & \frac{k_\parallel}{|k_\parallel|}\frac{n}{\sqrt{\beta}} Z_n\Lambda_n(\beta) \\ -in\Lambda_n'(\beta) & \frac{n^2}{\beta}\Lambda_n(\beta) - 2\beta\Lambda_n'(\beta) & -i\frac{k_\parallel}{|k_\parallel|}\sqrt{\beta}\, Z_n\Lambda_n'(\beta) \\ \frac{k_\parallel}{|k_\parallel|}\frac{n}{\sqrt{\beta}} Z_n\Lambda_n(\beta) & i\frac{k_\parallel}{|k_\parallel|}\sqrt{\beta}\, Z_n\Lambda_n'(\beta) & Z_n^2\Lambda_n(\beta) \end{bmatrix}, \tag{4.10}$$

$$Z_n = \frac{\omega - n\Omega}{|k_\parallel|\sqrt{T/m}}, \tag{4.11}$$

$$\beta \equiv \frac{k_\perp^2 T}{m\Omega^2}, \tag{4.12}$$

$$\Lambda_n(\beta) \equiv I_n(\beta)\exp(-\beta). \tag{4.13}$$

The functions $\Lambda_n(\beta)$ appear commonly in a theoretical treatment of Maxwellian plasmas in a magnetic field. It is instructive to note: (i) $0 \le \Lambda_n(\beta) \le 1$; (ii) $\Lambda_0(\beta)$ decreases monotonically from $\Lambda_0(0) = 1$; and (iii) $\Lambda_n(\beta)$ for $n \ne 0$ start from $\Lambda_n(0) = 0$, reach the maxima, and then decrease. For large values of β, those functions approach

$$\Lambda_n(\beta) \to \frac{1}{\sqrt{2\pi\beta}} \exp\left(-\frac{n^2}{2\beta}\right) \quad (\beta \to \infty). \tag{4.14}$$

These may be used as approximate representations for $\Lambda_n(\beta)$ when the harmonic number n is large. Equation (4.14) then indicates that the maxima of $\Lambda_n(\beta)$ occur approximately at $\beta \approx n^2$ or

$$|n\Omega| \approx k_\perp \sqrt{\frac{T}{m}}, \tag{4.15}$$

and that the maximum values of $\Lambda_n(\beta)$ decrease as $1/|n|$.

An important application of (4.14) takes place when the limit of $B \to 0$ is considered. In this limit, the cyclotron frequency Ω approaches zero. We may thus replace the summation over the harmonic numbers by an integration over a continuous variable ξ according to

$$\Lambda_n(\beta) \to \frac{|\Omega|}{k_\perp \sqrt{2\pi T/m}} \exp\left(-\frac{mn^2\Omega^2}{2k_\perp^2 T}\right),$$

$$n\Omega \to \xi ,$$

$$\sum_{n=-\infty}^{\infty} \to \frac{1}{|\Omega|}\int_{-\infty}^{\infty} d\xi . \tag{4.16}$$

Such a scheme of going over to a continuum limit offers a convenient device by which a connection between a calculation with a magnetic field and one without a magnetic field may be establilshed.

4.2 DYNAMIC SCREENING

The dielectric response function provides a direct measure of the extent to which an external test charge is screened by the induced space charge in the plasma, as we remarked in Section 3.2C. The screening is dynamic in that it depends, not only on the wave vector, but on the frequency variable as well. In this section we study a number of consequences stemming from such a dynamic screening.

A. Debye Screening and Effective Mass Correction

Let us revisit the cases treated in Section 3.2C where a test particle with electric charge Z_0e, mass M_0, and velocity $\mathbf{v}_0$ was introduced in the OCP. Substituting Eq. (4.1) into Eq. (3.44), we find that the induced potential field is expressed as

$$\Phi_{\text{ind}}(\mathbf{r},t) = -4\pi Z_0 e \sum_{\mathbf{k}} \left[\frac{1}{k^2} - \frac{1}{k^2 + k_{\text{D}}^2 W(\mu\beta_0)}\right] \exp\left[i\mathbf{k}\cdot(\mathbf{r}-\mathbf{v}_0 t)\right] , \tag{4.17}$$

where μ is the direction cosine between $\mathbf{k}$ and $\mathbf{v}_0$, and

$$\beta_0 \equiv \frac{v_0}{\sqrt{T/m}} . \tag{4.18}$$

When the test charge is stationary at the origin (i.e., $\mathbf{v}_0 = 0$), the static dielectric function,

$$\varepsilon(\mathbf{k},0) = 1 + \frac{k_D^2}{k^2}, \tag{4.19}$$

enters the evaluation of the induced potential field Eq. (4.17); it is now independent of time and is calculated as

$$\Phi_{\text{ind}}(\mathbf{r}) = -\frac{Z_0 e}{r}[1 - \exp(-k_D r)]. \tag{4.20}$$

Summation between Eqs. (4.20) and (1.15), the bare field produced by the test charge, reproduces the Debye-Hückel result Eq. (1.20). The Debye screening is therefore a consequence of the dielectric response function Eq. (4.1).

The interaction energy $E_{\text{int}}(v_0)$ between the test charge and the plasma may be calculated from Eq. (4.17) as

$$E_{\text{int}}(v_0) = Z_0 e \Phi_{\text{ind}}(\mathbf{r},t)$$

$$= -\frac{(Z_0 e)^2}{\pi} \int_{-1}^{1} d\mu \int_0^{\infty} dk \frac{k_D^2 W(\mu\beta)}{k^2 + k_D^2 W(\mu\beta_0)}. \tag{4.21}$$

For a slowly moving test charge such that $\beta_0 \ll 1$, an expansion of Eq. (4.21) with respect to β_0 yields

$$E_{\text{int}}(v_0) = 2\frac{U_{\text{int}}}{n} + \left(1 - \frac{\pi}{8}\right)\frac{(Z_0 e)^2 k_D m v_0^2}{6T} + \cdots \tag{4.22}$$

where U_{int} has been given by Eq. (1.29). An effective mass correction ΔM_0 of the test charge due to its interaction with the plasma is then calculated as

$$\Delta M_0 = \left[\frac{\partial^2 E_{\text{int}}(v_0)}{\partial v_0^2}\right]_{v_0=0} = \frac{1}{3}\left(1 - \frac{\pi}{8}\right)\frac{(Z_0 e)^2 k_D m}{T}. \tag{4.23}$$

The effective mass correction stems from a deformation of the screening cloud from a spherical shape (4.20) to an ellipsoidal one. As Eq. (4.17) illustrates, $k_D\sqrt{W(\mu\beta_0)}$ may be regarded as an effective Debye wave number; when $\beta_0 \neq 0$, this quantity, being a function of μ, leads to a screening length with a directional dependence.

B. Stopping Power and Cherenkov Emission of Plasma Waves

When the velocity of the test charge is large so that (3.48) is satisfied, the stopping power is given generally by Eq. (3.49).

Let us now pay attention to that part of the stopping power arising from the interaction of the test charge with collective modes $\omega \approx \pm\omega_p$ of the plasma. In reference to Eq. (3.60), we substitute

$$\mathrm{Im}\left[\frac{1}{\varepsilon(\mathbf{k},\omega)}\right] = -\frac{\pi}{2}\omega_p[\delta(\omega - \omega_p) - \delta(\omega + \omega_p)]\,,\,(\omega \approx \pm\omega_p) \quad (4.24)$$

in Eq. (3.47) to obtain

$$-\left(\frac{dW}{dt}\right)_{\mathrm{coll}} = 2\pi^2(Z_0 e\omega_p)^2 \sum_{k\,(<k_D)} \frac{1}{k^2}[\delta(\omega_p - \mathbf{k}\cdot\mathbf{v}_0) + \delta(\omega_p + \mathbf{k}\cdot\mathbf{v}_0)]\,. \quad (4.25)$$

The test particle loses its energy to the collective modes of the plasma at this rate; hence it must correspond to the rate at which plasma waves are emitted by a passage of the test particle in the plasma. Condition for the emission is derived from the arguments in the δ functions as

$$\omega_p = |\mathbf{k}\cdot\mathbf{v}_0|\,, \quad (4.26)$$

the relation known for the *Cherenkov emission* of the plasma waves. To satisfy both (4.26) and $k < k_D$ for plasma waves, the particle velocity must exceed the thermal velocity in the plasma.

It is instructive to note that this emission mechanism does not depend on the amplitude of the plasma wave already existent in the plasma. Such a process thus corresponds to a *spontaneous emission* of plasma waves, in contrast with the induced emission processes to be considered in Section 4.3B.

4.3 PLASMA OSCILLATIONS AND LANDAU DAMPING

In Section 3.2D, the dispersion relation (3.54) for the collective mode was derived. In this section we study the consequences of these equations for an OCP and for a two-component plasma consisting of electrons and ions in thermodynamic equilibrium.

A. Collective Mode

In light of Eq. (4.1), the dispersion relation reads

$$1+\left(\frac{k_{\mathrm{D}}}{k}\right)^2 W\left(\frac{\omega}{k\sqrt{T/m}}\right)=0\,. \tag{4.27}$$

This equation can be solved analytically in the high-frequency regime

$$\left|\frac{\omega}{k\sqrt{T/m}}\right| >> 1\,, \tag{4.28}$$

where the expansion (4.6) is applicable for the W function; Eq. (4.27) becomes

$$1-\frac{\omega_p^2}{\omega^2}-\frac{3\omega_p^2k^2(T/m)}{\omega^4}-\cdots+i\sqrt{\frac{\pi}{2}}\frac{\omega k_{\mathrm{D}}^2}{k^3\sqrt{T/m}}\exp\left[-\frac{\omega^2}{2k^2(T/m)}\right]=0\,. \tag{4.29}$$

We seek a solution to this equation in the form Eq. (3.56) under the assumption (3.57), to find

$$\omega_k^2=\omega_p^2+3\frac{T}{m}k^2+\cdots, \tag{4.30}$$

$$\gamma_k=\sqrt{\frac{\pi}{8}}\,\omega_p\left(\frac{k_{\mathrm{D}}}{k}\right)^3\exp\left(-\frac{m\omega_k^2}{2k^2T}\right). \tag{4.31}$$

In order that the solution (4.30) be consistent with (4.28), the values of k must be restricted to

$$k << k_{\mathrm{D}}\,. \tag{4.32}$$

This condition in turn guarantees the validity of (3.57). Consequently, Eqs. (4.30) and (4.31) represent a correct solution to Eq. (4.27) in the long-wavelength regime (4.32). As k increases, these expressions for ω_k and γ_k become inaccurate. The correct values of the W function must be used for a solution to Eq. (4.27) in such a short-wavelength regime.

Equation (4.30) obviously corresponds to the collective mode (the plasma oscillation) considered in Section 1.4A. For $k \neq 0$, the frequency is slightly shifted from ω_p due to thermal motion of the individual particles.

In the long-wavelength regime (4.32), the plasma oscillations (4.30) are well-defined elementary excitations with relatively long lifetimes. Let us now split the dielectric response function into the real and imaginary parts as in Eq. (3.55). In the vicinity of the collective modes $\omega = \pm\omega_k$, we may write

$$\varepsilon(\mathbf{k},\omega) = 1 - \left(\frac{\omega_p}{\omega}\right)^2 - \left(\frac{\omega_p}{\omega}\right)^2 \frac{\omega_k^2 - \omega_p^2}{\omega^2} - \cdots, \tag{4.33}$$

and then Eq. (3.60) gives

$$\mathrm{Im}\frac{1}{\varepsilon(\mathbf{k},\omega)} \approx -\frac{\pi\,\omega_k^5}{2\omega_p^2(3\omega_k^2 - 2\omega_p^2)}[\delta(\omega - \omega_k) - \delta(\omega + \omega_k)]\,. \tag{4.34}$$

This approximate expression is examined in light of the ~ sum rule Eq. (3.42). A calculation directly from Eq. (4.34) yields

$$\int_{-\infty}^{\infty} d\omega\, \omega\, \mathrm{Im}\frac{1}{\varepsilon(\mathbf{k},\omega)} = -\frac{\pi\,\omega_k^6}{\omega_p^2(3\omega_k^2 - 2\omega_p^2)}\,.$$

Although ω_k^2 exceeds ω_p^2 by a term proportional to k^2, we here find that Eq. (4.34) exhausts the sum rule Eq. (3.42) up to the contributions proportional to k^2. A simple approximation in accord with the sum rule may be obtained by adjusting the coefficients of the δ functions and thereby writing, instead of Eq. (4.34),

$$\mathrm{Im}\frac{1}{\varepsilon(\mathbf{k},\omega)} \approx -\frac{\pi}{2}\frac{\omega_p^2}{\omega_k}[\delta(\omega - \omega_k) - \delta(\omega + \omega_k)] \quad (k \ll k_{\mathrm{D}})\,. \tag{4.35}$$

The fact that the collective modes $\omega = \pm\omega_k$ completely exhaust the sum rule Eq. (3.42) for $k \ll k_{\mathrm{D}}$ implies that the values of the integrand $\omega\,\mathrm{Im}[1/\varepsilon(k,\omega)]$ are zero outside the domain near $\omega = \pm\omega_k$. Hence a spectrum with two-peak structure as in Fig. 1.4 is indicated.

To complete the comparison with the results of Section 1.4B, we extend the sum-rule analysis into the domain $k \gg k_{\mathrm{D}}$. Here we have $\varepsilon_1 = 1$, and

$$\mathrm{Im}\frac{1}{\varepsilon(\mathbf{k},\omega)} = -\sqrt{\frac{\pi}{2}}\frac{\omega k_{\mathrm{D}}^2}{k^3\sqrt{T/m}}\exp\left[-\frac{\omega^2}{2k^2(T/m)}\right] \qquad (k \gg k_{\mathrm{D}})\,. \tag{4.36}$$

Clearly, this spectrum represents the contribution arising from thermal motion of the individual particles, with the shape as depicted in Fig. 1.5. It is easy to confirm that Eq. (4.36) likewise satisfies the sum rule Eq. (3.42).

B. Wave-Particle Interaction

The appearance of the positive imaginary part (4.31) implies that the plasma oscillation cannot live forever. This phenomenon is called the *Landau damping*, since the existence of such a damping was first predicted by Landau [1946].

A question that may naturally arise in these connections is the relation between this damping and an apparent time reversibility of the Vlasov equation. According to Landau's original treatment, macroscopic quantities such as the electric field and charge density are damped exponentially, but perturbations in the phase-space distribution $f(\mathbf{r},\mathbf{v};t)$ of the particles oscillate indefinitely. Since the charge denstity is given by an integration such as Eq. (3.100), we may regard the damping as arising from the phase mixing between various parts of the distribution function. A macroscopic physical quantity might therefore reappear in the plasma if the direction of phase evolution in the microscopic elements can be reversed. Such a possibility has been clearly demonstrated in a plasma-wave echo process; this will be treated in Chapter 5.

In a quantum-theoretical approach, one can analyse the time rate of change in the wave intensities through calculations of the rates of emission and absorption for the wave quanta (the plasmons) by individual particles [Pines & Schrieffer 1962]. Thus, the Landau damping may be interpreted as arising from the balance between the rate of absorption and that of induced emission. The spontaneous emission, on the other hand, represents an effect to the first order in discreteness parameters, and hence outside the scope of the Vlasov equation; it was treated separately in Section 4.2B.

A classical counterpart to the foregoing explanation may be given in terms of the resonant coupling between the waves and those particles moving with approximately the same velocities as the phase velocity $v_p = \omega/k$ of the wave [Bohm & Gross 1949a, b]. We assume a plasma wave with wave number k and frequency ω, with an amplitude sufficiently small so that a linearized treatment is applicable. Figure 4.3 shows the potential variation associated with the space-charge distribution in the frame co-moving with the wave. Those charged particles with velocities almost the same as the phase velocity of the wave interact resonantly with the wave, continuously sensing the wave potential at a same phase. The condition for the wave-particle resonance is thus expressed as

$$\omega = \mathbf{k} \cdot \mathbf{v} . \tag{4.37}$$

If the amplitude of the wave is small but finite, particles (in the domain A of Fig. 4.3) moving slightly faster than the wave will be decelerated through the resonant interaction with the wave; their individual velocities decrease to

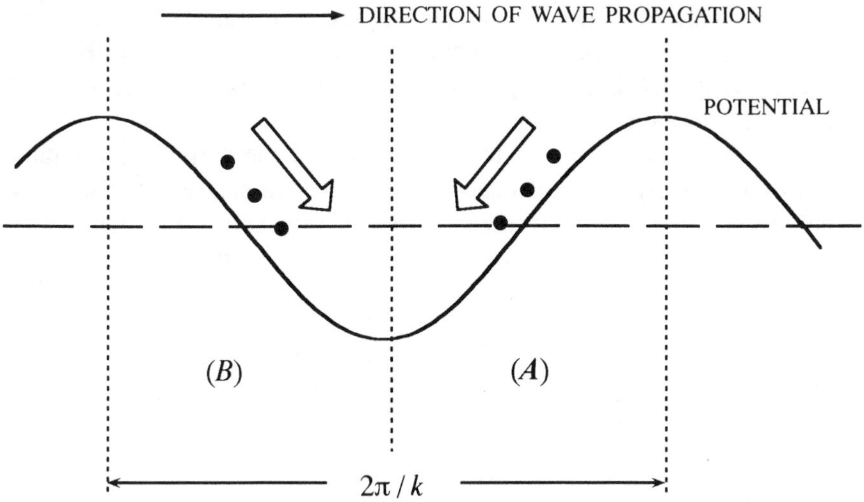

Figure 4.3 Resonant interaction between waves and particles.

v_p and their extra kinetic energies are transferred to the wave. Particles in B moving slightly slower than the wave will be accelerated and thereby absorb energy from the wave. Since the distribution decreases as the velocity increases, a damping takes place because there are more slow particles to absorb energy than fast particles to transfer energy to the wave. If, on the contrary, the distribution function increases with velocity near v_p, it is possible that the wave will grow in amplitude, an indication of a plasma-wave instability, which we shall study in Chapter 6.

C. Electron and Ion Two-Component Plasmas

We now consider the cases of two-component plasmas, where both electrons and ions participate in the dynamics. We shall use the subscripts $\sigma = e$ for the electrons and $\sigma = i$ for the ions. The number density of the electrons is denoted by n, so that the ion-number density is given by $n_i = n/Z$ due to the macroscopic charge neutrality. For simplicity, we may sometimes write $m_e = m$ and $m_i = M$.

The velocity distributions are expressed as

$$f_\sigma(\mathbf{v}) = \sqrt{\frac{m_\sigma}{2\pi T_\sigma}} \exp\left(-\frac{m_\sigma v^2}{2T_\sigma}\right) \qquad (\sigma = e, i)\,. \tag{4.38}$$

Thus we take into account a possibility that the temperatures of the electrons and of the ions may be different, while each species may be described by its own Maxwellian distribution Eq. (4.38). To begin, we examine the plausibility of such an assumption.

Approach to equilibrium or Maxwellization of the particle distribution is achieved through exchange of energies between particles via collisions. The relaxation times for the Maxwellization of electrons and of ions and for temperature equality between them are in the approximate ratios

$$\tau_{ee} : \tau_{ii} : \tau_{ei} \approx 1 : \sqrt{\frac{m_i}{m_e}} : \frac{m_i}{m_e} . \tag{4.39}$$

Because

$$\frac{m_i}{m_e} >> 1 \tag{4.40}$$

the use of the Maxwellians with unequal temperatures is reasonable.

The ordering (4.39) can be understood from a consideration of binary elastic collisions, as in Section 1.3A. We recall in particular that the cross sections for energy and momentum transfers are related by Eq. (1.14).

When $m_1 = m_2$, we find $\langle 2V/v \rangle \approx 1$, so that $Q_m \approx Q_E$. Since the rate of momentum transfer as calculated in Eq. (1.28) is proportional to $\mu^{-1/2}$, the relation

$$\frac{\tau_{ee}}{\tau_{ii}} \approx \sqrt{\frac{m_e}{m_i}}$$

follows.

When $m_1 << m_2$, the transfer of energy becomes quite inefficient, so that $Q_E \approx (m_1/m_2)Q_m$. Since the mass of the lighter species enters the rate of momentum transfer Eq. (1.28) in these circumstances, the remaining relation

$$\frac{\tau_{ee}}{\tau_{ei}} \approx \frac{m_e}{m_i}$$

is obtained.

Substitution of the distributions (4.38) in Eq. (3.136) yields the dielectric response function

$$\varepsilon(\mathbf{k},\omega) = 1 + \frac{k_e^2}{k^2} W\left(\frac{\omega}{k\sqrt{T_e/m_e}}\right) + \frac{k_i^2}{k^2} W\left(\frac{\omega}{k\sqrt{T_i/m_i}}\right) \tag{4.41}$$

where

$$k_\sigma^2 = \frac{4\pi n_\sigma (Z_\sigma e)^2}{T_\sigma} . \tag{4.42}$$

The properties of the collective modes may be investigated through the dispersion relation Eq. (3.54).

The high-frequency collective mode corresponding to the plasma oscillation of Eq. (4.30) may be obtained by expanding the two W functions in Eq. (4.41) through the asymptotic series (4.6). Assuming thus

$$|\omega| \gg k\sqrt{\frac{T_e}{m_e}} > k\sqrt{\frac{T_i}{m_i}} , \tag{4.43}$$

we find the solution to Eq. (3.54) as

$$\omega_k^2 = \omega_p^2 + \frac{3}{\omega_p^2}\left[\omega_e^2 \frac{T_e}{m_e} + \omega_i^2 \frac{T_i}{m_i}\right] k^2 , \tag{4.44}$$

$$\gamma_k = \sqrt{\frac{\pi}{8}}\frac{\omega_k^2}{k^3}\left\{\frac{k_e^2}{\sqrt{T_e/m_e}}\exp\left[-\frac{\omega_k^2}{2k^2(T_e/m_e)}\right] + \frac{k_i^2}{\sqrt{T_i/m_i}}\exp\left[-\frac{\omega_k^2}{2k^2(T_i/m_i)}\right]\right\} \tag{4.45}$$

where Eq. (3.122) has been used. The solution is consistent with the assumption (4.43) in the long-wavelength regime:

$$k^2 \ll k_e^2 . \tag{4.46}$$

The nature of the collective mode obtained here is essentially the same as that discussed in Section 4.2A. Both ω_k and γ_k are slightly modified, however, from the values in Eqs. (4.30) and (4.31) because both electrons and ions participate in the collective motions. The last term in Eq. (4.45) represents the Landau damping due to the ions.

Thus far we have not discovered anything particularly new in our study of a two-component system. Presence of the heavier ions, however, makes it possible to look for an additional, intermediate-frequency domain, other than (4.43), in which the imaginary parts of the dielectric response function may take on small values, that is,

$$k\frac{\sqrt{T_e}}{m_e} \gg |\omega| \gg k\frac{\sqrt{T_i}}{m_i} . \tag{4.47}$$

In this frequency domain, the small argument expansion (4.4) for the electrons and the large argument expansion (4.6) for the ions may be used in Eq. (4.41). Equation (3.54) may then be solved through the procedure of the previous subsection with assumption (3.57). In the long-wavelength regime such that $k^2 << k_e^2$, we find

$$\omega = \omega_k \left\{ \pm 1 - i\sqrt{\frac{\pi}{8}} \left[\sqrt{\frac{m_e}{m_i}} + \left(\frac{T_e}{T_i}\right)^{3/2} \exp\left(-\frac{T_e}{2T_i} - \frac{3}{2} \right) \right] \right\} \tag{4.48a}$$

where

$$\omega_k = \sqrt{\frac{T_e}{m_i} + \frac{3T_i}{m_i}}\, k . \tag{4.48b}$$

In light of this solution, we see that (3.57) is valid when both (4.40) and

$$\frac{T_e}{T_i} >> 1 \tag{4.49}$$

are satisfied; these conditions also guarantee the frequency domain (4.47).

The collective mode described by (4.48) is called the *ion-acoustic wave*. In the long-wavelength limit, it has a dispersion relation of sound-wave type:

$$\omega_k = sk . \tag{4.50a}$$

In the temperature range of (4.49), the propagation velocity is

$$s = \sqrt{\frac{T_e}{m_i}} \tag{4.50b}$$

and its damping rate is

$$\left| \frac{\gamma_k}{\omega_k} \right| = \sqrt{\frac{\pi}{8}} \sqrt{\frac{m_e}{m_i}} . \tag{4.51}$$

The ion-acoustic wave is essentially the collective mode associated with the dynamics of ions. If the electrons are smeared out, the resultant ion OCP has the collective modes with frequencies $\pm\omega_i$, the ion plasma frequency. This frequency is, however, so low that the mobile electrons can easily follow the motion of the ions. The electrons respond adiabatically to the ionic motion and thereby act to screen their electrostatic fields; the screening constant appropriate to these adiabatic circumstances is clearly (cf., Eq. 4.19)

$$\varepsilon_e(\mathbf{k}) = 1 + \left(\frac{k_e}{k}\right)^2 . \tag{4.52}$$

As a result, the collective mode of the ions is described by the screened plasma frequency

$$\omega_k^2 = \frac{\omega_i^2}{\varepsilon_e(\mathbf{k})} . \tag{4.53}$$

In the limit of long wavelengths, Eq. (4.53) approaches Eq. (4.50). As the wave number increases, the frequency deviates from the linear relation of Eq. (4.50) and tends to the ion plasma frequency; there the ion Landau damping becomes substantial. The earlier mathematical treatment based on (4.47) sustains the physical picture presented above.

4.4 EXCITATIONS IN MAGNETIZED PLASMAS

In the presence of a magnetic field, charged particles move in helical orbitals along the magnetic lines of force. Extents of motions in the perpendicular directions are thereby suppressed substantially. In the limit of a strong magnetic field, the plasma behaves as if it were a one-dimensional gas. These features are naturally reflected in the properties of the collective modes: Propagation characteristics of the plasma oscillations in a magnetic field are greatly modified from those without a magnetic field. Another notable feature is the periodicity brought about by the cyclotron motion of the particles. A plasma thus exhibits a new collective mode, propagating mainly in directions perpendicular to the magnetic field, with a frequency near the fundamental or a higher-harmonic frequency of the cyclotron motion [Bernstein 1958]. In this section we shall study these aspects of longitudinal excitations both for the OCP and for the two-component, electron-ion plasma.

The longitudinal dielectric function for the plasmas under present consideration may be obtained from the dielectric tensor (4.9) according to Eq. (3.135) as

$$\varepsilon(\mathbf{k},\omega) = 1 + \frac{k_{\mathrm{D}}^2}{k^2}\left\{1 + \sum_n \frac{\omega}{\omega - n\Omega}\left[W\left(\frac{\omega - n\Omega}{|k_\parallel|\sqrt{T/m}}\right) - 1\right]\Lambda_n(\beta)\right\} . \tag{4.54}$$

The nature of the collective modes may be investigated by substituting Eq. (4.54) into the dispersion relation (3.54).

A. Plasma Oscillations

We begin with the cases of an OCP in which the magnetic field is assumed to be strong, so that

$$\Omega^2 >> \omega_p^2 . \tag{4.55}$$

In the vicinity of the plasma frequency, the dielectric function (4.54) may be approximated as

$$\varepsilon(\mathbf{k},\omega) \approx 1 + \frac{k_D^2}{k^2}\left\{[1 - \Lambda_0(\beta)] + W\left(\frac{\omega}{|k_{||}|\sqrt{T/m}}\right)\Lambda_0(\beta)\right\}, \tag{4.56}$$

because the neglected terms are smaller by a factor on the order of $(\omega_p/\Omega)^2$. In view of Eq. (4.56), the plasma oscillation can be a well-defined excitation only in the long-wavelength regime, $k^2 << k_D^2$. Combination of this and (4.55) yields

$$\beta << 1 \tag{4.57}$$

for the plasma oscillation.

We now seek for a solution to Eq. (3.54) in the form

$$\omega = \pm\omega_0(\mathbf{k}) - i\gamma_0(\mathbf{k}) . \tag{4.58}$$

With the aid of expansion (4.6) as well as (4.55) and (4.57), we then obtain from Eq. (4.56)

$$\omega_0(\mathbf{k}) = \frac{|k_{||}|}{k}\omega_p\sqrt{\Lambda_0(\beta)} \approx \frac{|k_{||}|}{k}\omega_p , \tag{4.59a}$$

$$\frac{\gamma_0(\mathbf{k})}{\omega_0(\mathbf{k})} = \frac{\sqrt{\pi}}{8}\left(\frac{k_D}{k}\right)^3 [\Lambda_0(\beta)]^{3/2} \exp\left[-\frac{k_D^2\Lambda_0(\beta)}{2k^2}\right]. \tag{4.59b}$$

Equation (4.59a) clearly indicates a one-dimensional propagation of the plasma oscillation in the direction of the magnetic field. For longevity of such a wave, the condition

$$k^2 << k_D^2\Lambda_0(\beta) \tag{4.60}$$

needs to be satisfied.

Extension of the aforementioned considerations to the cases of the two-component, electron-ion plasmas is rather straightforward. Because of the large mass ratio (4.40), the properties of the electron plasma oscillation are little affected by the presence of the ions. We shall therefore be concerned only with the low-frequency collective modes associated with the ions.

We look for a solution corresponding to the ion-acoustic wave in the frequency domain such that

$$|k_{||}|\sqrt{\frac{T_i}{m_i}} << |\omega| << |k_{||}|\sqrt{\frac{T_e}{m_e}} \tag{4.61a}$$

and in the long-wavelength limit such that

$$|k_{||}|\sqrt{\frac{T_e}{m_e}} < \Omega_i . \tag{4.61b}$$

The dielectric response function is then simplified as

$$\varepsilon(\mathbf{k},\omega) = 1 + \frac{k_e^2}{k^2} + \frac{k_i^2}{k^2}[1 - \Lambda_0(\beta_i)] - \frac{k_{||}^2}{k^2}\frac{\omega_i^2}{\omega^2}\Lambda_0(\beta_i) + i\sqrt{\frac{\pi}{2}}$$

$$\times\left\{\frac{\omega}{|k_{||}|\sqrt{T_e/m_e}}\frac{k_e^2}{k^2} + \frac{\omega}{|k_{||}|\sqrt{T_i/m_i}}\frac{k_i^2}{k^2}\Lambda_0(\beta_i)\exp\left[-\frac{\omega^2}{2k_{||}^2(T_i/m_i)}\right]\right\}. \tag{4.62}$$

Adopting the form of (4.58), we find the solution to Eq. (3.54) as

$$\omega_0(\mathbf{k}) = \frac{|k_{||}|}{k}\omega_i\sqrt{\frac{\Lambda_0(\beta_i)}{\varepsilon_e(\mathbf{k}) + [1 - \Lambda_0(\beta_i)]k_i^2/k^2}} .$$

In the limit of long wavelengths, this expression reduces to

$$\omega_0(\mathbf{k}) = |k_{||}|s . \tag{4.63}$$

Again, Eq. (4.63) shows a one-dimensional propagation of the ion-acoustic wave in the direction of the magnetic field.

The frequency domain (4.61a) implies (4.49). In the long-wavelength limit, we then obtain a simplified expression for the damping rate

$$\frac{\gamma_0(\mathbf{k})}{\omega_0(\mathbf{k})} = \sqrt{\frac{\pi}{8}}\sqrt{\frac{m_e}{m_i}} . \tag{4.64}$$

This formula is essentially the same as Eq. (4.51).

B. The Bernstein Modes

In addition to the plasma oscillation (4.59), the dielectric function (4.54) permits a propagation of collective modes with frequencies in the vicinity of the harmonics of the cyclotron frequency. In such a frequency domain, we may write

$$\varepsilon(\mathbf{k},\omega) = 1 + \frac{k_D^2}{k^2}[1 - \Lambda_0(\beta)] - \frac{k_D^2}{k^2}\frac{\omega}{\omega - n\Omega}\left[1 - W\left(\frac{\omega - n\Omega}{|k_{||}|\sqrt{T/m}}\right)\right]\Lambda_n(\beta)$$

$$(\omega \approx n\Omega;\ n = \pm 1, \pm 2, \dots)\,. \tag{4.65}$$

We again seek a solution to the dispersion relation in the form

$$\begin{aligned}\omega &= \omega_n(\mathbf{k}) - i\gamma_n(\mathbf{k}) \\ &= n\Omega[1 + \Delta_n(\mathbf{k})] - i\gamma_n(\mathbf{k})\end{aligned} \tag{4.66}$$

with an assumption

$$|k_{||}|\frac{\sqrt{T/m}}{|n\Omega|} << |\Delta_n(\mathbf{k})| << 1\,. \tag{4.67}$$

We thus obtain

$$\Delta_n(\mathbf{k}) = \frac{k_D^2\Lambda_n(\beta)}{k^2 + k_D^2[1 - \Lambda_0(\beta)]}\,, \tag{4.68a}$$

$$\left|\frac{\gamma_n(\mathbf{k})}{\omega_n(\mathbf{k})}\right| = \sqrt{\frac{\pi}{2}}\frac{|n\Omega|[\Delta_n(\mathbf{k})]^2}{|k_{||}|\sqrt{T/m}}\exp\left\{-\frac{[n\Omega\Delta_n(\mathbf{k})]^2}{2k_{||}^2(T/m)}\right\}. \tag{4.68b}$$

Roughly, $\Lambda_n(\beta)$ takes on the maximum value $1/|n|$ at (4.15). Hence we estimate

$$[\Delta_n(\mathbf{k})]_{\max} \approx \frac{1}{|n|^3}\left(\frac{\omega_p}{\Omega}\right)^2. \tag{4.69}$$

In light of (4.55), the second half of (4.67) is thus guaranteed.

The first half of (4.67) amounts to the condition $|\gamma_n(\mathbf{k})/\omega_n(\mathbf{k})| << 1$. With the aid of (4.15), we may then rewrite this condition approximately as

$$|k_{||}| << k_\perp\Delta_n(\mathbf{k})\,. \tag{4.70}$$

In view of (4.69), we find that the propagation directions of these collective modes are confined near the directions perpendicular to the magnetic field. In particular, when $k_{\parallel}=0$, we have $\gamma_n(\mathbf{k})=0$; undamped waves propagate across the magnetic field.

Physically these features can be understood in terms of the wave-particle resonance picture of Section 4.3B. When a wave propagates with an infinite phase velocity in a direction parallel to the magnetic field (i.e., $|k_{\parallel}| \to 0$), no particles are available to interact resonantly with the wave, and hence no Landau damping. When $k_{\parallel}$ becomes finite, some particles can satisfy the resonance condition, $\omega_n \approx k_{\parallel}v_{\parallel}$; the waves may be damped.

Finally let us investigate the dispersion relation in the vicinity of the harmonics $n\Omega_i$ of the ion cyclotron frequency in the two-component plasmas. The dielectric response function may be approximated in these cases as

$$\varepsilon(\mathbf{k},\omega) = 1 + \frac{k_e^2}{k^2} + \frac{k_i^2}{k^2}[1-\Lambda_0(\beta_i)] - \frac{k_i^2}{k^2}\frac{\omega}{\omega - n\Omega_i}\left[1 - W\left(\frac{\omega - n\Omega_i}{|k_{\parallel}|\sqrt{T_i/m_i}}\right)\right]\Lambda_n(\beta_i)$$

$$(\omega \approx n\Omega_i;\ n=\pm1,\pm2,\ldots)\,. \tag{4.71}$$

Equation (4.71) differs from Eq. (4.65) by a static screening stemming from the electrons. Again setting as in Eq. (4.66) for the ions, we obtain

$$\Delta_n(\mathbf{k}) = \frac{k_i^2\Lambda_n(\beta_i)}{k^2\varepsilon_e(\mathbf{k}) + k_i^2[1-\Lambda_0(\beta_i)]}\,, \tag{4.72a}$$

$$\left|\frac{\gamma_n(\mathbf{k})}{n\Omega_i}\right| = \sqrt{\frac{\pi}{2}}\frac{|n\Omega_i|[\Delta_n(\mathbf{k})]^2}{|k_{\parallel}|\sqrt{T_i/m_i}}\exp\left\{-\frac{[n\Omega_i\Delta_n(\mathbf{k})]^2}{2k_{\parallel}^2(T_i/m_i)}\right\}. \tag{4.72b}$$

In many practical cases, it is reasonable to assume

$$\omega_i^2 >> \Omega_i^2 \tag{4.73}$$

even with a substantially strong magnetic field. In these circumstances, we may approximate

$$\Delta_n(\mathbf{k}) \approx \Lambda_n(\beta_i)\frac{T_e}{T_e+T_i}\,. \tag{4.74}$$

The wave vectors are restricted within the regime

$$|k_{\parallel}| << k_{\perp}\Delta_n(\mathbf{k})\,. \tag{4.75}$$

Propagations nearly perpendicular to the magnetic field are therefore indicated in these cases as well.

4.5 ELECTROMAGNETIC WAVES IN PLASMAS WITHOUT EXTERNAL MAGNETIC FIELD

We now consider the propagation of electromagnetic waves in plasmas near thermodynamic equilibrium, beginning with the simple cases of an isotropic plasma without an external magnetic field.

A. The Transverse Dielectric Function

The transverse dielectric function is calculated in accord with Eq. (3.96). Substituting the Maxwellian Eq. (1.1) in Eq. (3.132) we obtain

$$\varepsilon_T(k,\omega) = 1 - \frac{\omega_p^2}{\omega^2}\left[1 - W\left(\frac{\omega}{k\sqrt{T/m}}\right)\right]. \tag{4.76}$$

Since the phase velocity ω/k of the resultant transverse wave will exceed light velocity c in vacuum, the W function should vanish in Eq. (4.76); no particles can match the condition (4.37). Hence

$$\varepsilon_T(k,\omega) = 1 - \left(\frac{\omega_p}{\omega}\right)^2. \tag{4.77}$$

The propagation of an electromagnetic wave in the plasma under consideration is described by the dispersion relation

$$\left(\frac{ck}{\omega}\right)^2 = 1 - \left(\frac{\omega_p}{\omega}\right)^2, \tag{4.78a}$$

or

$$\omega^2 = \omega_p^2 + c^2k^2. \tag{4.78b}$$

In the long-wavelength regime, $k \ll \omega_p/c$, the phase velocity of the wave substantially exceeds light velocity in vacuum.

Equation (4.78a) or (4.78b) accounts for the dispersion and broadening of radio signals traveling through an interstellar space. Since ω_p^2 is proportional to the density of charged particles, density fluctuations in interstellar space create local variations in the velocities and directions of the radio-wave propagation.

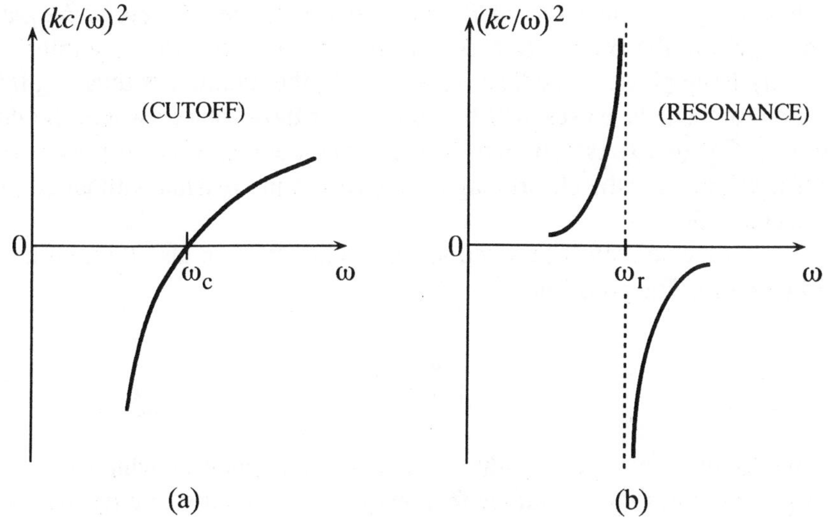

Figure 4.4 (a) Cutoff; (b) resonance.

The effective path length of the signal from a radio source to an observer on the Earth would thus be modulated by these fluctuations; the signal would be broadened in shape and scintillate in time. Equation (4.78a) implies that such effects would be more emphasized for radio signals with lower frequencies.

B. Cutoffs and Resonances

According to Eqs. (4.78), the electromagnetic wave propagates in a plasma only when its frequency exceeds the plasma frequency. As the wave frequency decreases, the transverse dielectric function $\varepsilon_T(k,\omega)$ also decreases; it vanishes at $\omega = \omega_p$. The square of the phase velocity would diverge to $+\infty$ at this point. If the frequency further decreases, both $\varepsilon_T(k,\omega)$ and $(\omega/k)^2$ become negative; the wave cannot propagate in the frequency domain, $|\omega| < \omega_p$.

As the preceding argument has indicated, a point at which $\varepsilon_T(k,\omega)$ or $(ck/\omega)^2$ (cf. Eq. 3.63) changes its sign marks a boundary between propagating and nonpropagating frequency domains. Generally, two cases exist for such a possibility, as Fig. 4.4 illustrates: One is the case in which $\varepsilon_T(k,\omega)$ changes its sign through

$$\varepsilon_T(k,\omega) = 0 . \tag{4.79}$$

The frequency at which Eq. (4.79) takes place defines the *cutoff frequency*. Those waves in the propagating domain near the cutoff frequency travel with extremely large phase velocities; dissipation in the medium is thus negligible. The electromagnetic waves will be reflected at those places where the cutoff condition Eq. (4.79) is satisfied. The plasma frequency corresponds to the cutoff frequency for the electromagnetic waves in the plasma without an external magnetic field.

The transverse dielectric function can change its sign also through infinity (see Fig. 4.4b). The equation

$$\left(\frac{\omega}{ck}\right)^2 = 0 \tag{4.80}$$

thus marks the resonance condition, and the frequency at which Eq. (4.80) takes place defines the *resonance frequency*. In the vicinity of a resonance, the waves travel with infinitesimally small phase velocities; dissipation in the medium thus becomes substantial. The electromagnetic waves will be strongly absorbed by the medium under these resonance conditions. We shall see that various resonances take place in magnetized plasmas.

4.6 WAVE PROPAGATION IN MAGNETIZED PLASMAS

We next consider the electromagnetic waves in a plasma with a constant uniform magnetic field. The system is no longer isotropic; all the elements of the dielectric tensor must be taken into account. The wave propagation is analyzed through the dispersion relation Eq. (3.92), or

$$\det\left|\boldsymbol{\varepsilon}(\mathbf{k},\omega) - \left(\frac{ck}{\omega}\right)^2 \left(\mathbf{I} - \frac{\mathbf{kk}}{k^2}\right)\right| = 0\,. \tag{4.81}$$

In what follows, the consequences of this dispersion relation will be discussed in two special cases: when **k** is parallel to **B** and when k is perpendicular to **B**.

A. Transformation to Circularly Polarized Waves

When the wave propagates parallel to the magnetic field, that is, when $k_\perp = 0$ and $|k_{\parallel}| = k$, the tensor $\Pi(\beta, Z_n; n)$ given by Eq. (4.10) now takes a simplified form in the limit of $\beta \to 0$ as

$$\Pi(0,Z_n;n)=\begin{bmatrix} \frac{1}{2}[\delta(n,1)+\delta(n,-1)] & \frac{i}{2}[\delta(n,1)-\delta(n,-1)] & 0 \\ -\frac{i}{2}[\delta(n,1)-\delta(n,-1)] & \frac{1}{2}[\delta(n,1)+\delta(n,-1)] & 0 \\ 0 & 0 & Z_0^2\delta(n,0) \end{bmatrix} \tag{4.82}$$

where $\delta(n,n')$ is Kronecker's delta,

$$\delta(n,n')=\begin{cases} 1 & (n=n'), \\ 0 & (n\neq n'). \end{cases} \tag{4.83}$$

A vectorial quantity such as the electric field has been described thus far in terms of its Cartesian components; this representation is convenient when the linearly polarized waves are eigenmodes of the wave propagation. In the present case, however, the circularly polarized waves turn out to be the proper solution to the dispersion relation. It is therefore useful to introduce a scheme of transformation from the linearly polarized waves to the circularly polarized waves.

The transformation may be achieved with the aid of a unitary matrix

$$\mathbf{U}=\begin{pmatrix} \frac{1}{\sqrt{2}} & -\frac{i}{\sqrt{2}} & 0 \\ -\frac{i}{\sqrt{2}} & \frac{1}{\sqrt{2}} & 0 \\ 0 & 0 & 1 \end{pmatrix}. \tag{4.84}$$

It satisfies the condition for unitarity

$$\mathbf{U}\cdot\mathbf{U}^{\dagger}=\mathbf{U}^{\dagger}\cdot\mathbf{U}=\mathbf{I}$$

where $\mathbf{U}^{\dagger}$ is the Hermitian conjugate matrix to $\mathbf{U}$ (cf. Problem 3.10). An electric-field vector expressed in terms of the Cartesian components is transformed as

$$\begin{pmatrix} E_r \\ E_l \\ E_z \end{pmatrix}=\mathbf{U}\cdot\begin{pmatrix} E_x \\ E_y \\ E_z \end{pmatrix}$$

or

$$E_r = \frac{1}{\sqrt{2}}(E_x - iE_y)$$

$$E_l = -\frac{i}{\sqrt{2}}(E_x + iE_y)\,. \tag{4.85}$$

It is clear that the new components, E_r and E_l, given by Eqs. (4.85) represent the right- and left-circularly polarized components, respectively, if the frequencies are restricted to the domain $\omega > 0$.

A tensorial quantity such as the dielectric tensor (4.9) is transformed as $\mathbf{U} \cdot \boldsymbol{\varepsilon} \cdot \mathbf{U}^\dagger$. With the aid of Eq. (4.82), the dielectric tensor after such a transformation is now expressed in a diagonal form:

$$\mathbf{U} \cdot \boldsymbol{\varepsilon} \cdot \mathbf{U}^\dagger = \begin{pmatrix} \varepsilon_r & 0 & 0 \\ 0 & \varepsilon_l & 0 \\ 0 & 0 & \varepsilon \end{pmatrix}.$$

The diagonal elements for a multicomponent plasma may thus be calculated as

$$\varepsilon_r(k,\omega) = 1 - \sum_\sigma \frac{\omega_\sigma^2}{\omega(\omega + \Omega_\sigma)}\left[1 - W\left(\frac{\omega + \Omega_\sigma}{k\sqrt{T_\sigma/m_\sigma}}\right)\right], \tag{4.86a}$$

$$\varepsilon_l(k,\omega) = 1 - \sum_\sigma \frac{\omega_\sigma^2}{\omega(\omega - \Omega_\sigma)}\left[1 - W\left(\frac{\omega - \Omega_\sigma}{k\sqrt{T_\sigma/m_\sigma}}\right)\right], \tag{4.86b}$$

$$\varepsilon(k,\omega) = 1 + \sum_\sigma \frac{k_\sigma^2}{k^2} W\left(\frac{\omega}{k\sqrt{T_\sigma/m_\sigma}}\right). \tag{4.86c}$$

Equation (4.86c) is the longitudinal dielectric function with $k_\perp = 0$, which was investigated in Section 4.3.

The dispersion tensor Eq. (3.88) is likewise diagonalized after the unitary transformation. The dispersion relation for the right- (or left-) circularly polarized wave is thus obtained as

$$\left(\frac{ck}{\omega}\right)^2 = \varepsilon_{r(l)}(k,\omega)\,. \tag{4.87}$$

Consequences of these dispersion relations are now investigated for a number of specific cases.

B. Helicon, Alfvén Wave, and Whistler

We first take up the cases of an OCP consisting of electrons. The dielectric function for the right-circularly polarized wave is

$$\varepsilon(k,\omega)=1-\frac{\omega_e^2}{\omega(\omega-|\Omega_e|)}\left[1-W\left(\frac{\omega-|\Omega_e|}{k\sqrt{T_e/m_e}}\right)\right]. \tag{4.88}$$

The imaginary part of this function will be negligible if

$$||\Omega_e|-\omega|>>k\sqrt{\frac{T_e}{m_e}}. \tag{4.89}$$

The dispersion relation (4.87) for the right-circularly polarized wave under the assumption (4.89) becomes

$$\left(\frac{ck}{\omega}\right)^2=1-\frac{\omega_e^2}{\omega(\omega-|\Omega_e|)}. \tag{4.90}$$

This relation is illustrated by the solid lines in Fig. 4.5.

The cutoff frequency is calculated from Eq. (4.79) as

$$\omega_{cr}=\frac{1}{2}\left(\sqrt{\Omega_e^2+4\omega_e^2}+|\Omega_e|\right). \tag{4.91}$$

The resonance occurs at the frequency

$$\omega_R=|\Omega_e|. \tag{4.92}$$

In the magnetic field, the electrons gyrate in the same direction as the right-circularly polarized wave; a resonance is expected when the wave frequency matches the cyclotron frequency.

When the temperature is finite, however, the conditions (4.89) may not be satisfied in the vicinity of the resonance frequency Eq. (4.92). Thus in the domain

$$|\Omega_e|-k\sqrt{\frac{T_e}{m_e}}<\omega<|\Omega_e|+k\sqrt{\frac{T_e}{m_e}} \tag{4.93}$$

the imaginary part of Eq. (4.88) and therefore the decay of the wave will be substantial. The physical origin of such a decay can be easily understood:

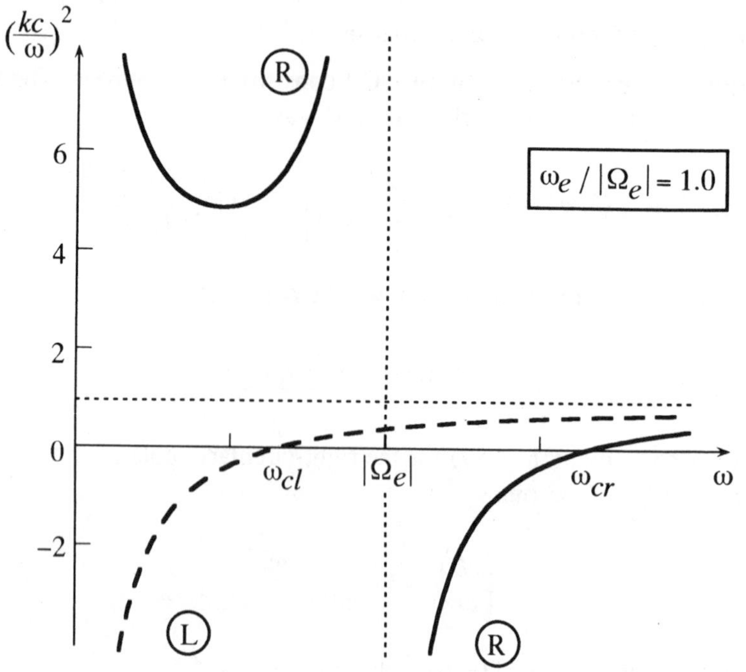

Figure 4.5 Right- and left-circularly polarized waves in the electron plasma.

When a wave with a frequency ω and a wave number k propagates along the magnetic field in a plasma, an electron with velocity $v_{||}$ along the magnetic field experiences, because of the usual Doppler effect, a shifted frequency

$$\omega' = \omega \pm k v_{||} .$$

If ω' coincides with the electron cyclotron frequency, then the resonant absorption of the wave energy by electrons will take place; Eq. (4.93) represents the condition for such a resonance. This phenomenon, called the *Doppler-shifted cyclotron resonance*, is thus related closely to the Landau damping of the plasma oscillations. It has been investigated for both solids [Chambers 1956; Millers & Haering 1962] and plasmas [Drummond 1958a,b]. The use of this Doppler-shifted absorption for a point-by-point mapping of the Fermi surface has been considered by Stern [1963].

A most remarkable feature in the dispersion relation Eq. (4.90) is the emergence of a new propagation mode at low frequencies. In the low-frequency domain, the first term, unity, may be ignored on the right-hand side of Eq. (4.90); the solution then is

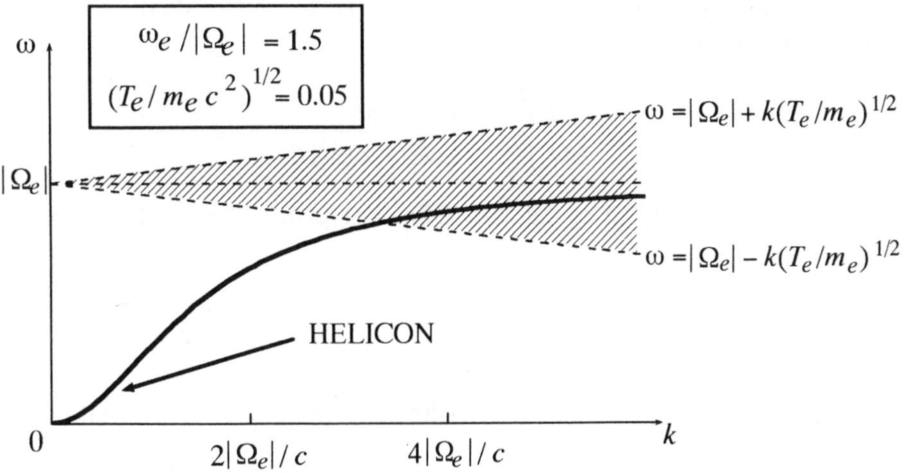

Figure 4.6 Low-frequency mode in electron plasma.

$$\omega = \frac{|\Omega_e|(ck)^2}{\omega_e^2 + (ck)^2}. \tag{4.94}$$

In the limit of long wavelengths, this relation becomes

$$\omega = |\Omega_e| \left(\frac{ck}{\omega_e}\right)^2. \tag{4.95}$$

For short wavelengths, Eq. (4.94) approaches $|\Omega_e|$. Figure 4.6 illustrates features of Eq. (4.94). In the vicinity of $|\Omega_e|$, the Doppler-shifted cyclotron resonance becomes important; the shaded area in Fig. 4.6 depicts the domain of strong absorption (4.93). Such a possibility of electromagnetic wave propagation along the magnetic field in a plasma, at a frequency much lower than both the plasma frequency and the cyclotron frequency, was first predicted by Aigrain [1961] and by Konstantinov and Perel prime [1960]; the term "*helicon*" was coined by Aigrain to describe the propagation mode Eq. (4.95).

The appearance of the helicon mode in the low-frequency limit is in fact a unique feature of an uncompensated plasma like an electron gas. Such a situation occurs frequently in solids, in which space charges of the oppositely charged mobile carriers may not exactly compensate each other. For this reason, the helicon was investigated extensively in solid state plasmas [Bowers, Legendy & Rose 1961; Kaner & Skobov 1966, 1968; Glicksman 1971].

The left-circularly polarized mode may be treated similarly. In place of Eq. (4.90), we now have

$$\left(\frac{ck}{\omega}\right)^2 = 1 - \frac{\omega_e^2}{\omega(\omega + |\Omega_e|)}. \tag{4.96}$$

This relation is depicted by the dashed line in Fig. 4.5. The cutoff occurs at the frequency

$$\omega_{cl} = \frac{1}{2}\left[\sqrt{\Omega_e^2 + 4\omega_e^2} - |\Omega_e|\right]. \tag{4.97}$$

No resonance takes place for a left-circularly polarized wave in an electron gas.

As Fig. 4.5 illustrates, a substantial difference generally exists in propagation characteristics between the right- and left-circularly polarized waves in magnetized plasmas. When a linearly polarized wave is injected toward a plasma along the magnetic field, its right- and left-circular components thus propagate differently in the plasma: In those cases where the propagation velocities are different, the plane of linear polarization will rotate after a passage of a finite length in the plasma; a phenomenon called *Faraday rotation* takes place. In some other cases, only one of the two circularly polarized components may be able to propagate in the plasma; in these circumstances, a linearly polarized wave may be converted into a circularly polarized wave.

Let us next consider a two-component plasma consisting of electrons and ions where their space charges are compensated and vanish on the average, i.e.,

$$\sum_\sigma Z_\sigma n_\sigma = 0. \tag{4.98}$$

The dielectric function for the right-circularly polarized wave is

$$\varepsilon_r(k,\omega) = 1 - \frac{\omega_e^2}{\omega(\omega - |\Omega_e|)}\left[1 - W\left(\frac{\omega - |\Omega_e|}{k\sqrt{T_e/m_e}}\right)\right]$$
$$- \frac{\omega_i^2}{\omega(\omega + \Omega_i)}\left[1 - W\left(\frac{\omega + \Omega_i}{k\sqrt{T_i/m_i}}\right)\right]. \tag{4.99}$$

Assuming the conditions for little absorption

$$||\Omega_e| - \omega| >> k\sqrt{\frac{T_e}{m_e}}, \; \Omega_i + \omega >> k\sqrt{\frac{T_i}{m_i}}, \tag{4.100}$$

we find the dispersion relation for the right-circularly polarized wave

$$\left(\frac{ck}{\omega}\right)^2 = 1 - \frac{\omega_e^2}{\omega(\omega - |\Omega_e|)} - \frac{\omega_i^2}{\omega(\omega + \Omega_i)}. \tag{4.101}$$

The behavior of this function is illustrated by solid lines in Fig. 4.7. The cutoff frequency is located at

$$\omega_{cr} = \frac{1}{2}\left[\sqrt{(|\Omega_e| + \Omega_i)^2 + 4(\omega_e^2 + \omega_i^2)} + |\Omega_e| - \Omega_i\right] \tag{4.102}$$

and the resonance frequency is given again by

$$\omega_R = |\Omega_e|. \tag{4.103}$$

In the vicinity of this resonance, the first condition of (4.100) becomes violated; strong absorption ensues due to the Doppler-shifted cyclotron resonances of the electrons. The second condition of (4.100), on the other hand, is satisfied in ordinary circumstances.

The feature to be noted here is again the appearance of a low-frequency mode of propagation. Neglecting the first term on the right-hand side of Eq. (4.101) and assuming the mass relation (4.40), we find in the low-frequency domain

$$\omega = \frac{|\Omega_e|(ck)^2}{\omega_e^2 + (ck)^2}\left[\frac{1}{2} + \sqrt{\frac{\omega_i^2}{(ck)^2} + \frac{1}{4}}\right]. \tag{4.104}$$

In the long-wavelength limit such that

$$(ck)^2 << 4\omega_i^2 << \omega_e^2, \tag{4.105}$$

Eq. (4.104) tends to

$$\omega = \frac{\Omega_i}{\omega_i} ck, \tag{4.106}$$

where the charge neutrality condition (4.98) has been used. The propagation mode described by Eq. (4.106) is the *Alfvén wave* [Alfvén 1942], and its propagation velocity

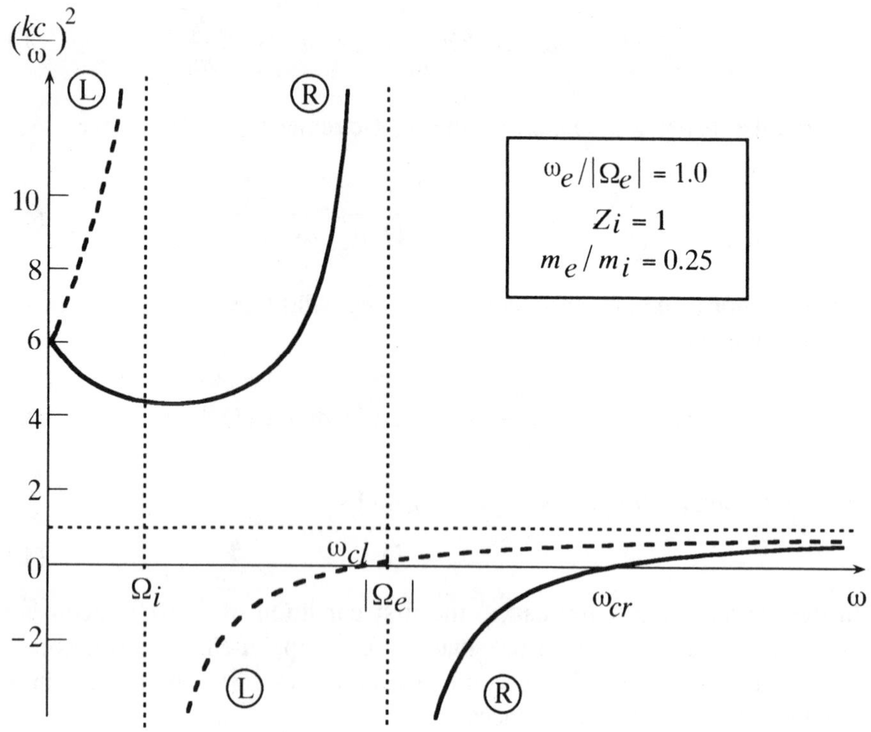

Figure 4.7 Right- and left-circularly polarized waves in a compensated two-component plasma.

$$c_A = \frac{B}{\sqrt{4\pi\rho_m}} \tag{4.107}$$

is called the *Alfvén velocity*, where

$$\rho_m = \sum_\sigma m_\sigma n_\sigma \tag{4.108}$$

is the mass density. The Alfvén wave has been investigated in various substances, including gaseous-discharge plasmas [Allen et al. 1959; Jephcott 1959; Wilcox, Boley & DeSilva 1960], liquid metals [Lundquist 1949; Lehnert 1954], and solids [Buchsbaum & Galt 1961; Williams & Smith 1964; Kaner & Skobov 1968; Glicksman 1971].

It is instructive to examine the values of the propagation velocity for the Alfvén wave. For such a purpose we introduce an effective dielectric constant via

$$\varepsilon_{\mathrm{A}} = \left(\frac{c}{c_{\mathrm{A}}}\right)^2 = \frac{4\pi\rho_{\mathrm{m}}c^2}{B^2} . \tag{4.109}$$

In a hydrogen plasma, this dielectric constant takes on a value

$$\varepsilon_{\mathrm{A}} = 1.9 \times 10^{-2} \frac{n}{B^2} \tag{4.110}$$

where n and B are measured in units of cm^{-3} and gauss. When $n = 10^{12}$ and $B = 10^4$, $\varepsilon_{\mathrm{A}} = 2 \times 10^2$; it may be reasonable to assume $\varepsilon_{\mathrm{A}} >> 1$ even in a magnetic field with substantial strength.

In the intermediate-wavelength regime such that

$$4\omega_i^2 << (ck)^2 << \omega_e^2 , \tag{4.111}$$

Eq. (4.104) becomes

$$\omega = |\Omega_e| \left(\frac{ck}{\omega_e}\right)^2 . \tag{4.112}$$

Note that this dispersion relation is identical to Eq. (4.95) for the helicon; only the domains of wave numbers appear different. The propagation mode described in Eq. (4.112) is called a *whistler*. It has particularly become notorious in connection with the ionospheric radio noises with decreasing pitches which originate from disturbances created by lightning in the polar areas of the earth.

The physical origin for a heliconlike mode in a compensated plasma is easy to understand: In the wave-number regime as specified by (4.111), the corresponding frequencies are too high for the ions to be able to follow dynamically; hence the ions act as if they were immobile space charges. On the other hand, the frequencies involved are still low so that the electrons can follow the wave motion. The combined result should resemble the low-frequency mode obtained in an uncompensated electron plasma.

Overall behavior of the low-frequency right-circularly polarized mode (4.104) together with the domain of strong absorption due to the Doppler-shifted electron cyclotron resonances is shown in Fig. 4.8.

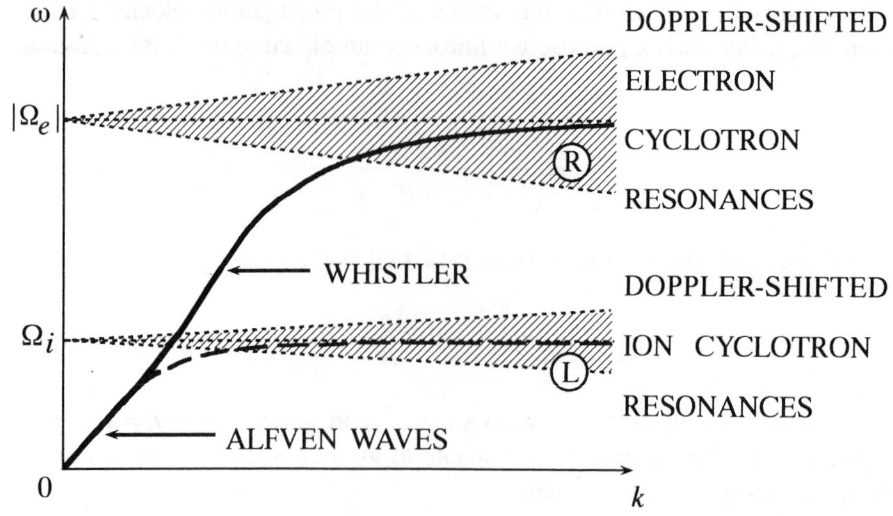

Figure 4.8 Low-frequency mode in a compensated electron-ion plasma.

The propagation characteristics of the left-circularly polarized wave may be analyzed analogously. Assuming

$$|\Omega_i - \omega| >> k\sqrt{\frac{T_i}{m_i}}\,, \, |\Omega_e| + \omega >> k\sqrt{\frac{T_e}{m_e}}\,, \tag{4.113}$$

we obtain the dispersion relation

$$\left(\frac{ck}{\omega}\right)^2 = 1 - \frac{\omega_e^2}{\omega(\omega + |\Omega_e|)} - \frac{\omega_i^2}{\omega(\omega - \Omega_i)}\,. \tag{4.114}$$

This relation is illustrated by the dashed lines in Fig. 4.7. The cutoff frequency is calculated to be

$$\omega_{cl} = \frac{1}{2}\left[\sqrt{(|\Omega_e| + \Omega_i)^2 + 4(\omega_e^2 + \omega_i^2)} - |\Omega_e| + \Omega_i\right] \tag{4.115}$$

and the resonance frequency is

$$\omega_L = \Omega_i\,. \tag{4.116}$$

In the vicinity of this resonance, the first condition of (4.113) is violated; the Doppler-shifted cyclotron resonances of the ions now cause strong attenuation.

In the low-frequency domain, Eq. (4.114) becomes

$$\omega = \frac{|\Omega_e|(ck)^2}{\omega_e^2 + (ck)^2}\left[-\frac{1}{2} + \sqrt{\frac{\omega_i^2}{(ck)^2} + \frac{1}{4}}\right]. \tag{4.117}$$

This dispersion relation reduces to that of the Alfvén wave, Eq. (4.106), in the limit of long wavelengths. The Alfvén wave in a compensated magnetized plasma can propagate as either a right- or a left-circularly polarized wave, and hence also as a linearly polarized wave. As the wave number increases, Eq. (4.117) approaches Ω_i. The behavior of Eq. (4.117) is shown in Fig. 4.8 together with the domain of strong absorption due to the Doppler-shifted ion cyclotron resonances.

C. Wave Propagation Perpendicular to the Magnetic Field

In this case, $k_{\parallel} = 0$ and $k = k$; the wave propagates in the x direction (cf., Fig. 3.2). The dielectric tensor (4.9) is now written as

$$\boldsymbol{\varepsilon}(k,\omega) = \begin{pmatrix} \varepsilon_1 & -i\varepsilon_x & 0 \\ i\varepsilon_x & \varepsilon_2 & 0 \\ 0 & 0 & \varepsilon_3 \end{pmatrix} \tag{4.118}$$

where for an OCP

$$\varepsilon_1(k,\omega) = 1 - \frac{k_D^2}{k^2}\sum_n \frac{(n\Omega)^2}{\omega(\omega - n\Omega)}\Lambda_n(\beta), \tag{4.119}$$

$$\varepsilon_2(k,\omega) = 1 - \frac{k_D^2}{k^2}\sum_n \frac{(n\Omega)^2}{\omega(\omega - n\Omega)}\left[\Lambda_n(\beta) - \frac{2\beta^2}{n^2}\Lambda_n'(\beta)\right], \tag{4.120}$$

$$\varepsilon_3(k,\omega) = 1 - \frac{\omega_p^2}{\omega^2}\sum_n \frac{\omega}{\omega - n\Omega}\Lambda_n(\beta), \tag{4.121}$$

$$\varepsilon_x(k,\omega) = \frac{\omega_p^2}{\omega^2}\sum_n \frac{\omega}{\omega - n\Omega} n\Lambda_n'(\beta). \tag{4.122}$$

The dispersion relation (4.81) splits itself into two equations,

$$\left(\frac{ck}{\omega}\right)^2 = \frac{\varepsilon_1\varepsilon_2 - \varepsilon_x^2}{\varepsilon_1}, \tag{4.123}$$

$$\left(\frac{ck}{\omega}\right)^2 = \varepsilon_3 . \tag{4.124}$$

The former case, Eq. (4.123), is called the *extraordinary wave.* From an examination of Eq. (3.91), we find that $E_z = 0$ and the remaining components satisfy

$$\frac{E_x}{E_y} = i\frac{\varepsilon_x}{\varepsilon_1} . \tag{4.125}$$

Since E_x corresponds to the longitudinal component of the electric field vector while E_y is a transverse one, the extraordinary wave represents a *hybrid mode.* The cutoff frequencies are determined from

$$\varepsilon_1\varepsilon_2 = \varepsilon_x^2 . \tag{4.126}$$

The *hybrid resonances* take place at

$$\varepsilon_1 = 0 . \tag{4.127}$$

Under these resonance conditions, we find from Eq. (4.125) that E_y also vanishes; the extraordinary wave turns into a longitudinal mode at a hybrid resonance. In fact, this longitudinal wave may be identified as one of the Bernstein modes considered in Section 4.4B. It is possible to show with the aid of the identity

$$\sum_n \Lambda_n(\beta) = 1 \tag{4.128}$$

that the dielectric functions Eqs. (4.54) and (4.119) are the same when $k_{||} = 0$.

The latter mode, Eq. (4.124), is called the *ordinary wave.* In this case, $E_z \neq 0$; the electric field of the wave is parallel to the external magnetic field. The ordinary wave is obviously a transverse wave.

The elements of the dielectric tensor, Eqs. (4.119)–(4.122), exhibit strong spatial dispersion arising from the cyclotron motion of charged particles with finite Larmor radii. For a plasma at a finite temperature, *nonlocal effects* brought about by the thermal motion of charged particles sometimes play an important role in determining the propagation characteristics of the electromagnetic waves. The finiteness of the Larmor radii of the charged particles makes it possible to transport the disturbances created by the waves in the directions perpendicular to the magnetic field. When the average Larmor radius is comparable to the wavelength, such a nonlocal effect may become particularly substantial.

If, however, the wavelength under consideration is much longer than the average Larmor radius of the particles, we may approximately treat the wave propagation by neglecting the effects of thermal motion and thus by going over to the limit $T \to 0$. In such a cold-plasma limit, the spatial dispersion is suppressed; Eqs. (4.119)–(4.122) are expressed in substantially simplified forms as

$$\varepsilon_1 = \varepsilon_2 = 1 - \frac{\omega_p^2}{\omega^2 - \Omega^2}, \tag{4.129}$$

$$\varepsilon_3 = 1 - \frac{\omega_p^2}{\omega^2}, \tag{4.130}$$

$$\varepsilon_x = \frac{\omega_p^2 \Omega}{\omega(\omega^2 - \Omega^2)}. \tag{4.131}$$

It is clear from Eq. (4.130) that the ordinary wave (4.124) is not affected by the presence of the magnetic field in this limit; Eq. (4.78a) is applicable in this case.

The dispersion relation for the extraordinary wave in a compensated electron-ion plasma at $T=0$ is shown in Fig. 4.9. In the limit of low frequencies, one finds under the condition Eq. (4.98)

$$\varepsilon_1 = \varepsilon_2 \to 1 + \frac{4\pi \rho_m c^2}{B^2}, \tag{4.132}$$

$$\varepsilon_x \to 0. \tag{4.133}$$

The extraordinary wave thus starts in the low-frequency domain as an Alfvén wave (cf. Eq. 4.109). The first resonance at ω_{LH} is called the *lower hybrid resonance*; the frequency is given by

$$\omega_{LH}^2 = |\Omega_e \Omega_i| \frac{\omega_e^2 + |\Omega_e \Omega_i|}{\omega_e^2 + \Omega_e^2}. \tag{4.134}$$

The first cutoff frequency ω_{c1} appears at

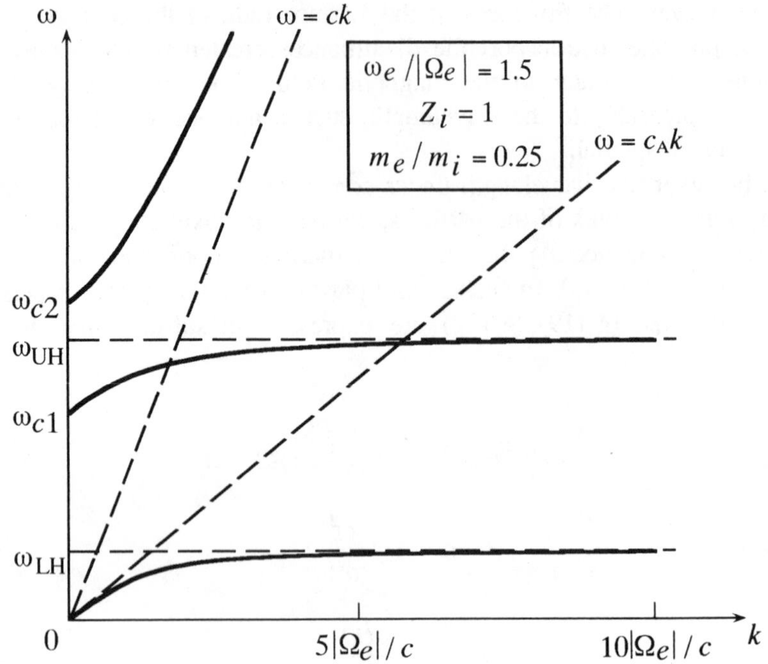

Figure 4.9 Extraordinary wave in a cold electron-ion plasma.

$$\omega_{c1} = \frac{1}{2}\left[\sqrt{\Omega_e^2 + 4\omega_e^2} - |\,\Omega_e\,|\right], \tag{4.135}$$

and then the *upper hybrid resonance* takes place at

$$\omega_{UH} = \sqrt{\Omega_e^2 + \omega_e^2}\,. \tag{4.136}$$

Finally, the second cutoff frequency ω_{c2} appears at

$$\omega_{c2} = \frac{1}{2}\left[\sqrt{\Omega_e^2 + 4\omega_e^2} + |\,\Omega_e\,|\right], \tag{4.137}$$

and this branch turns into the free propagation mode $\omega = ck$.

PROBLEMS

4.1 Derive Eq. (4.3).

4.2 Show the expansions (4.4) and (4.6).

4.3 Carry out the calculations leading to Eq. (4.9). In the calculations, it may be helpful to note in Eq. (3.129)

$$\sum_{n=-\infty}^{\infty} \Pi(v_\perp, v_\parallel; n) = \begin{pmatrix} \frac{1}{2}v_\perp^2 & 0 & 0 \\ 0 & \frac{1}{2}v_\perp^2 & 0 \\ 0 & 0 & v_\parallel^2 \end{pmatrix},$$

which is derived with the aid of the formulas in Problem 3.13, and

$$\int_0^\infty dx x^2 J_n(x) J_n'(x) \exp\left(-\frac{x^2}{2\beta}\right) = \beta^2 \Lambda_n'(\beta),$$

$$\int_0^\infty dx x^3 [J_n(x)]^2 \exp\left(-\frac{x^2}{2\beta}\right) = n^2 \beta \Lambda_n(\beta) - 2\beta^3 \Lambda_n'(\beta),$$

which result from differentiations of Eq. (4.8) with respect to p, q, or both.

4.4 Derive Eq. (4.22).

4.5 Obtain the solution (4.30) and (4.31).

4.6 For the ion-acoustic mode (4.53), show that

$$\left.\frac{\partial \varepsilon_1(\mathbf{k},\omega)}{\partial \omega}\right]_{\omega=\omega_k} = \frac{2}{\omega_k}\varepsilon_e(\mathbf{k})$$

where $\varepsilon_e(\mathbf{k})$ is the static screening constant (4.52) of the electrons.

4.7 Derive Eq. (4.54).

4.8 Show that in the limit $B \to 0$, Eq. (4.54) turns into Eq. (4.1).

4.9 Consider a test particle with electric charge $Z_0 e$ and velocity $\mathbf{v}_0$ introduced in a plasma with the dielectric response function given by Eq. (4.54); $\mathbf{v}_0$ is parallel to $\mathbf{B}$. Calculate that part of the interaction energy in the limit of a strong magnetic field which is proportional to v_0^2 for the slowly moving test charge.

4.10 Following the procedure of Section 3.2D, derive an expression for $\mathrm{Im}[1/\varepsilon(\mathbf{k},\omega)]$ approximate in the vicinity of the Bernstein mode Eq. (4.66).

4.11 Consider the effects of particle collisions on the low-frequency modes of electromagnetic wave propagation in plasmas along the magnetic field, using the dielectric tensor obtained in Problem 3.12; we ignore the thermal effects by assuming $T=0$ in that expression.

(a) Show that the helicon mode now suffers attenuation due to the collisions as described by

$$\omega = |\Omega_e| \left(\frac{ck}{\omega_e}\right)^2 \left[1 - \frac{i}{\tau_e |\Omega_e|}\right].$$

(b) Show that for the Alfvén wave the dispersion relation becomes

$$\omega = c_A k - \frac{i}{2} \frac{m_e/\tau_e + m_i/\tau_i}{m_e + m_i}.$$

4.12 Derive Eqs. (4.134)–(4.137).

CHAPTER 5

Transient Processes

In the preceding chapter, we studied the fundamental properties of plasmas in thermodynamic equilibrium and in a stationary state. Another class of problems, related rather closely to the experimental aspects, is the transient processes in plasmas. It is the purpose of the present chapter to study some of the problems in this category, based on the Vlasov description of plasmas.

We begin with a treatment of the temporal and spatial propagation of small-amplitude waves. The distinction between temporal and spatial propagation stems from the experimental boundary conditions: When a spatial distribution of potential field is set initially at time $t = 0$, the subsequent evolution of the potential field in the plasma constitutes a temporal-propagation problem; it involves a temporal decay due to Landau damping. Instead of such a time-dependent situation, one can alternatively formulate the problem so as to single out a spatial evolution of potential disturbance in the plasma. Thus, we shall consider a case in which a time-dependent potential disturbance is applied to a plasma at a plane $x = 0$; the potential field will then propagate and decay in space. Both of these problems will be formulated in terms of the Vlasov dielectric functions.

An important question associated with such propagation characteristics is the physical relation between the decay of the potential field and the reversibility of the Vlasov equation. The plasma-wave echo will thus be taken up as an example demonstrating such a microscopic reversibility in the phase evolutions vis-à-vis Landau damping.

We shall then treat the problems associated with the behavior of large-amplitude waves in a plasma. Here the trapping of those particles traveling with almost the same velocities as the phase velocity of the wave plays an essential part; the plasma wave thereby exhibits an amplitude oscillation in its decay process.

Finally we shall consider other nonlinear processes associated with propagations of large-amplitude waves; these include a formation of solitary waves, self-trapping, and self-focusing. Relevant nonlinear mechanisms are the hydrodynamic flows and the ponderomotive forces.

5.1 PROPAGATION OF SMALL-AMPLITUDE PLASMA WAVES

Consider the application of a weak external potential field

$$\Phi_{\rm ext}(\mathbf{r},t) = \sum_{\mathbf{k}} \int \frac{d\omega}{2\pi} \Phi_{\rm ext}(\mathbf{k},\omega) \exp\left[i(\mathbf{k}\cdot\mathbf{r} - \omega t) + \eta t\right] \; (\eta \to +0) \tag{5.1}$$

to an OCP. In the dielectric formulation of Section 3.2, the Fourier components $\Phi_{\rm ind}(\mathbf{k},\omega)$ of the induced potential field are given by

$$\Phi_{\rm ind}(\mathbf{k},\omega) = -\left[1 - \frac{1}{\varepsilon(\mathbf{k},\omega)}\right] \Phi_{\rm ext}(\mathbf{k},\omega) \,. \tag{5.2}$$

The distribution function $f(\mathbf{r},\mathbf{v};t)$ deviates from its equilibrium value $f(\mathbf{v})$ so that

$$\begin{aligned} f(\mathbf{r},\mathbf{v};t) &= f(\mathbf{v}) + \delta f(\mathbf{r},\mathbf{v};t) \\ &= f(\mathbf{v}) + \sum_{\mathbf{k}} \int \frac{d\omega}{2\pi} \delta f_{\mathbf{k},\omega}(\mathbf{v}) \exp\left[i(\mathbf{k}\cdot\mathbf{r} - \omega t) + \eta t\right] \\ &\quad (\eta \to +0) \,. \end{aligned} \tag{5.3}$$

The Fourier components $\delta f_{\mathbf{k},\omega}(\mathbf{v})$ of the deviation may be determined from a calculation analogous to Eq. (3.108) as

$$\delta f(\mathbf{v})_{\mathbf{k},\omega} = -\frac{Ze}{m} \frac{\Phi_{\rm ext}(\mathbf{k},\omega)}{(\omega - \mathbf{k}\cdot\mathbf{v} + i\eta)\, \varepsilon(\mathbf{k},\omega)} \mathbf{k} \cdot \frac{\partial f(\mathbf{v})}{\partial \mathbf{v}} \,. \tag{5.4}$$

Equations (5.2) and (5.4) are the basic equations for the present study of a small-amplitude wave propagation in a plasma.

We choose the x axis arbitrarily in the direction of the wave vector $\mathbf{k}$. The Vlasov dielectric function of a plasma without a magnetic field calculated in Eq. (3.136) now reads as

$$\varepsilon(k,\omega) = 1 - \frac{\omega_p^2}{k^2}\int dv \frac{f'(v)}{v - (\omega/k) - i(\eta/k)} \tag{5.5}$$

where $v \equiv \mathbf{v}\cdot\mathbf{k}/k$ is the velocity component in the direction of the wave propagation (i.e., in the x direction),

$$f(v) \equiv \int dv_y \int dv_z f(\mathbf{v}) \tag{5.6}$$

is a one-dimensional velocity distribution in the x direction, and the prime denotes a differentiation with respect to v.

Allowing for the possibilities that ω and k may be complex variables, we examine the symmetry properties of the complex function (5.5): Thus when k is a real variable,

$$\varepsilon(-k, -\omega^*) = \varepsilon^*(k,\omega)\,, \tag{5.7}$$

while when ω is real,

$$\varepsilon(-k^*, -\omega) = \varepsilon^*(k,\omega)\,. \tag{5.8}$$

For a Maxwellian plasma, the dispersion relation for the plasma oscillation has been given by Eq. (4.29). This equation can likewise be investigated for the two cases: (1) when k is real, and (2) when ω is real. These separate cases correspond, respectively, to the temporal and spatial propagation of the plasma oscillation.

When k is real and $k^2 \ll k_D^2$, Eq. (4.29) may be solved for ω as

$$\begin{aligned}\omega &\equiv \pm\omega(k) - i\gamma(k)\\ &= \pm\omega_p\left(1 + \frac{3k^2}{2k_D^2}\right) - i\sqrt{\frac{\pi}{8}}\,\omega_p \frac{k_D^3}{|k|^3}\exp\left[-\left(\frac{k_D^2}{2k^2} + \frac{3}{2}\right)\right].\end{aligned} \tag{5.9}$$

Figure 5.1 schematically depicts the locations of these solutions on the complex ω plane; this figure applies for both positive and negative values of k.

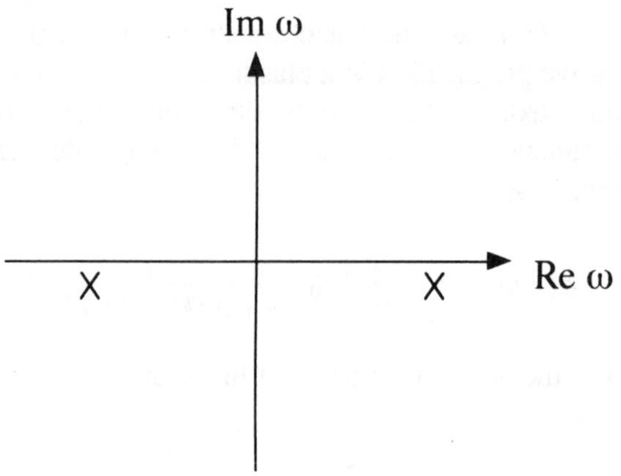

Figure 5.1 Locations of the plasma-wave solutions Eq.(5.9) for a real value of k.

When ω is real and $\omega^2 \le \omega_p^2$, Eq. (4.29) may be solved for k as

$$k = \pm \frac{\omega^2}{\omega_p \sqrt{3T/m}} \sqrt{1 - \frac{\omega_p^2}{\omega^2}}$$

$$\times \left\{ 1 + i\sqrt{\frac{27\pi}{8}} \frac{\omega_p^5}{\omega^5} \left(1 - \frac{\omega_p^2}{\omega^2} \right)^{-5/2} \exp\left[-\frac{3\omega_p^2}{2(\omega^2 - \omega_p^2)} \right] \right\} . \tag{5.10}$$

Figure 5.2 shows the locations of these solutions on the complex k plane; in this case $\omega > 0$ and $\omega < 0$ need to be distinguished.

A remarkable difference exists between the characteristics of the solutions described on the ω plane (Fig. 5.1) and those on the k plane (Fig. 5.2). As a consequence, the plasma wave exhibits a causal behavior in time, but not in space.

A. Temporal Propagation

Let us now specify the form of the external disturbance (5.1) as

$$\Phi_{\rm ext}(\mathbf{r},t) = \Phi_0 \delta(\omega_p t)[\exp(ik_0 x) + {\rm cc}] \qquad (k_0 > 0) . \tag{5.11}$$

This form implies a case in which a sinusoidal potential distribution with a fixed wavelength $2\pi/k_0$ is impulsively applied to the plasma at $t = 0$. The Fourier components of the potential field are

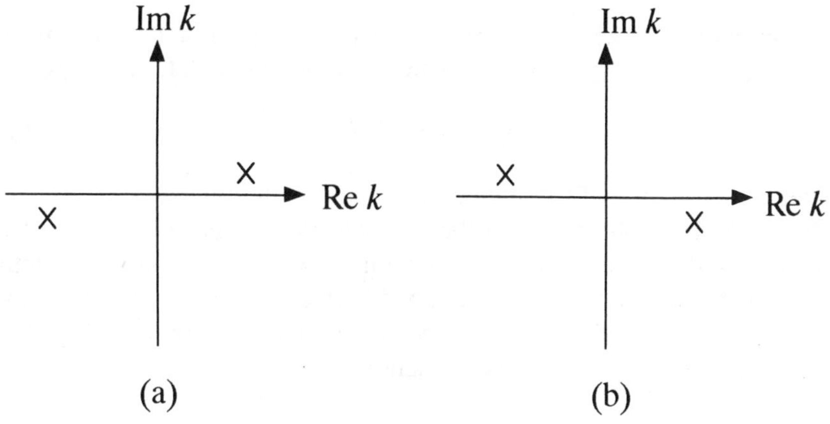

Figure 5.2 Locations of the plasma-wave solutions Eq. (5.10) for positive and negative values of ω: (a) ω>0, (b) ω<0.

$$\Phi_{\text{ext}}(\mathbf{k},\omega) = \frac{\Phi_0}{\omega_p}[\delta(k,k_0) + \delta(k,-k_0)]\delta(k_y,0)\delta(k_z,0) \tag{5.12}$$

where $k = k_x$ and Kronecker's delta (4.83) has been used.

The induced potential field is obtained by substituting Eq. (5.12) into Eq. (5.2); carrying out the inverse transformations, we find

$$\Phi_{\text{ind}}(x,t) = -\frac{\Phi_0}{2\pi\omega_p}\int_{-\infty}^{\infty} d\omega \left[1 - \frac{1}{\varepsilon(k_0,\omega)}\right] \exp[i(k_0x - \omega t)]$$
$$-\frac{\Phi_0}{2\pi\omega_p}\int_{-\infty}^{\infty} d\omega \left[1 - \frac{1}{\varepsilon(-k_0,\omega)}\right] \exp[-i(k_0x + \omega t)]\,. \tag{5.13}$$

For $t > 0$, we may close the contour of the ω integration by an infinite semicircle in the lower half-plane. The residue at each of the poles of the inverse dielectric response function contributes to the integration. The location of a pole on the complex ω plane determines the frequency and the lifetime of each collective mode. Obviously, those poles lying closest to the real axis represent the collective modes with the longest lifetimes; contributions from such modes eventually predominate. In view of Eq. (5.9) the field decays as

$$\Phi_{\text{ind}}(x,t) \approx \exp[-\gamma(k_0)t] \qquad (t > 0)\,, \tag{5.14}$$

a manifestation of Landau damping. For $t < 0$, we may close the contour by an infinite semicircle in the upper half-plane. In view of Fig. 5.1, one finds

$$\Phi_{\mathrm{ind}}(x,t) = 0 \qquad (t < 0)\,, \tag{5.15}$$

another demonstration of causality built in the response function.

Thus, we have observed that the potential field induced in the plasma by the application of an impulsive perturbation (5.11) will die away exponentially due to Landau damping. The situation differs significantly, however, for an induced perturbation of the phase-space distribution given by Eq. (5.4). To see this, we carry out the inverse transformation

$$\delta f(x,v;t) = -\frac{Ze\Phi_0 k_0}{2\pi m\omega_p}\int_{-\infty}^{\infty} d\omega \frac{f'(v)}{\omega - k_0 v + i\eta}\,\frac{1}{\varepsilon(k_0,\omega)}\exp\left[i(k_0 x - \omega t)\right]$$

$$+\frac{Ze\Phi_0 k_0}{2\pi m\omega_p}\int_{-\infty}^{\infty} d\omega \frac{f'(v)}{\omega + k_0 v + i\eta}\,\frac{1}{\varepsilon(-k_0,\omega)}\exp\left[-i(k_0 x + \omega t)\right]\,. \tag{5.16}$$

The ω integration can be performed in the same way as before: For $t < 0$, Eq. (5.16) vanishes. For $t > 0$, various poles of the integrand contribute. Note that the integrand now has poles at $\omega = \pm k_0 v - i\eta$ in addition to the usual poles arising from the inverse dielectric function. Hence Eq. (5.16) is written as

$$\delta f(x,v;t) = -\frac{2Ze\Phi_0 k_0}{m\omega_p}\mathrm{Im}\left\{\frac{f'(v)}{\varepsilon(k_0,k_0 v)}\exp\left[ik_0(x - vt)\right]\right\}$$

$$+\,[\text{contributions from the plasma–wave poles}] \qquad (t > 0)\,. \tag{5.17}$$

The last term of Eq. (5.17) behaves in the same way as Eq. (5.14); it decays exponentially. The first term on the right-hand side of Eq. (5.17), however, does not show such a decay; it oscillates indefinitely with a frequency $k_0 v$. Thus in the Vlasov description, where particle collisions are absent, the phase-space distribution $f(\mathbf{r},\mathbf{v};t)$ will keep the memory of the initial perturbation permanently.

This difference in the time-dependent behavior between Eqs. (5.14) and (5.17) may be understood in the following way: A macroscopic quantity such as the potential field involves a velocity space integration of a distribution function such as Eq. (5.17). In the process of such an integration, the phases $k_0 vt$ of the microscopic elements will now be mixed, because of added contributions from various elements in the velocity space. Consequently, the velocity integration of the first term on the right-hand side of Eq. (5.17) will also decay in time.*

On the basis of the foregoing considerations, we can then design the following *gedanken* experiment to illustrate the reversible nature of the microscopic processes: Apply a pulsed disturbance of the form (5.11) to the plasma and wait for a time interval τ. If $\tau >> |\gamma(k_0)|^{-1}$ is chosen, the induced potential field (5.14) as well as the last term of Eq. (5.17) will have decayed away; macroscopically, it will appear that the plasma has returned to a state without perturbations. Actually, however, the first term of Eq. (5.17) still remains. We now reverse the directions of phase evolution in those microscopic elements. After passage of another time interval τ, the phase relations of those elements will be restored to their original states at $t=0$; a macroscopic quantity will thereby reappear in the plasma.

In reality it would be impossible, of course, to perform an experiment exactly in the way described above. It is possible, however, to produce an effect physically equivalent to the foregoing through a plasma-wave echo experiment; we shall study such a process after a brief investigation of spatial propagation of a plasma wave.

B. Spatial Propagation

Its conceptual simplicity notwithstanding, the external disturbance in the form of Eq. (5.11) may not be easy to create in a plasma. Instead, it may not be difficult to apply a potential in the form

$$\Phi_{\text{ext}}(\mathbf{r},t) = \Phi_0 \delta(k_D x)[\exp(-i\omega_0 t) + \text{cc}] \qquad (\omega_0 > 0) \tag{5.18}$$

to the plasma through an insertion of a grid at $x=0$ whose potential is sinusoidally modulated by an external source. Let us therefore consider a consequence of such an experiment.

Since the Fourier components of Eq. (5.18) are

$$\Phi_{\text{ext}}(\mathbf{k},\omega) = \frac{\Phi_0}{k_D}[\delta(\omega - \omega_0) + \delta(\omega + \omega_0)]\delta(k_y,0)\delta(k_z,0) , \tag{5.19}$$

the induced potential field is calculated as

---

* In actually carrying through the velocity integration for Eq. (5.17), a contour integration may be required in the complex v plane. Closing such a contour by an infinite semicircle would ordinarily encounter an essential singularity associated with $f(v)$ at $v=+i\infty$ or $v=-i\infty$. Such a singularity, however, is canceled in the integrand by a similar singularity occurring in $\varepsilon(k_0,k_0 v)$ or $\varepsilon(-k_0,-k_0 v)$. Hence, the contour of integration may be closed by an infinite semicircle in the appropriate half-plane.

$$\Phi_{\text{ind}}(x,t) = -\frac{\Phi_0}{2\pi k_{\text{D}}}\int_{-\infty}^{\infty} dk\left[1-\frac{1}{\varepsilon(k,\omega_0)}\right]\exp\,[i(kx-\omega_0 t)]$$

$$-\frac{\Phi_0}{2\pi k_{\text{D}}}\int_{-\infty}^{\infty} dk\left[1-\frac{1}{\varepsilon(k,-\omega_0)}\right]\exp\,[-i(kx+\omega_0 t)]\,. \quad (5.20)$$

For $x > 0$, the contour of integration is closed by an infinite semicircle in the upper half of the complex k plane. Expressing Eq. (5.10) as

$$k = k_{\text{r}}(\omega) + ik_{\text{i}}(\omega)\,, \quad (5.21)$$

we find from Fig. 5.2 that those poles located in the upper half-plane contribute to Eq. (5.20) through such an integration. Each mode has a phase velocity directed toward the positive x direction (i.e., $\omega/k_{\text{r}} > 0$) and decays exponentially as $\exp(-|k_{\text{i}}|x)$. For $x < 0$, the contour of the k integration may be closed by an infinite semicircle in the lower half-plane. The poles in this half-plane now contribute to Eq. (5.20). Each mode has a phase velocity directed toward the negative x direction (i.e., $\omega/k_{\text{r}} < 0$) and decays exponentially as $\exp(|k_{\text{i}}|x)$. Contrary to the case of a temporal propagation, the spatial propagation of the plasma wave takes place symmetrically with regard to the positive and negative x directions (unless of course the plasma itself has its own asymmetry).

The induced perturbation of the phase-space distribution can be calculated similarly from Eq. (5.4):

$$\delta f(x,v;t) = -\frac{Ze\Phi_0}{2\pi mk_{\text{D}}}\int_{-\infty}^{\infty} dk\frac{kf'(v)}{\omega_0 - kv + i\eta}\frac{1}{\varepsilon(k,\omega_0)}\exp\,[i(kx-\omega_0 t)]$$

$$+\frac{Ze\Phi_0}{2\pi mk_{\text{D}}}\int_{-\infty}^{\infty} dk\frac{kf'(v)}{\omega_0 + kv - i\eta}\frac{1}{\varepsilon(k,-\omega_0)}\exp\,[i(kx+\omega_0 t)]\,. \quad (5.22)$$

The poles at $k = \pm(\omega_0/v) + i(\eta/v)$ contribute to a propagation in the $+x$ direction when $v > 0$; naturally, the particles with $v > 0$ convey the effects of disturbance imparted at $x = 0$ into the region $x > 0$. By the same token, the particles with $v < 0$ carry information into the negative x region. As with the temporal-propagation cases of the preceding subsection, these effects propagate in space without decay.

One can use relations such as (5.20) to investigate the propagation characteristics of a collective mode in a plasma. The propagation velocity and the spatial decay rate may thus be measured through an interferometric technique;

both the electron plasma oscillation [Malmberg & Wharton 1964,1966; Van Hoven 1966; Derfler & Simonen 1966] and the ion-acoustic wave [Wong, D'Angelo & Motley 1962; Wong, Motley & D'Angelo 1964] have been thereby investigated.

5.2 PLASMA-WAVE ECHOES

The reversible nature of the Vlasov equation is reflected in the microscopic phase evolution of the particle distribution function after application of a disturbance to the plasma. A macroscopic quantity such as the electric-field potential, on the other hand, will decay through Landau damping. It has been recognized that a macroscopic quantity might then reappear in the plasma if the direction of the phase evolution in the microscopic elements could be reversed. Such a reversal is in fact possible through the application of a second disturbance; a macroscopic field will reappear in the plasma many Landau-damping periods after application of the second disturbance, in the form of an echo [Gould, O'Neil & Malmberg 1967; O'Neil & Gould 1968]. The plasma echo is related to other known echo phenomena, such as spin echo [Hahn 1950], cyclotron resonance echo [Hill & Kaplan 1965], and photon echo [Abella, Kurnit & Hartmann 1966].

As the examples in the previous section have shown, plasma-wave propagation may be analyzed as a one-dimensional problem through an appropriate choice of the external perturbation. The Vlasov equation in these circumstances may be expressed as

$$\frac{\partial f}{\partial t}+v\frac{\partial f}{\partial x}-\frac{Ze}{m}\frac{\partial \Phi}{\partial x}\frac{\partial f}{\partial v}=0 \tag{5.23}$$

where f is a one-dimensional distribution function defined according to Eq. (5.6). The potential field is the sum of the external and induced potentials determined from the Poisson equation

$$\frac{\partial^2 \Phi}{\partial x^2}=\frac{\partial^2 \Phi_{\text{ext}}}{\partial x^2}-4\pi Zen\left[\int dvf-1\right]. \tag{5.24}$$

The last term on the left-hand side of Eq. (5.23), therefore, represents the nonlinear term of the Vlasov equation. When two separate disturbances are applied to the plasma, this nonlinear term acts to mix the effects of such disturbances. The plasma echo appears as a consequence of such mixing.

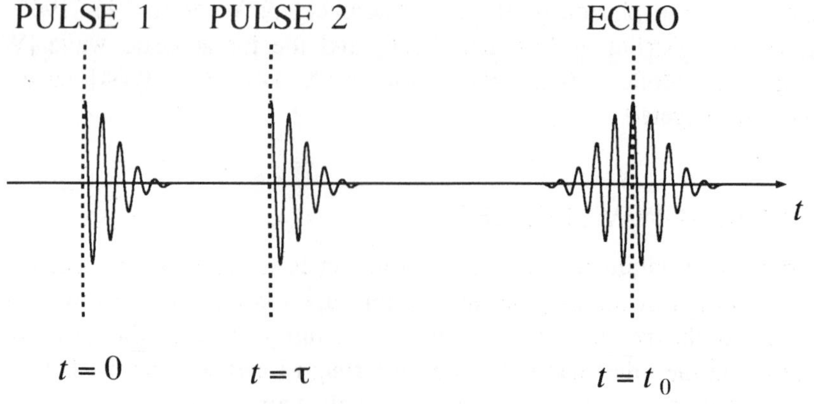

Figure 5.3 Temporal plasma-wave echo.

A. Fundamental Processes

Consider an external disturbance

$$\Phi_1(x,t) = \Phi_1 \delta(\omega_p t)[\exp(ik_1 x) + \mathrm{cc}] \tag{5.25}$$

applied to the plasma at $t = 0$. According to the calculations in Section 5.1A, the induced potential field decays at a rate $|\gamma(k_1)|$; for $t >> |\gamma(k_1)|^{-1}$ the induced field vanishes (see Fig. 5.3). The perturbation $\delta f_1(\mathbf{x},\mathbf{v};t)$ induced in the distribution function still remains, however; the phase evolution in the first term on the right-hand side of Eq. (5.17) is

$$\delta f_1(x,v;t) \approx \exp[\pm ik_1(x - vt)]\,. \tag{5.26}$$

We now apply a second pulse

$$\Phi_2(x,t) = \Phi_2 \delta[\omega_p(t-\tau)][\exp(ik_2 x) + \mathrm{cc}] \tag{5.27}$$

at $t = \tau >> |\gamma(k_1)|^{-1}$. Through the nonlinear term in Eq. (5.23), this pulse couples with the distribution function $f(x,v;t)$ and creates new disturbances in the plasma. The distribution function now consists of two separate contributions: the unperturbed part $f(v)$ and a perturbation $\delta f_1(\mathbf{x},\mathbf{v};t)$ due to the application of the first pulse.

That part of the disturbances stemming from coupling between (5.27) and $f(v)$ behaves in the same manner as the disturbance created by the first pulse: The induced potential field decays at a rate $|\gamma(k_2)|$; the long-time behavior of the perturbed part of the distribution function will be

$$\delta f_2(x,v;t) \approx \exp\{\pm ik_2[x - v(t-\tau)]\} . \tag{5.28}$$

Macroscopically, therefore, the pulse will disappear in time $t >> \tau + |\gamma(k_2)|^{-1}$.

The plasma echo arises as a result of coupling between $\Phi_2(x,t)$ and $\delta f_1(\mathbf{x},\mathbf{v};t)$. The phase evolution of the distribution function perturbed through such a coupling may then be described by a product of Eqs. (5.26) and (5.28) as

$$\delta f_{12}(x,v;t) \approx \exp\{i[(k_1 \pm k_2)x - k_1 vt + k_2 v(t-\tau)]\} + \text{cc} . \tag{5.29}$$

Calculation of a macroscopic quantity generally involves an integration of such a distribution function over the velocity space, which would ordinarily result in phase mixing and thereby the disappearance of a macroscopic quantity. The phase mixing can be avoided, however, if the phases are independent of the particle velocities. Let us assume that such a condition is realized for Eq. (5.29) at $t = t_0$; it sets a relation, $k_1 v t_0 \pm k_2 v(t_0 - \tau) = 0$, or $t_0 = [\pm k_2/(k_1 \pm k_2)]\tau$. Since $t_0 > \tau$, we must choose the lower signs; hence

$$t_0 = \frac{k_2}{k_2 - k_1}\tau . \tag{5.30}$$

Consequently, when $k_2 > k_1$, a macroscopic field will reappear around the time $t = t_0$, varying sinusoidally in space with wave number

$$k_3 = k_2 - k_1 . \tag{5.31}$$

The phenomenon described above is called a *temporal plasma-wave echo,* since it involves a temporal propagation of plasma waves.

We can consider similar echo phenomena arising from a spatial propagation of plasma waves as well. Thus, we apply a first signal

$$\Phi_1(x,t) = \Phi_1 \delta(k_D x)[\exp(-i\omega_1 t) + \text{cc}] \tag{5.32}$$

at $x = 0$ in the plasma; the field damps away for $|x| >> |k_i(\omega_1)|^{-1}$. A second signal

$$\Phi_2(x,t) = \Phi_2 \delta[k_D(x-l)][\exp(-i\omega_2 t) + \text{cc}] \tag{5.33}$$

is then applied at $x = l >> |k_i(\omega_1)|^{-1}$; it will also vanish for $|x-l| >> |k_i(\omega_2)|^{-1}$. When $\omega_2 > \omega_1$, an echo will appear at

$$x_0 = \frac{\omega_2}{\omega_2 - \omega_1} l \tag{5.34}$$

with frequency

$$\omega_3 = \omega_2 - \omega_1 \ . \tag{5.35}$$

This is, therefore, a *spatial plasma-wave echo.*

The two examples considered above are referred to as second-order plasma echoes; these derive from the second-order perturbation contributions of the nonlinear term with respect to the external disturbances—first order in the disturbance $\Phi_1(x,t)$ and first order in another disturbance $\Phi_2(x,t)$. One can likewise extend the arguments and consider higher-order echoes. The nth-order effect of the disturbance Eq. (5.25) on the distribution is thus given by

$$\delta f_1^{(n)}(x,v;t) = \Phi_1^n \exp\left[\pm ink_1(x - vt)\right] \tag{5.36}$$

and the mth-order effect of the disturbance Eq. (5.27) is

$$\delta f_2^{(m)}(x,v;t) = \Phi_2^m \exp\left\{\pm imk_2[x - v(t-\tau)]\right\} . \tag{5.37}$$

The $(m+n)$th-order temporal echo appears as a result of coupling between (5.36) and (5.37) at time

$$t_0 = \frac{mk_2}{mk_2 - nk_1} \tau \tag{5.38}$$

with wave number

$$k_3 = mk_2 - nk_1 \ . \tag{5.39}$$

The amplitude of the resulting echo is proportional to $\Phi_1^n \Phi_2^m$. Similarly, the $(m+n)$th spatial echo appears at the coordinate

$$x_0 = \frac{m\omega_2}{m\omega_2 - n\omega_1} l \tag{5.40}$$

with frequency

$$\omega_3 = m\omega_2 - n\omega_1 \ . \tag{5.41}$$

These plasma-wave echo phenomena have been investigated experimentally both for the electron plasma oscillation [Malmberg et al. 1968] and for the ion-acoustic wave [Ikezi & Takahashi 1968; Baker, Ahern & Wong 1968].

B. Detailed Calculation

Based on the ideas outlined in the previous section, one calculates the detailed features of the plasma-wave echo from the Vlasov-Poisson equations (5.23) and (5.24). The calculations described in this subsection are due mainly to O'Neil and Gould [1968].

We write the distribution function as a summation of an unperturbed part and a perturbation; the latter is Fourier decomposed so that

$$f(x,v;t) = f(v) + \sum_{\mathbf{k}}{}' \int \frac{d\omega}{2\pi} \delta f_{k\omega}(v) \exp\left[i(kx - \omega t) + \eta t\right] . \qquad (5.42)$$

The prime means omission of the term with $k = 0$, which has been included in the unperturbed distribution. The potential field $\Phi(x,t)$ may likewise be Fourier decomposed; the term with $k = 0$ does not contribute because of the presence of neutralizing space-charge background.

These expressions are substituted into the Vlasov equation (5.23) and the Poisson equation (5.24). Retaining all the effects of the nonlinear term in Eq. (5.23), we find

$$\delta f_{k,\omega}(v) = -\frac{Ze}{m}\frac{kf'(v)}{\omega - kv + i\eta}\Phi(k,\omega) - \frac{Ze}{m}\frac{1}{\omega - kv + i\eta}$$

$$\times \sum_{\mathbf{k}'}{}' \int \frac{d\omega'}{2\pi}(k-k')\Phi(k-k',\omega-\omega')\frac{\partial}{\partial v}\delta f_{k',\omega'}(v) . \qquad (5.43)$$

This equation differs from Eq. (5.4) by the last term on its right-hand side. The potential field is then obtained from the Poisson equation as

$$\Phi(k,\omega) = \frac{\Phi_{\text{ext}}(k,\omega)}{\varepsilon(k,\omega)} - \frac{\omega_p^2}{k^2\varepsilon(k,\omega)}\int dv \frac{1}{\omega - kv + i\eta}$$

$$\times \sum_{\mathbf{k}'}{}' \int \frac{d\omega'}{2\pi}(k-k')\Phi(k-k',\omega-\omega')\frac{\partial}{\partial v}\delta f_{k',\omega'}(v) \qquad (5.44)$$

where the dielectric function $\varepsilon(k,\omega)$ has been defined by Eq. (5.5). Equations (5.43) and (5.44) are the basic equations for a theoretical treatment of the plasma-wave echo phenomena. With the aid of perturbation-theoretical calculations, one can in principle analyze a temporal or spatial echo to any order of external disturbances.

For the sake of definiteness, we shall henceforth consider the case of a temporal plasma echo. The external disturbance is given by the summation of (5.25) and (5.27); its Fourier components are

$$\Phi_{\text{ext}}(k,\omega) = \frac{\Phi_1}{\omega_p}[\delta(k,k_1) + \delta(k,-k_1)]$$

$$+ \frac{\Phi_2}{\omega_p}[\delta(k,k_2) + \delta(k,-k_2)] \exp(i\omega\tau) . \tag{5.45}$$

We substitute this expression into Eqs. (5.43) and (5.44), and carry out the perturbation calculations with respect to Φ_1 and Φ_2.

To the first order, Eqs. (5.43) and (5.44) yield standard results from a linear response analysis,

$$\Phi^{(\mathrm{I})}(k,\omega) = \frac{\Phi_1}{\omega_p \varepsilon(k,\omega)}[\delta(k,k_1) + \delta(k,-k_1)]$$

$$+ \frac{\Phi_2}{\omega_p \varepsilon(k,\omega)}[\delta(k,k_2) + \delta(k,-k_2)] \exp(i\omega\tau) , \tag{5.46a}$$

$$\delta f_k^{(\mathrm{I})}(v) = -\frac{Ze}{m}\frac{kf'(v)}{\omega - kv + i\eta}\Phi^{(\mathrm{I})}(k,\omega) . \tag{5.46b}$$

No echoes are expected in this order.

To the second order in the perturbation, the potential field is given by

$$\Phi^{(\mathrm{II})}(k,\omega) = -\frac{Ze\omega_p^2}{mk\varepsilon(k,\omega)}\sum_{\mathbf{k}'}{}' \int\frac{d\omega'}{2\pi}\int dv \frac{k'(k-k')f'(v)}{(\omega - kv + i\eta)^2(\omega' - k'v + i\eta)}$$

$$\times \Phi^{(\mathrm{I})}(k',\omega')\Phi^{(\mathrm{I})}(k-k',\omega-\omega') . \tag{5.47}$$

Equation (5.46a) is substituted into this equation; a second-order temporal echo will appear from those terms containing cross products between Φ_1 and Φ_2. Retaining such terms only, we obtain

$$\Phi^{(\mathrm{II})}(k,\omega) = -\frac{Ze\Phi_1\Phi_2 k_1 k_2}{mk\varepsilon(k,\omega)} \int \frac{d\omega'}{2\pi} \int dv \frac{f(v)}{(\omega - kv + i\eta)^2}$$

$$\times \left\{ \frac{\delta(k,k_3) \exp(i\omega'\tau)}{(\omega' - k_2 v + i\eta)\varepsilon(k_2,\omega')\varepsilon(-k_1,\omega-\omega')} \right.$$

$$+ \frac{\delta(k,-k_3) \exp(i\omega'\tau)}{(\omega' + k_2 v + i\eta)\varepsilon(-k_2,\omega')\varepsilon(k_1,\omega-\omega')}$$

$$+ \frac{\delta(k,k_3) \exp[i(\omega-\omega')\tau]}{(\omega' + k_1 v + i\eta)\varepsilon(-k_1,\omega')\varepsilon(k_2,\omega-\omega')}$$

$$\left. + \frac{\delta(k,-k_3) \exp[i(\omega-\omega')\tau]}{(\omega' - k_1 v + i\eta)\varepsilon(k_1,\omega')\varepsilon(-k_2,\omega-\omega')} \right\} \tag{5.48}$$

where k_3 has been given by Eq. (5.31).

The ω' integration for the first two terms on the right-hand side may be carried out by closing the contour in the upper half-plane. The only poles that can contribute to such integrations are the zeros of $\varepsilon(\pm k_1,\omega - \omega')$; they give rise to exponential decay $\exp(-|\gamma(k_1)|\tau)$. Since $\tau >> |\gamma(k_1)|^{-1}$, those terms may be neglected. The integration for the last two terms may be calculated by closing the contour in the lower half-plane. Two different kinds of poles are now involved: the poles from the zeros of $\varepsilon(\pm k_1,\omega')$ and those at $\omega' = \pm k_1 v - i\eta$. The contributions from the former poles behave in the same way as the first two terms of Eq. (5.48); these may be neglected. The only significant contributions, therefore, come from the latter poles at $\omega' = \pm k_1 v - i\eta$.

Taking account of the simplifications described above, we now carry out inverse Fourier transformation of Eq. (5.48). Expressing the result as

$$\Phi^{(\mathrm{II})}(x,t) = \Phi^{(\mathrm{II})}(t) \exp(ik_3 x) + \mathrm{cc}\,,$$

we find

$$\Phi^{(\mathrm{II})}(t) = -i\frac{Ze\Phi_1\Phi_2 k_1 k_2}{mk_3} \int \frac{d\omega}{2\pi} \int dv \frac{f(v)}{(\omega - k_3 v + i\eta)^2}$$

$$\times \frac{\exp\{i[k_1 v\tau - \omega(t-\tau)]\}}{\varepsilon(-k_1,-k_1 v)\varepsilon(k_2,\omega + k_1 v)\varepsilon(k_3,\omega)}\,.$$

The ω integration can be calculated by closing the contour in the lower half-plane. Three kinds of poles are involved: the double pole at $\omega = k_3 v - i\eta$, the zeros of $\varepsilon(k_2, \omega + k_1 v)$, and the zeros of $\varepsilon(k_3, \omega)$. The contributions from the last two kinds of poles produce exponentially decaying factors $\exp[-|\gamma(k_2)|(t-\tau)]$ and $\exp[-|\gamma(k_3)|(t-\tau)]$; assuming $|\gamma(k_2)|(t-\tau) >> 1$ and $|\gamma(k_3)|(t-\tau) >> 1$, we neglect those contributions. The equation thus becomes

$$\Phi^{(\mathrm{II})}(t) = -\frac{Ze\Phi_1\Phi_2 k_1 k_2}{mk_3}\int dv \frac{f(v)}{\varepsilon(-k_1, -k_1 v)}$$

$$\times \left\{ \frac{\partial}{\partial \omega}\left[\frac{\exp\{i[k_1 v\tau - \omega(t-\tau)]\}}{\varepsilon(k_2, \omega + k_1 v)\varepsilon(k_3, \omega)} \right] \right\}_{\omega = k_3 v} . \quad (5.49)$$

Equation (5.49) will be split into three terms when the differentiation with respect to ω is actually carried out. Among them, the terms involving the derivatives of the dielectric functions simply oscillate in time with constant amplitudes. The term arising from the derivative of the exponential function, on the other hand, exhibits a secular growth in time proportional to $t-\tau$. In the vicinity of the echo peak

$$t - \tau = t_0 - \tau = \frac{k_1}{k_3}\tau , \quad (5.50)$$

this secular term represents the major contribution. Keeping only this term and using Eq. (5.50) approximately for $t-\tau$ appearing in its coefficient, we find that Eq. (5.49) becomes

$$\Phi^{(\mathrm{II})}(t_0 + t') = -\frac{Ze\Phi_1\Phi_2 k_1^2 k_2 \tau}{imk_3^2}\int dv \frac{f(v)\exp(-ik_3 vt')}{\varepsilon(-k_1, -k_1 v)\varepsilon(k_2, k_2 v)\varepsilon(k_3, k_3 v)} , \quad (5.51)$$

where $t' = t - t_0$ is the time measured from the peak position t_0. The shape of an echo profile is thus found by performing the velocity-space integration in Eq. (5.51). Contour of the integration may be closed in the upper half-plane for $t' < 0$ and in the lower half-plane for $t' > 0$; the singularities of $f(v)$ at $v = \pm i\infty$ will be canceled by similar singularities associated with the dielectric functions (see footnote, p.149). The zeros of the dielectric functions in Eq. (5.51) therefore determine the shape of the echo as it builds up and decays.

For a Maxwellian plasma, we may use the result of Eq. (5.9). The echo builds up as $\exp[(k_3/k_1)|\gamma(k_1)|(t-t_0)]$; it decays as $\exp[-(k_3/k_2)|\gamma$

$(k_2)|(t-t_0)]$ and $\exp[-|\gamma(k_3)|(t-t_0)]$. Hence the echo shape is not symmetric with respect to the echo peak.

C. Effects of Collisions

As the arguments in the preceding sections have demonstrated, the existence of an echo depends critically on the preservation of delicate phase memory during the evolution of microscopic elements. We would, therefore, anticipate that deflections of particle trajectories caused by collisions, however small they may be, will have sensitive effects on the echo through destruction of the delicate phase memories. In this section, such collisional effects on plasma-wave echoes will be treated; for a more detailed investigation, the reader is referred to [Karpman 1966; Su & Oberman 1968; Hinton & Oberman 1968; O'Neil 1968; Jensen, Malmberg & O'Neil 1969].

For analysis of collisional effects, we add to the right-hand side of Eq. (5.23) the collision term considered in Section 2.4. The solution to such a kinetic equation for plasma-wave propagation is ordinarily a complicated problem. The situation is simplified, however, since we are here interested in a change in the distribution function arising from relatively small momentum transfers. Relaxation processes may then be described by the model collision term proposed by Lenard and Bernstein [1958]:

$$\left.\frac{\partial f}{\partial t}\right]_c = \nu \frac{\partial}{\partial v}\left[vf + \frac{T}{m}\frac{\partial f}{\partial v}\right], \tag{5.52}$$

where v is the velocity component in the direction of wave propagation and ν is an effective collision frequency. Equation (5.52) retains two of the most important properties of a Coulomb collision term such as Eq. (2.67): it describes a diffusive behavior and vanishes if a one-dimensional Maxwellian distribution reduced from Eq. (1.1) is substituted. The collision term assumes a general form of the Fokker-Planck equation, which will be discussed later in Chapter 8.

Substituting in Eq. (5.52) the increment (5.26) created by the first pulse (5.25), we obtain

$$\left.\frac{\partial}{\partial t}\delta f_1(x,v;t)\right]_c = \nu\left[1 - ik_1 vt - \frac{T}{m}(k_1 t)^2\right]\delta f_1(x,v;t)$$

$$\to \nu\left[ik_1 vt + \frac{T}{m}(k_1 t)^2\right]\delta f_1(x,v;t)\,,$$

where the upper sign of (5.26) has been chosen; a long-time behavior has been singled out in the last step. Hence with collisions,

$$\delta f_1(x,v;t)_c = \delta f_1(x,v;t)\exp\left[-\nu\left(\frac{ik_1vt^2}{2}+\frac{Tk_1^2t^3}{3m}\right)\right]. \tag{5.53}$$

The amount of phase advance during the time interval τ is

$$-k_1v\tau\left(1+\frac{\nu\tau}{2}\right); \tag{5.54}$$

the decay factor due to the collisions is

$$\exp\left[-\nu\left(\frac{Tk_1^2\tau^3}{3m}\right)\right]. \tag{5.55}$$

At $t=\tau$, the second pulse (5.27) is applied to the plasma. Without collisions, the increment of distribution would propagate after this time with a phase advance

$$\delta f_3(x,v;t) \approx \exp\{-ik_3[x-v(t-\tau)]\}\,.$$

With collisions, however, this will be modified as

$$\delta f_3(x,v;t)_c = \delta f_3(x,v;t)\exp\left\{\nu\left[i\frac{k_3v(t-\tau)^2}{2}-\frac{Tk_3^2(t-\tau)^3}{3m}\right]\right\}. \tag{5.56}$$

The phase advances during the time interval $t-\tau$ by the amount

$$k_3v(t-\tau)\left(1+\nu\frac{t-\tau}{2}\right); \tag{5.57}$$

the amplitude decays by a factor

$$\exp\left[-\nu\frac{Tk_3^2(t-\tau)^3}{3m}\right]. \tag{5.58}$$

At the echo peak $t=t_0$, the total phase advance, (5.54) plus (5.57), must vanish. Assuming $\nu\tau \ll 1$ and $\nu(t_0-\tau) \ll 1$, we calculate the peak position from this condition as

$$t_0 = \frac{k_2}{k_3}\tau\left[1-\frac{\nu\tau k_1}{2k_2}\left(1-\frac{k_1}{k_3}\right)\right]. \tag{5.59}$$

Hence collisions act to shift the position of the echo peak slightly. From Eqs. (5.55) and (5.58), we also find that the echo amplitude decreases by a factor

$$\exp\left(-\nu\frac{Tk_1^2\tau^2t_0}{3m}\right) \tag{5.60}$$

because of the interparticle collisions.

5.3 LARGE-AMPLITUDE PLASMA WAVES

The treatment of plasma-wave propagation in Section 5.1 assumed the case with a small amplitude so that the main body of the distribution function remained invariant during the process of wave evolution. The wave showed a decay by Landau damping at the rate

$$\gamma_L = -\frac{\pi\omega_p^2}{(\partial\varepsilon_1/\partial\omega)k^2}f'\left(\frac{\omega}{k}\right) \tag{5.61}$$

obtainable from a comparison between Eqs. (3.58) and (5.5).

Alternatively, one can consider the other limiting cases [Al'tshul' & Karpman 1965; O'Neil 1965]: A large-amplitude plasma wave is applied to the plasma; its amplitude may then remain almost constant. The distribution function, on the other hand, is modulated by the wave potential; it now becomes time dependent.

The distinction between these two cases is related to the ways in which the nonlinear term of the Vlasov equation (5.23) is linearized. Generally the linearization may be performed according to

$$\frac{\partial\Phi}{\partial x}\frac{\partial f}{\partial v} = \frac{\partial\Phi}{\partial x}\frac{\partial f_0}{\partial v} + \frac{\partial\Phi_0}{\partial x}\frac{\partial f}{\partial v} \tag{5.62}$$

where the quantities with the subscript 0 are to be regarded as time independent. Landau's treatment of a small-amplitude plasma wave corresponds to retaining the first term only in the linearization. For a large-amplitude wave, however, the second term may be dominant.

The criterion for deciding which of the two treatments is more applicable to a given circumstance, therefore, depends on the wave amplitude. Such a criterion may be obtained by looking into the behavior of those particles trapped in the potential troughs of the large-amplitude plasma wave.

A. Trapped Particles

Consider a monochromatic wave in a plasma whose potential field is given by

$$\Phi(x,t) = \Phi_0 \cos(kx - \omega t) . \tag{5.63}$$

A Galilean transformation

$$X = x - \frac{\omega}{k} t \tag{5.64}$$

from the laboratory frame (x,t) to a frame (X,t) comoving with the wave transforms Eq. (5.63) to

$$\Phi(X,t) = \Phi_0 \cos(kX) . \tag{5.65}$$

Since the particle velocity v is likewise transformed to

$$V = v - \frac{\omega}{k} , \tag{5.66}$$

the total energy of a particle in this moving frame of reference is

$$W = \frac{1}{2} m V^2 + Ze\Phi_0 \cos(kX) . \tag{5.67}$$

In the phase space (X,V), a particle thus moves along a trajectory $W =$ constant. (See Fig. 5.4.) Those particles with $W < 0$ are thereby trapped by the wave potential, bouncing back and forth in the trough; they perform a periodic motion in the phase space.

The period of such a bouncing motion may be estimated from consideration of those particles near the bottom of the trough. Since the electric field associated with Eq. (5.65) is $\Phi_0 k \sin(kX)$, the equation of motion for those particles with $|kX| \ll 1$ is

$$m\ddot{X} = Ze\Phi_0 k^2 X .$$

Hence the *bounce frequency* ω_B is determined as

$$\omega_B = \sqrt{\left| \frac{Ze\Phi_0 k^2}{m} \right|} . \tag{5.68}$$

As the wave amplitude increases, so does the bounce frequency.

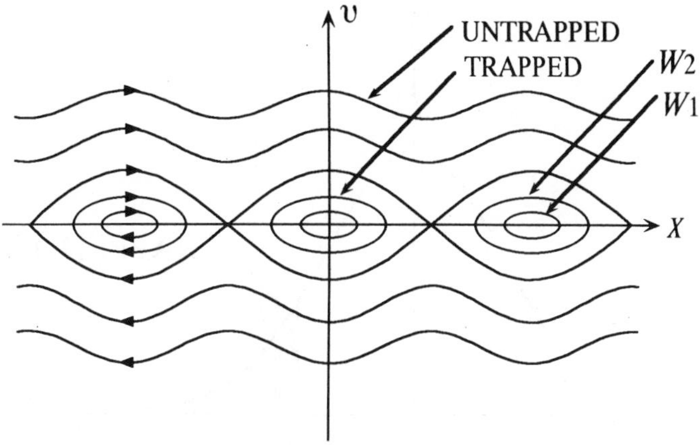

Figure 5.4 The phase trajectories of trapped and untrapped particles.

The criterion for the validity of Landau's treatment of small-amplitude plasma waves may be found from a comparison between this bounce frequency and Landau's decay rate (5.61). Thus, when the wave amplitude is small so that

$$\omega_B << | \gamma_L | , \tag{5.69}$$

the wave will be Landau-damped considerably by the time a *trapped particle* has completed a cycle of the bounce motion. The trapping is not a significant effect here; a small-amplitude treatment neglecting the last term of Eq. (5.62) is justified in these circumstances.

On the other hand, if the wave amplitude is large so that

$$\omega_B >> | \gamma_L | , \tag{5.70}$$

then the trapped particles will have completed many cycles of their bounce motion before the wave starts to decay. The second term on the right-hand side of Eq. (5.62) now represents the major effects. In the next subsection, we consider qualitative consequences arising from such trapping effects.

B. Amplitude Oscillation

Suppose that the distribution function $f(v)$ initially has the form depicted by the solid line in Fig. 5.5, together with the wave field Eq. (5.63) or (5.65) in

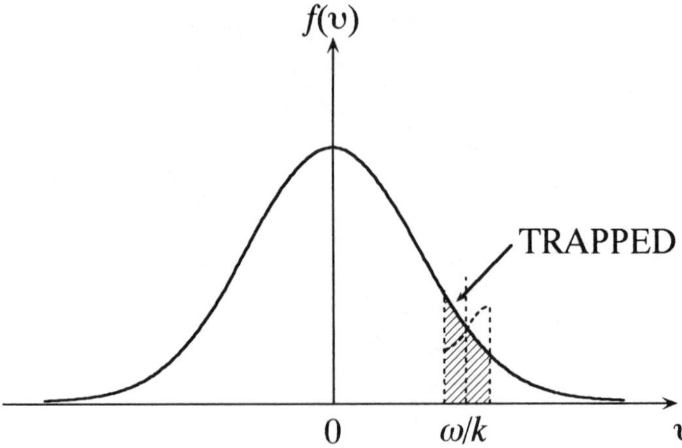

Figure 5.5 Initial distribution of the particles and the trapped part.

the plasma. The particles in the shaded area are then trapped in the wave potential. Since $f'(\omega/k) < 0$, the γ_L calculated according to Eq. (5.61) takes on a positive value at $t = 0$.

Let us now follow the motion of the trapped particles for a half period, π/ω_B. The particles will have completed a half cycle of their bounce motion; their distribution may assume a shape inverted with respect to the axis $v = \omega/k$ in the trapped domain. This is not exactly true, however, for those particles close to the outer boundary of the trapped domain; the bounce frequency Eq. (5.68) has been calculated only for those particles near the bottom of the trough. The equation of motion for those particles close to the periphery of the trapped domain deviates substantially from that of a harmonic motion. As a consequence, the shape of the distribution for the trapped particles at $t = \pi/\omega_B$ will be slightly dulled from the exact mirror image of the distribution at $t = 0$, as the broken line in Fig. 5.5 illustrates. Let us nonetheless note that $f'(\omega/k)$ is now expected to be positive; hence $\gamma_L < 0$ at $t = \pi/\omega_B$.

If we wait another half period, most of the trapped particles will return to their original phases at $t = 0$; γ_L will become positive again. Thus, the value of γ_L is expected to oscillate around $\gamma_L = 0$ at the frequency ω_B, as the distribution in the trapped domain is modulated by the wave at the same frequency.

It is apparent, however, that this sort of oscillation cannot continue indefinitely. For one thing, the effects of anharmonicity act to smooth out the distribution in the trapped domain. In addition, we must note that ample time is available for those particles trapped in the same potential trough to interact

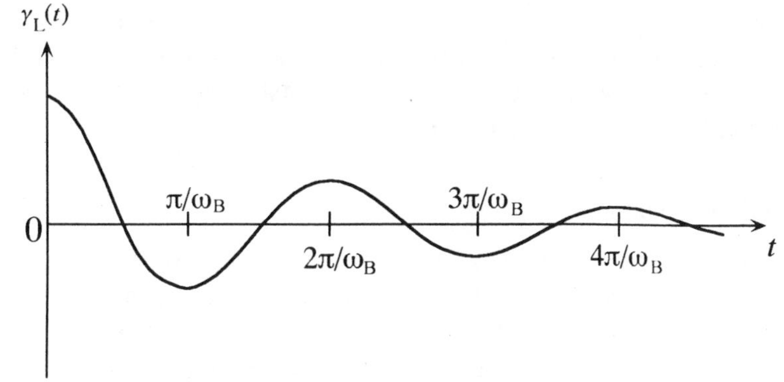

Figure 5.6 Schematic behavior of γ_L.

with each other. The particle interactions, however weak, will eventually phase-mix between those trapped particles on the same trajectory (W = constant) as well as between those on the slightly different trajectories W_1 and W_2 in Fig. 5.4. All these effects will contribute to a flattening of the distribution or a formation of a plateau in the trapped domain. Consequently, the time-dependent behavior of γ_L should resemble that of a damped oscillation, asymptotically approaching $\gamma_L = 0$ (see Fig. 5.6).

Thus far we have not taken into account the change in the wave amplitude; it has been assumed constant, large enough to satisfy (5.70). In the limit as $|\gamma_L|/\omega_B \to 0$, the wave amplitude may in fact remain constant. With first-order effects of $|\gamma_L|/\omega_B$, however, the wave amplitude Φ_0 in Eq. (5.63) no longer remains constant. Instead, it varies in accord with the equation

$$\frac{d|\Phi_0|^2}{dt} = -2\gamma_L(t)|\Phi_0|^2, \tag{5.71}$$

which may be solved as

$$|\Phi_0(t)|^2 = |\Phi_0(0)|^2 \exp\left[-2\int^t dt'\gamma_L(t')\right]. \tag{5.72}$$

In light of Fig. 5.6, we may thus expect that $|\Phi_0(t)|$ likewise contains an oscillatory component with frequency $\sim \omega_B$.

This phenomenon is called the *amplitude oscillation*; it reflects the significance of the particle-trapping effects in large-amplitude plasma waves. The period of the amplitude oscillation as well as the criterion, (5.69) versus (5.70), has been investigated experimentally for both the electron plasma oscillation [Malmberg & Wharton 1967; Wharton, Malmberg & O'Neil 1968] and the ion-acoustic wave [Sato et al. 1968].

C. The BGK Solutions

In the previous subsection, a *plateau formation* was predicted as a time asymptotic solution for the distribution function in the trapped domain of a large-amplitude plasma wave. It should also be noted that when such a plateau is formed, $f'(\omega/k) = 0$; the wave amplitude in turn remains stationary. Based on such an observation, one can construct an exact stationary solution to the Vlasov-Poisson equations containing an arbitrary electrostatic wave with finite amplitude.

We work in the phase space (X,V) defined in Eqs. (5.64) and (5.66), the frame comoving with the electrostatic wave Eq. (5.63). In that part of the phase space where the trapping takes place (i.e., $W < 0$), the distribution function may be set as a constant

$$f(X,V) = f_p \qquad (W < 0) . \tag{5.73}$$

A plateau of height f_p has thus been formed; f_p is related to the number of particles trapped in the potential troughs, which should be determined later in a self-consistent way. In the untrapped domain, we may assume that $f(X,V)$ is a function of W and the sign of V only:

$$f(X,V) = \begin{cases} F_1(W) & (W \geq 0,\ V > 0) , \\ F_2(W) & (W \geq 0,\ V < 0) . \end{cases} \tag{5.74}$$

The distribution function must be continuous, so that

$$F_1(0) = F_2(0) = f_p . \tag{5.75}$$

The time-independent version of the Vlasov equation in the phase space (X,V) is

$$V\frac{\partial f}{\partial X} - \frac{Ze}{m}\frac{\partial \Phi}{\partial X}\frac{\partial f}{\partial V} = 0 . \tag{5.76}$$

In light of Eqs. (5.65) and (5.67), it is clear that the distribution function (5.73) and (5.74) indeed satisfies the Vlasov equation (5.76). The existence of the plateau (5.73) guarantees the stationarity of the wave.

The remaining task, therefore, is to examine the solution in terms of the Poisson equation, which reads

$$\frac{\partial^2 \Phi}{\partial X^2} = -4\pi Zen \left[\int_{-\infty}^{\infty} dV f(X,V) - 1 \right]. \tag{5.77}$$

Substitution of (5.73) and (5.74) into Eq. (5.77) yields an integral equation which determines the shapes of $F_1(W)$ and $F_2(W)$ and the level of f_p. One thus obtains an exact solution to the Vlasov-Poisson equations for arbitrary stationary waves. It is in fact possible to extend such an approach and to construct waves of quite an arbitrary shape, such as isolated pulses and sinusoidal waves. The existence and the properties of the solutions to such nonlinear problems have been investigated by Bernsein, Greene, and Kruskal [1957], and are thus called the *BGK solutions.*

5.4 NONLINEAR EFFECTS

In addition to echoes and trapped particles, varieties of nonlinear effects have been observed in conjunction with the propagation of plasma waves. In this section, we treat two such nonlinear effects: solitary waves and ponderomotive forces.

A. Solitary Waves

Ion-acoustic wave solitons stem from a nonlinear effect associated with a flow of the plasma. In a fluid-dynamic description of plasmas (cf., Problem 3.12), a continuity equation for the ion fluid reads

$$\frac{\partial n_i}{\partial t} + \frac{\partial (n_i u_i)}{\partial x} = 0\,, \tag{5.78}$$

where n_i and u_i are the number density and flow velocity of the ions; a one-dimensional propagation in the x–direction has been assumed.

The continuity equation for the flow of momentum is

$$\frac{\partial (n_i u_i)}{\partial t} + \frac{\partial (n_i u_i^2)}{\partial x} = -\frac{e}{m_i} n_i \frac{\partial \Phi}{\partial x}\,. \tag{5.79}$$

Here $\Phi(x)$ refers to the electrostatic potential associated with the ion-acoustic waves and $Z_i = 1$ is assumed. In the derivation of Eq. (5.79) the contributions of viscosity (i.e., ion-ion collisions) and the partial pressure of ions are neglected. The partial pressure of the ions is proportional to the ion temperature T_i; as we examined in Section 4.3C, $T_e >> T_i$ must be satisfied for the existence of ion-acoustic waves. Under these circumstances the effect of the partial pressure of ions can be ignored. Since we are dealing with a weakly coupled plasma, the viscosity term arising from interparticle collisions can likewise be ignored in light of the earlier study in Section 1.4C. Substitution of Eq. (5.78) in Eq. (5.79) yields

$$\frac{\partial u_i}{\partial t} + u_i \frac{\partial u_i}{\partial x} = -\frac{e}{m_i} \frac{\partial \Phi}{\partial x} . \tag{5.80}$$

The ion-acoustic wave is a low-frequency phenomenon as far as the dynamics of electrons is concerned. Hence one expresses the electron density n_e in a quasistatic approximation

$$n_e = n \exp\left(\frac{e\Phi}{T}\right) \tag{5.81}$$

according to the Boltzmann statistics. Here n refers to the average number density of electrons or ions. One can thus close the set of equations by Eqs. (5.78), (5.80), (5.81) and the Poisson equation

$$\frac{\partial^2 \Phi}{\partial x^2} = 4\pi e(n_e - n_i) , \tag{5.82}$$

which determines the electrostatic potential.

That the collective motion described by this set of equations corresponds to the ion-acoustic wave discussed in Section 4.3C can be confirmed in the following way: Write the electron and ion densities as sums of their average value n and deviations therefrom, δn_e and δn_i. The set of equations, (5.78) and (5.80)–(5.82), are then linearized with respect to variations δn_e, δn_i, Φ, and u_i. Assuming that these variations have a sinusoidal space-time variation $\cos(kx - \omega_k t)$, we arrive at the dispersion relation (4.53) between k and ω_k (Problem 5.6).

In the confines of the linear analysis mentioned above, the second term on the left-hand side of Eq. (5.80) plays no part. This term, being a second-order nonlinear term with respect to the variation u_i, becomes significant at a large wave amplitude. To extract a relevant nonlinear effect, we consider an equation

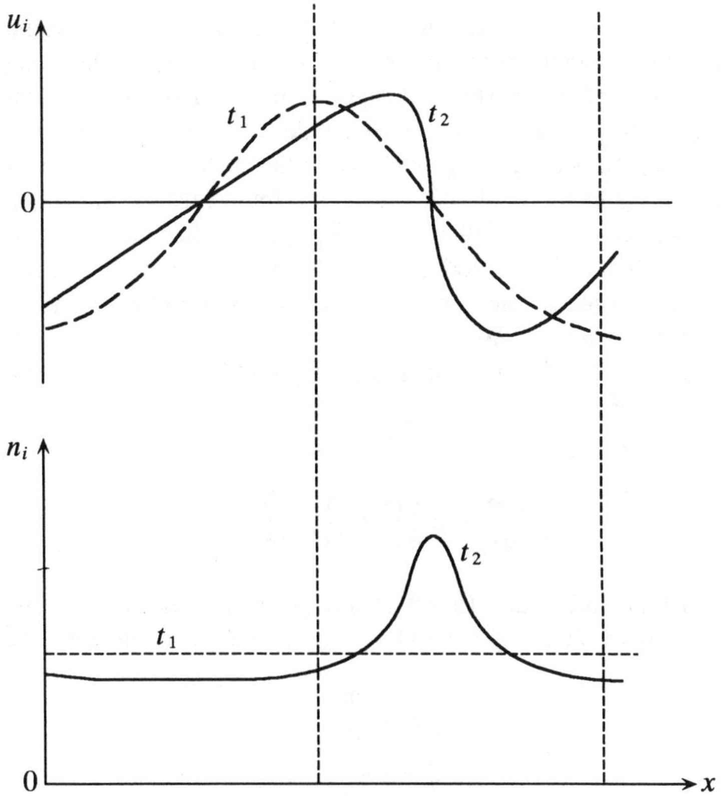

Figure 5.7 Steepening of wave front due to the nonlinear effect of plasma flow.

$$\frac{\partial u_i}{\partial t} + u_i \frac{\partial u_i}{\partial x} = 0 \tag{5.83}$$

where the Coulomb-force term on the right-hand side of Eq. (5.80) is neglected. Let us note that the "coefficient" u_i in the second term on the left-hand side plays the part of "propagation velocity" for the variation. Assume at time t_1 that u_i and n_i have distributions as shown by the dashed curves in Fig. 5.7. The portion where u_i is large moves fast in accordance with Eq. (5.83); at t_2 ($> t_1$) the distributions will become as depicted by solid curves in Fig. 5.7. As a consequence the wave front of u_i steepens sharply; a spatial variation of density is induced in accord with the continuity equation (5.78).

As the wave front of the ion-acoustic wave steepens and the variations of density and flow velocity become abrupt spatially, a dispersion effect of the wave sets in. In the long-wavelength regime, the ion-acoustic wave obeys the dispersion relation (4.50a) of the acoustic type; no changes in the wave form will result along the wave propagation. When the steepening of the wave front takes place owing to the nonlinear effect, the intensities of short-wavelength components increase. Those obey the propagation characteristics in accord with Eq. (4.53); the phases and group velocities are now functions of the wavelength. A change in the wave form results as a consequence of such a dispersion effect. Solitary waves may be generated through balance between the steepening effect of nonlinearity and the dispersion effect in the short-wavelength regime.

One can in fact derive

$$\frac{1}{s}\frac{\partial \varphi}{\partial t}+\varphi\frac{\partial \varphi}{\partial \xi}+\frac{1}{2k_e^2}\frac{\partial^3 \varphi}{\partial \xi^3}=0 \tag{5.84}$$

starting with Eqs. (5.78) and (5.80)–(5.83) [Washimi and Taniuti 1966]. An approximate way of obtaining Eq. (5.84) is described in Problem 5.7. Here

$$\varphi=\frac{e\Phi}{T}, \tag{5.85}$$

$$\xi=x-st. \tag{5.86}$$

The second term on the left-hand side of Eq. (5.84) corresponds to the second term on the left-hand side of Eq. (5.80), the nonlinear effect in the plasma flow. The third term on the left-hand side of Eq. (5.84) represents the dispersion effect, derived from the second term in the expansion

$$\omega_k=sk\left(1-\frac{k^2}{k_e^2}+\cdots\right)$$

of Eq. (4.53) with respect to k^2. The leading term, sk, of the expansion has been taken care of by the transformation (5.86).

Equation (5.84) takes the form of the *Korteweg-de Vries equation* derived in 1895 on the shallow-water wave; it has a special solution representing a soliton

$$\varphi=\varphi_0\,\text{sech}^2\left(\frac{x-ut}{\Delta}\right), \tag{5.87}$$

$$u = s\left(1 + \frac{\varphi_0}{3}\right), \tag{5.88}$$

$$\Delta = \frac{\sqrt{6/\varphi}}{k_e}. \tag{5.89}$$

According to this solution, a soliton exists only in the positive direction of the potential $\varphi = (T/e)\varphi$; the propagation velocity u increases from s and the width Δ decreases as the amplitude φ_0 increases. These features have been clearly exhibited in the ion-acoustic wave soliton experiments performed by Ikezi [1973].

B. Ponderomotive Forces

Pressure of a longitudinal wave or a transverse wave in a plasma implies an energy density proportional to the square of the wave amplitude. When this energy density varies spatially, a force proportional to its gradient is exerted onto the plasma. Generally such a force is referred to as a *ponderomotive force* [e.g., Landau & Lifshitz 1960]; it is a nonlinear effect essential in the description of large-amplitude waves in a plasma.

A heuristic example of the ponderomotive force may be offered by consideration of an electric field, $\mathbf{E} = E\hat{\mathbf{x}}$, applied to a dielectric medium, containing n "molecules" per unit volume. The field strength is assumed to vary slowly (relative to the molecular size) in the x direction, in which the field is directed. If a dipole moment $\mathbf{d} = \mu\mathbf{E}$ is induced in a molecule with a polarizability μ under the action of the electric field, then the force on the molecule is calculated as

$$\mathbf{f} = \left(\mathbf{d}\cdot\frac{\partial}{\partial\mathbf{r}}\right)\mathbf{E} = \hat{\mathbf{x}}\left(\frac{\mu}{2}\right)\frac{\partial}{\partial x}E^2. \tag{5.90}$$

This force is thus exerted in the x direction.

The force per unit volume acting on the medium with the dielectric constant, $\varepsilon = 1 + 4\pi n\mu$, is then expressed as

$$F_x = \frac{\varepsilon - 1}{8\pi}\frac{\partial}{\partial x}E^2. \tag{5.91}$$

It is important here to note that the direction of this force depends on the sign of $\varepsilon - 1$.

We have thus calculated the force acting on a dielectric medium when a static electric field varies slowly in space. A force analogous to it arises in the presence of a high-frequency electric field associated with, for example, the plasma oscillation. Let us assume an oscillatory electric field (longitudinal wave) in a plasma with frequency ω and wave number k, whose amplitude varies slowly in the x direction. The field component in the x direction is expressed as

$$E(x,t) = E_{\mathrm{L}}(x) \cos (kx - \omega t) \tag{5.92}$$

where we assume that $E_{\mathrm{L}}(x)$ remains almost constant over a distance of the wavelength $2\pi/k$.

In the case of the high-frequency electric field Eq. (5.92), we replace the dielectric constant ε in Eq. (5.91) with the dielectric response function $\varepsilon(k,\omega)$, and E^2 with an average of the square of $E(x,t)$. This average is carried out over the high-frequency component, so that

$$\langle E(x,t)^2 \rangle = \frac{E_{\mathrm{L}}(x)^2}{2} .$$

Hence the slow spatial variation contained in $E_{\mathrm{L}}(x)$ remains after the average. For a high-frequency oscillation, the expansion (4.33) is applicable. Retaining up to the terms proportional to ω^{-2}, we find an expression for the ponderomotive force of plasma oscillation acting on the plasma as

$$F_x = -\frac{\omega_p^2}{16\pi\,\omega^2} \frac{\partial}{\partial x} E_{\mathrm{L}}(x)^2 . \tag{5.93}$$

According to Eq. (5.93), the ponderomotive force acts from the region of large-amplitude plasma wave to the region of small-amplitude plasma wave; the plasma is thus pushed away from the domain of the large amplitude wave. As the plasma density decreases, the local frequency of the plasma oscillation also decreases. The plasma oscillation is consequently confined in a domain where plasma density is "cavitated," as Fig. 5.8 illustrates. *Self-trapping* of a large-amplitude plasma wave thus takes place owing to the action of the ponderomotive force and subsequent localization of the plasma wave [Zakharov 1972; Kim, Stenzel & Wong 1974; Wong & Cheung 1984].

A ponderomotive force operates as a nonlinear effect also when a large-amplitude electromagnetic wave (transverse wave) propagates in a plasma. The ponderomotive force is mainly operative in the direction of the oscillating electric field, as Eq. (5.90) illustrates; for the transverse wave it is perpendicu-

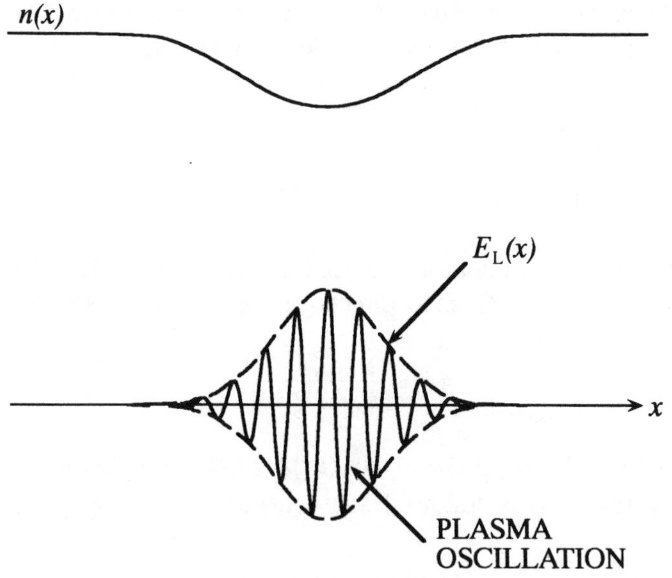

Figure 5.8 Self-trapping of large-amplitude plasma oscillations and cavitation of the plasma.

lar to the propagation. An electromagnetic field propagating in the x direction with the electric field

$$E_y(\mathbf{r},t) = E_T(\mathbf{r}) \cos (kx - \omega t) \tag{5.94}$$

directed in the y direction gives the ponderomotive force in the y direction

$$F_y = \frac{\varepsilon_T(\omega) - 1}{16\pi} \frac{\partial}{\partial y} E_T^2 . \tag{5.95}$$

Here $\varepsilon_T(\omega)$ is a transverse dielectric function of the plasma considered in Sections 4.5A and 4.6A.

Since the directional dependence in the case of a transverse wave is different from that in the case of the longitudinal wave, a nonlinear effect associated with the ponderomotive force appears in the form of *self-focusing* [e.g., Zakharov 1967].

The ponderomotive force effects and particularly the self-trapping processes as described in Fig. 5.8 play important parts in the formulation of a strong-turbulence theory for plasma waves. The reader is referred to a review on developments in those fields given by Goldman [1984].

PROBLEMS

5.1 Derive Eq. (5.10).

5.2 Consider an OCP without an external magnetic field whose one-dimensional velocity distribution (in the x direction) is given by a Lorentzian

$$f(v) = \frac{1}{\pi} \frac{\zeta}{v^2 + \zeta^2} .$$

At $t = 0$, a pulse field $\Phi_{\text{ext}}(x,t) = \Phi_1 \cos(kx)\delta(\omega_p t)$ is applied to the plasma, and its linear response is followed subsequently; the pulse shape satisfies $(k\zeta)^2 << \omega_p^2$.

(a) Investigate the time-dependent behavior of the induced field $\Phi_{\text{ind}}(x,t)$.

(b) Calculate the perturbed part $\delta f(x,v;t)$ of the distribution function for $t >> 1/k\zeta$.

5.3 Derive Eqs. (5.43) and (5.44).

5.4 Following the treatment described in Section 5.2C, investigate effects of collisions on the temporal plasma-wave echo when the collision term assumes the form

$$\left.\frac{\partial f}{\partial t}\right]_c = -\nu(f - f_0) .$$

Here ν is an effective collision frequency and f_0 is an equilibrium distribution function.

5.5 When the spatial propagation (cf. Section 5.1B) of an amplitude oscillation (cf. Section 5.1B) is considered, show that the wave number for the spatial variation of the amplitude oscillation is given by

$$k_{\text{osc}} = \frac{\omega_B k_r(\omega)}{\omega} ,$$

and that the condition for the amplitude oscillation is given by

$$\frac{k_{osc}}{k_i(\omega)} \gtrsim \sqrt{2\pi} .$$

5.6 Carry out linearization of Eqs. (5.78) and (5.80)–(5.82) as described in the text and derive the dispersion relation Eq. (4.53).

5.7 A simple way of deriving Eq. (5.84) is the following. In addition to Eq. (5.85), we introduce dimensionless variables n' and u' through $n_i = n(1+n')$ and $u_i = su'$. We then carry out coordinate transformations via Eq. (5.86), and expand Eqs. (5.78) and (5.80)–(5.82) up to the quadratic terms with respect to the perturbations. The equations are combined and simplified because we may take $\varphi = n' = u'$ in the linear approximation. Carry out these calculations and derive Eq. (5.84).

CHAPTER 6

Instabilities

In plasmas near thermodynamic equilibria, the collective modes are stable elementary excitations, as we have seen in Chapter 4. Those exciations suffer damping through resonant interactions with individual particles; their lifetimes are finite.

When a plasma is away from thermodynamic equilibrium, collective modes may become unstable; amplitude of such an excitation tends to grow exponentially. Most of the plasmas in nature are significantly far from thermodynamic equilibrium. A current carrying plasma, for example, involves a relative drift motion between the electrons and the ions. A spatial inhomogeneity in the physical quantities, such as density, temperature, and strength of the magnetic field, will likewise create flows of particles, momenta, and energies in the plasma. Possibilities of plasma instabilities exist in any of these nonequilibrium circumstances.

In this chapter, we investigate the conditions under which onsets of various plasma instabilities may be expected. The physical mechanisms responsible for the onset of a plasma instability may be sought in the excess of free energy due to a departure from thermodynamic equilibrium. Hence it is meaningful to classify instabilities into two groups: instabilities in homogeneous plasmas and those in inhomogeneous plasmas.

In the former cases, one deals with nonequilibria in the velocity distributions. The kinetic energies associated with a stream or an anisotropic velocity distribution of the particles, for example, are the major sources of the extra free energy. Instabilities in this category are treated in the first half of this chapter.

In the second half, we shall consider those instabilities characteristic of inhomogeneous plasmas. Additional sources of free energy associated with the spatial variations of physical quantities are available in these circumstances. A nonuniform system therefore has a natural tendency to release this extra amount of free energy and thereby to approach a uniform equilibrium state. An external magnetic field applied to such a plasma usually places constraints on the motion of charged particles. The ordinary relaxations by Coulomb collisions are relatively inefficient processes in weakly coupled, high-temperature plasmas, as we have seen in Section 1.4C. The onset of an instability may thus be looked upon as an alternative avenue through which the plasma finds it preferable to release the extra amount of free energy; an anomalous relaxation would take place in such a plasma.

The nature of the instabilities should therefore depend critically on the interplay between the magnetic field and the spatial inhomogeneities. Both spatial inhomogenities and external force fields can induce drift motions in the charged particles. As in the cases of instabilities in homogeneous plasmas, such particle drifts play an essential part in the interpretation and understanding of instabilities in inhomogeneous plasmas.

6.1 PENROSE CRITERION FOR PLASMA-WAVE INSTABILITY

The stability of a plasma wave may be investigated through examination of the imaginary part γ_k of its characteristic frequency. When $\gamma_k > 0$, the plasma wave decays; the plasma is stable against the excitation of such a collective mode. When $\gamma_k < 0$, on the other hand, the plasma wave grows exponentally; the plasma is said to be unstable against such an excitation.

For a given one-dimensional velocity distribution $f(v)$ of a plasma, the properties of the plasma oscillation can be investigated through the dielectric response function (5.5). With the aid of Eq. (3.15), the dispersion relation in the vicinity of the real axis (i.e., $|\gamma_k/\omega_k| \ll 1$) becomes

$$1-\frac{\omega_p^2}{k^2}\mathcal{P}\int_{-\infty}^{\infty} dv\frac{f'(v)}{v-\omega/k}-i\pi\frac{\omega_p^2}{k^2}f'\left(\frac{\omega}{k}\right)=0 . \tag{6.1}$$

Analogously to Eq. (3.36), the decay rate may be determined as

$$\gamma_k=-\frac{\pi}{2}\frac{\omega_p^2\omega_k}{k^2}f'\left(\frac{\omega_k}{k}\right). \tag{6.2}$$

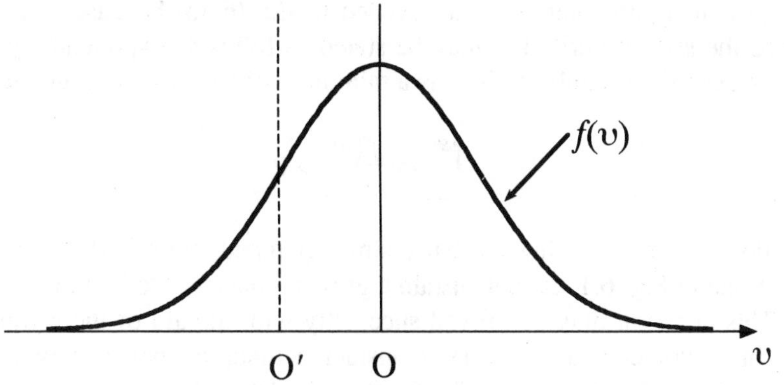

Figure 6.1 A velocity distribution function viewed from two inertial frames of reference, O and O'. The vertical lines indicate the lines at which $\upsilon = 0$ in the respective frames of reference.

It then follows that the plasma wave is stable when

$$\frac{\omega_k}{k} f'\left(\frac{\omega_k}{k}\right) < 0 . \tag{6.3}$$

This criterion has a merit of simplicity and transparency; it can be understood in terms of resonant coupling between the wave and particles.

In the form of (6.3), however, the criterion suffers a drawback in that it is not invariant under a Galilean transformation. Consider, for example, the situation as depicted in Fig. 6.1: A velocity distribution is observed from two different frames of reference, O and O′. In the frame O, $vf'(v) < 0$ everywhere. In the frame O′, however, $vf'(v)$ takes on positive values in the domain between O and O′. The magnitude and the sign of $vf'(v)$, therefore, depend on the frame of reference.

A stability criterion free of such a shortcoming has been obtained by Penrose [1960]. Consider again the dispersion relation (6.1) along the real axis on the complex ω plane; it represents the condition for marginal stability, expressed as

$$f'\left(\frac{\omega}{k}\right) = 0 , \tag{6.4a}$$

$$\int_{-\infty}^{\infty} dv \frac{f(v)}{v - \omega/k} = \left(\frac{k}{\omega_p}\right)^2 . \tag{6.4b}$$

The principal-part integral is not needed in Eq. (6.4b) because of Eq. (6.4a). Hence the stability criterion may be stated as follows: Exponentially growing modes exist if, and only if, there is a minimum of $f(v)$ at $v = v_m$ such that

$$\int_{-\infty}^{\infty} dv \frac{f(v)}{v - v_m} > 0 . \tag{6.5}$$

It follows from this criterion that a single-humped velocity distribution, such as the one in Fig. 6.1, cannot sustain a growing plasma oscillation.

This criterion may be proved succinctly with the aid of the Nyquist criterion in servomechanism theory. Consider a mapping between two complex planes ω and Z, specified by the functional relationship

$$Z(\omega) = \int_{\bar{C}} dv \frac{f(v)}{v - \omega / k} , \tag{6.6}$$

where $\bar{C}$ is the contour defined in Fig. 4.1. The dispersion relation $\varepsilon(k,\omega) = 0$ is then expressed as

$$\left(\frac{k}{\omega_p} \right)^2 = Z(\omega) . \tag{6.7}$$

Hence one concludes that, if the shape of $f(v)$ is such that a portion of the positive real axis in the Z plane is mapped through Eq. (6.6) into the upper half of the ω plane, then excitations with positive γ_k can exist for those values of k corresponding to that portion of the real Z axis. If, on the other hand, no portions of the positive real Z axis fall into the upper half of the ω plane through the mapping, then no solutions of ω from Eq. (6.7) have positive imaginary parts; plasma waves will be stable.

When $f(v)$ is a Maxwellian,

$$Z(\omega) = -\frac{m}{T} W \left(\frac{\omega}{k\sqrt{T/m}} \right) .$$

The upper half of the ω plane is mapped through this function onto the Z plane as in Fig. 6.2. The plasma is stable, since no portion of the positive real Z axis is enclosed by the contour. If we deform the shape of $f(v)$ from the Maxwellian so that the mapping now goes as shown in Fig. 6.3(a) or Fig. 6.3(b), for example, then a portion of the positive real Z axis is found inside the mapped contour. Exponentially growing plasma oscillations are possible in these circumstances.

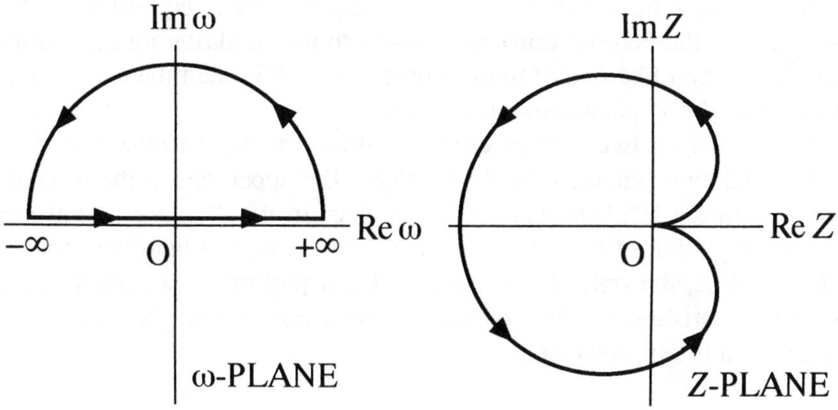

Figure 6.2 Mapping between ω and Z according to Eq. (6.6) for a Maxwellian plasma.

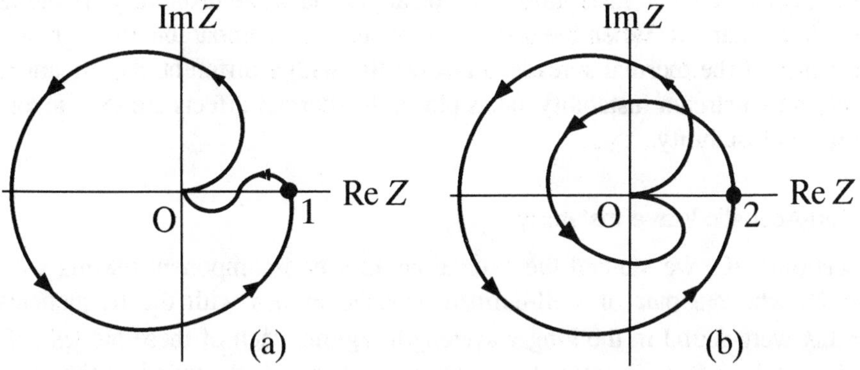

Figure 6.3 Examples of mapping indicating instabilities.

The important point to be noted in Fig. 6.3 is the intersection between the contour and the positive real Z axis at 1 or 2. The features common to all the intersections are: (a) $\mathrm{Im}Z(\omega)=0$; (b) the contour crosses such an intersection from $\mathrm{Im}Z(\omega)<0$ to $\mathrm{Im}Z(\omega)>0$; and (c) $\mathrm{Re}Z(\omega)>0$. The conditions (a) and (b) imply that $f(\omega/k)=0$ and $f'(\omega/k)>0$ at the intersection; hence it corresponds to a minimum of $f(v)$. The condition (c) is equivalent to (6.5). Consequently, the Penrose criterion amounts to the condition for the existence of an intersection like one of those shown in Fig. 6.3 when the real ω axis is mapped onto the Z plane through Eq. (6.6).

When such an intersection exists, a portion of the positive real Z axis should certainly be enclosed by the contour. (The upper half of the ω plane is mapped onto the left-hand side of the contour in the Z plane.) Conversely, when some portion of the positive real Z axis is mapped onto the upper half of the ω plane, the right-side boundary of that portion must correspond to a point which satisfies the three conditions listed earlier. The proof of the Penrose criterion is thus completed.

6.2 CURRENT-CARRYING PLASMAS

A drift motion between electrons and ions produces an electric current in a plasma. When the drift velocity is smaller than a thermal velocity of the particles, the velocity distributions significantly overlap between the two species. The ion-acoustic wave instability is a typical plasma-wave instability arising in these circumstances. When the drift velocity becomes comparable to or greater than either of the thermal spreads, an instability with a different classification, namely a two-stream instability, takes place; the thermal effects are less important in this instability.

A. Ion-Acoustic Wave Instability

In Section 4.3C, we studied the excitations in a two-component plasma with $T_e >> T_i$, where a pair of well-defined acoustic modes with the frequencies $\omega = \pm sk$ were found in the long-wavelength regime; each of these modes suffered a small and equal amount of damping. When a drift motion of the electrons as a whole is imparted relative to the ions, one of the acoustic modes acquires a longer lifetime; the other mode, on the contrary, suffers a stronger damping. When the drift velocity is increased further, the former mode becomes unstable. The existence of such an ion-acoustic wave instability under the action of an applied electric field has been considered by many investigators both for gaseous plasmas [Bernstein et al. 1960; Jackson 1960; Fried & Gould 1961] and for solid-state plasmas [Pines & Schrieffer 1961; Harrison 1962].

On the rest frame of the ions, the one-dimensional velocity distributions along the wave vector **k** may be expressed as

$$f_e(v) = \sqrt{\frac{m_e}{2\pi T_e}} \exp\left[-\frac{m_e(v - v_d)^2}{2T_e}\right], \tag{6.8a}$$

$$f_i(v) = \sqrt{\frac{m_i}{2\pi T_i}} \exp\left(-\frac{m_i v^2}{2T_i}\right), \tag{6.8b}$$

where v_d is the component of the drift velocity in the { roman bold k } direction. The dielectric response function for such a plasma is given by

$$\varepsilon(k,\omega) = 1 + \frac{k_e^2}{k^2} W\left(\frac{\omega - kv_d}{k\sqrt{T_e/m_e}}\right) + \frac{k_i^2}{k^2} W\left(\frac{\omega}{k\sqrt{T_i/m_i}}\right). \tag{6.9}$$

For the ion-acoustic wave, we consider the frequency domain (4.47) under an additional constraint,

$$v_d << \sqrt{\frac{T_e}{m_e}}, \tag{6.10}$$

so that the solution to the dispersion relation Eq. (3.54) is obtained as

$$\omega = \pm\omega_k = \pm\frac{\omega_i k}{\sqrt{k^2 + k_e^2}}, \tag{6.11a}$$

$$\gamma_k = \frac{\sqrt{\pi}}{8} \frac{\omega^3}{k^3 \omega_i^2}\left[\frac{k_e^2(\omega - kv_d)}{\sqrt{T_e/m_e}} + \frac{k_i^2 \omega}{\sqrt{T_i/m_e}} \exp\left(-\frac{\omega^2}{2k^2 T_i/m_i}\right)\right]_{\omega = \pm\omega_k}. \tag{6.11b}$$

Equation (6.11a) is the same as Eq. (4.53); Eq. (6.11b) differs from the imaginary part of Eq. (4.48a) by the presence of the drift velocity. Of the two solutions described by Eqs. (6.11), the branch $\omega = -\omega_k$ is always damped; its decay rate is greater than that in the absence of the drift motion.

To investigate a possible instability, we thus substitute $\omega = +\omega_k$ in Eq. (6.11b). In the limit of long wavelengths, we obtain

$$\gamma_k = \sqrt{\frac{\pi m_e}{8m_i}}\, k\left[(s - v_d) + s\sqrt{\frac{m_i T_e^3}{m_e T_i^3}} \exp\left(-\frac{T_e}{2T_i}\right)\right]. \tag{6.12}$$

When $T_e >> T_i$, the last term in Eq. (6.12), due to the ionic Landau damping, may be negligible. In these circumstances, the condition for the onset of the ion-acoustic wave instability is that the drift velocity exceeds the sound velocity.

The instability of the ion-acoustic wave is understood in term of the Cherenkov condition for wave excitations,

$$\mathbf{k} \cdot \mathbf{v}_{\mathrm{d}} > \omega_k \,, \tag{6.13}$$

which determines the apex angle of the *Cherenkov cone* for the wave vector **k**. The waves propagating in the directions satisfying (6.13) can be excited by the drifting electrons.

The ion-acoustic wave instability has been intensively investigated by experiments [e.g., Yamada & Raether 1975]. It has been looked upon as one of the principal sources of plasma-wave turbulence and the associated anomalous resistivity [Hamberger & Jancarik 1972], which we shall study in Chapter 9.

B. Two-Stream Instability

When the drift velocity of the electrons exceeds the condition (6.10) and reaches the domain,

$$v_{\mathrm{d}} \geq \sqrt{\frac{T_e}{m_e}} \,, \tag{6.14}$$

the electrons and ions are decoupled in the velocity space. The electronic polarizability in Eq. (6.9) can now be expanded as in Eq. (4.6), to yield the dispersion relation

$$1 - \frac{\omega_e^2}{(\omega - \mathbf{k} \cdot \mathbf{v}_{\mathrm{d}})^2} + \frac{\omega_i^2}{\omega^2} = 0 \,. \tag{6.15}$$

This leads to another typical class of plasma-wave instabilities, perhaps the simplest from a mathematical point of view, which takes place in a beam-plasma system [e.g., Briggs 1964]. Consider a monochromatic beam of charged particles with number density n_{b} injected into an OCP with number density n_0; for simplicity, we assume that the beam and the OCP consist of like particles. The velocity distribution function is then

$$f(\mathbf{v}) = \frac{n_0}{n}\delta(\mathbf{v}) + \frac{n_{\mathrm{b}}}{n}\delta(\mathbf{v} - \mathbf{v}_{\mathrm{d}}) \,, \tag{6.16}$$

and $n = n_0 + n_{\mathrm{b}}$, with the dispersion relation,

$$1 - \frac{\omega_b^2}{(\omega - \mathbf{k} \cdot \mathbf{v}_{\mathrm{d}})^2} + \frac{\omega_0^2}{\omega^2} = 0 \,, \tag{6.17}$$

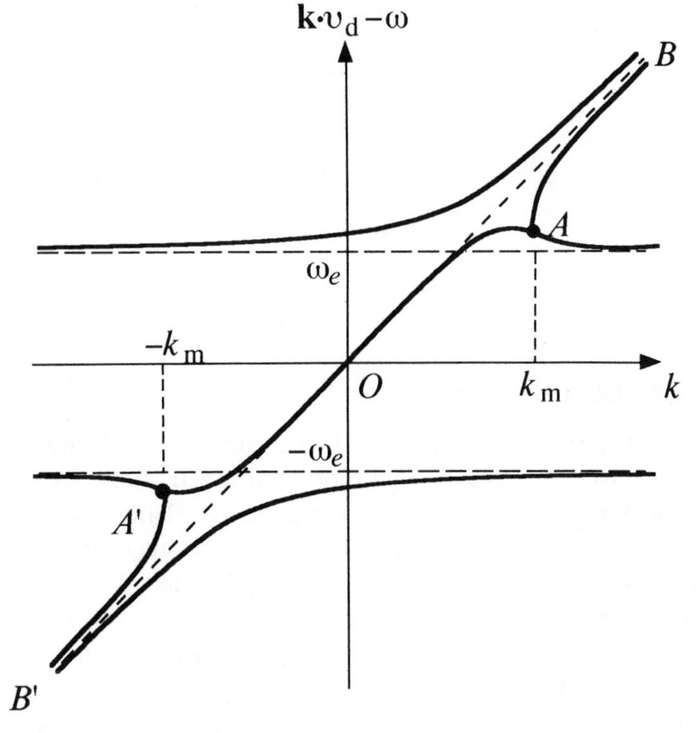

Figure 6.4 Dispersion relation of a beam-plasma system (schematic) with $\omega_b^2 \ll \omega_0^2$. The broken line *BOB'* represents the relation $\omega = k\upsilon_d$.

homologous to Eq. (6.15). Since these are quadruple equations with respect to ω, one can solve them exactly by an algebraic means [Buneman 1959]. Stability of the system can be analyzed by graphical means (cf., Problem 6.3).

Figure 6.4 schematically describes the solutions to the dispersion relation Eq. (6.15) for the electron-ion system. The branch AOA' represents the real part of the complex solutions, or the unstable branch. The maximum wave number for the unstable branch is given by

$$k_m = \frac{\omega_e}{v_d}\left[1 + \left(\frac{m_e}{m_i}\right)^{1/3}\right]^{3/2}. \tag{6.18}$$

The maximum growth rate,

$$\mathrm{Im}\,\omega_{\max} = \frac{\sqrt{3}}{2}\left(\frac{m_e}{2m_i}\right)^{1/3}\omega_e\,, \tag{6.19}$$

takes place at

$$\mathrm{Re}\,\omega \approx \left(\frac{m_e}{m_i}\right)^{1/3}\frac{\omega_e}{2} \tag{6.20}$$

where $\mathbf{k}\cdot\mathbf{v}_{\mathrm{d}} = \omega_e$.

6.3 MAGNETIC INSTABILITIES

In the presence of a strong magnetic field, the nature of collective modes differs remarkably from that in the absence of the magnetic field. As we have noted in Section 4.4A, the high-frequency plasma oscillation and the ion-acoustic wave prefer to propagate in the direction of the magnetic field. The energy source of such an oscillation is therefore sought in the kinetic energy of the particle motion parallel to the magnetic field. We have also noted in Section 4.4B the emergence of Bernstein modes, which propagate in directions perpendicular to the magnetic field. These are collective modes associated with the cyclotron motion of the particles, driven by the perpendicular elements in kinetic energies; when an imbalance in the energy contents between two collective modes exists, there arises a possibility of instability through coupling between those two modes. We begin this section with a discussion of instabilities from the point of view of such a coupling. These discussions will be then extended to the cases of electromagnetic instabilities.

A. Cyclotron Instabilities

When the frequencies and the wave vectors of two collective modes match each other, these can interact resonantly. When an imbalance in the energy contents exists between the two modes, the excessive amount of energy in one mode may be fed into the other through such a coupling. If the rate of the energy influx exceeds the decay rate inherent in the wave (e.g., Landau damping), then the wave may tend to grow in the form of the onset of an instability [Rosenbluth 1965].

One can in fact interpret the ion-acoustic wave instability of the previous section in terms of such a coupling between two collective modes. Generally, when a group of particles moves as a whole with a velocity $\mathbf{v}_{\mathrm{d}}$, a collective mode

$$\omega = \mathbf{k} \cdot \mathbf{v}_{\mathrm{d}} \tag{6.21}$$

associated with such a drift motion may be defined. The condition for Cherenkov excitation is that the frequency and the wave vector of this collective mode match those of the ion-acoustic wave. When the drift velocity increases further, free energy is fed from the drift mode (6.21) into the ion-acoustic waves, satisfying (6.13).

An instability quite analogous to the above may take place in a magnetized plasma when the Bernstain modes, on the one hand, and the plasma oscillation or the ion-acoustic wave, on the other, couple each other. The comparison to be made here is between the energy contained in the cyclotron motion of the particles and that associated with the parallel motion. When the former exceeds the latter, for example, a possibility arises that the energy of cyclotron motion is fed into the plasma oscillation or the ion-acoustic wave in the vicinity of the harmonics of the cyclotron frequency. Such instabilities were investigated by Harris [1959, 1961; Soper & Harris 1965].

An energy imbalance between the parallel and perpendicular degrees of freedom may be described by assigning two different temperatures, $T_{||}$ and $T_{\perp}$, in the velocity distributions:

$$f(v_{\perp}, v_{||}) = \left(\frac{m}{2\pi T_{\perp}}\right)\sqrt{\frac{m}{2\pi T_{||}}} \exp\left[-\frac{v_{\perp}^2}{2(T_{\perp}/m)} - \frac{v_{||}^2}{2(T_{||}/m)}\right]. \tag{6.22}$$

In light of relation (4.39), such an anisotropic distribution is more likely to occur in the ions rather than in the electrons. For this reason, we assume Eq. (6.22) only for the ions. Instabilities in the vicinity of the harmonics of the ion cyclotron frequency are thus to be considered. The electron distribution is assumed to be isotropic, however, with a drift velocity v_{d} along the magnetic field. The dielectric response function in these cases takes the form

$$\varepsilon(k,\omega) = 1 + \frac{k_e^2}{k^2} W\left(\frac{\omega}{|k_{||}|\sqrt{T_e/m_e}}\right) + \frac{k_i^2}{k^2}$$

$$\times \left\{1 + \sum_n \left(1 + \frac{T_{||}}{T_{\perp}} \frac{n\Omega}{\omega - n\Omega}\right) \times \left[W\left(\frac{\omega - n\Omega}{|k_{||}|\sqrt{T_{||}/m}}\right) - 1\right] \Lambda_n(\beta)\right\}. \tag{6.23}$$

For simplicity we here omit the subscript i except for ω_i and $k_i \equiv \sqrt{4\pi n_i (Z_i e)^2 / T_{||}}$; the quantities without subscript refer to those associated with the ions. In Eq. (6.23) the zero Larmor-radius approximation, $\beta_e \to 0$, has been adopted for the electrons.

In certain cases of plasma experiments, ion temperatures may be maintained at a level substantially higher than those of the electrons. Such may be expected when the major source of energy to the plasma comes through the kinetic energies of the ions. In an ion-injection machine, for example, energetic ions are injected into a plasma confinement system. Ion energies are thermalized in the system; a high ionic temperature may be attained. Transfer of ion energies to the electrons is a process much slower than establishment of the ion temperature. In a stationary situation, the electron temperature may thus remain relatively low.

Let us therefore consider a cold-electron case without a drift ($v_d = 0$), such that

$$\frac{T_e}{m_e} << \frac{T_{||}}{m} . \tag{6.24}$$

Since $k_e^2 >> k_i^2$, the real part $\varepsilon_1(\mathbf{k},\omega)$ of the dielectric function (6.23) can be expressed as

$$\varepsilon_1(\mathbf{k},\omega) = 1 - \frac{\omega_e^2 k_{||}^2}{\omega^2 k^2} .$$

Hence the frequencies of the collective mode are

$$\omega_k = \omega_e \frac{k_{||}}{k} = \omega_e \cos\theta \tag{6.25}$$

where θ is the angle between the wave vector and the magnetic field. This is the high-frequency plasma oscillation of Section 4.4A.

The imaginary part of the dielectric function can be calculated likewise from Eq. (6.23). The growth rate $-\gamma_k$ is thus obtained as

$$\gamma_k = \sqrt{\frac{8}{\pi}} \frac{\omega_k^2 k_e^2}{|k_{||}| k^2 \sqrt{T_e/m_e}} \exp\left[-\frac{\omega_k^2}{2k_{||}^2 T_e/m_e} \right]$$
$$+ \sqrt{\frac{\pi}{8}} \frac{k_i^2 \omega_k}{k^2} \sum_n \left(\frac{\omega_k - n\Omega}{|k_{||}|\sqrt{T_{||}/m}} + \frac{T_{||}}{T_\perp} \frac{n\Omega}{|k_{||}|\sqrt{T_{||}/m}} \right)$$
$$\times \exp\left[-\frac{(\omega_k - n\Omega)^2}{2k_{||}^2 T_{||}/m} \right] \Lambda_n(\beta) . \tag{6.26}$$

The first term on the right-hand side is positive definite, representing the electronic Landau damping. Since the wave frequencies (6.25) in the vicinity of the cyclotron harmonics, $n\Omega$, for the ions are of interest here, the Landau damping term becomes negligibly small when

$$(n\Omega)^2 >> k_{||}^2 \frac{T_e}{m_e} . \tag{6.27}$$

We assume this condition, consistently with the cold electron approximation (6.24).

We investigate the possibility of an instability in the vicinity of $\omega_k = n\Omega$, a condition for coupling between the plasma oscillations and the ion Bernstein modes. In light of Eq. (6.25), such a coupling can take place only when

$$\omega_e > n\Omega ,$$

thus setting the minimum electron density for the instability as

$$n_e > \frac{n^2 Z^2 B^2 m_e}{4\pi m^2 c^2} . \tag{6.28}$$

Having thus established the minimum density, we next examine the sign of the decay rate (6.26). Its first term is negligible on account of (6.27). The third (i.e., the last) term on the right-hand side is positive definite, contributing to the damping of the waves. The second term changes its sign as $\omega_k - n\Omega$ does. In the domain $\omega_k > n\Omega$, this term is positive and hence contributes to an additional damping. In the domain $n\Omega > \omega_k$, on the other hand, it becomes negative and thus indicates a possibility of instability. When the growth rate arising from this second term exceeds the decay rate due to the third, an instability sets in.

The second (middle) term of Eq. (6.26) brings about a maximum growth rate at $\omega_k = n\Omega - |k_{||}|\sqrt{T_{||}/m}$. This frequency must be greater than $(n - 1/2)\Omega$; otherwise the growth rate would be masked by the strong decay stemming from the $(n - 1)$th Bernstein mode. Hence we have the condition

$$|k_{||}| < \sqrt{\frac{m}{T_{||}}} \frac{\Omega}{2} . \tag{6.29}$$

Note that combination of (6.24) and (6.29) yields (6.27) unless n is extremely large. The maximum growth rate is thus given by

$$-\gamma_k|_{\max} \approx \sqrt{\frac{\pi}{8}}\frac{k_i^2 n\Omega}{k^2}\left(1-\frac{T_{\parallel}}{T_{\perp}}\frac{n\Omega}{|k_{\parallel}|\sqrt{T_{\parallel}/m}}\right)\exp\left(-\frac{1}{2}\right)\Lambda_n(\beta) . \tag{6.30}$$

For an onset of instability, this quantity must be positive; combining this requirement with (6.29), we finally obtain

$$\frac{T_{\parallel}}{T_{\perp}} < \frac{|k_{\parallel}|\sqrt{T_{\parallel}/m}}{n\Omega} < \frac{1}{2n} \tag{6.31}$$

as the condition for a plasma-wave instability in the vicinity of the cyclotron harmonic frequencies of the ions. When the cyclotron motions contain an excessive amount of kinetic energy so that $T_{\perp} > T_{\parallel}$, a possibility exists that this portion of energy may be fed into the plasma oscillations through coupling at a frequency slightly less than the nth harmonic frequency of the ionic Bernstein mode.

B. Loss-Cone Instability

In addition to the cases with the two-temperature Maxwellian of Eq. (6.22), an anisotropic velocity distribution may be realized in a plasma due to a deviation away from the Maxwellian. Such a situation may arise, for example, when only those particles located in a certain domain of the phase space can be confined, while other particles are allowed to escape from the system. The distribution function resulting from such a special constraint in the phase space will naturally have a shape different from a Maxwellian.

A typical example of such a non-Maxwellian distribution is found when plasma particles are confined in a magnetic mirror field. The field configuration tends to retain those particles which have more kinetic energy in the perpendicular direction than in the parallel direction with respect to the magnetic lines of force. The resulting imbalance in energy content will then induce a plasma-wave instability, that is, the *loss-cone instability* of Rosenbluth and Post [1965; Post & Rosenbluth 1966].

A charged particle in a magnetic field performs a helical motion around a magnetic line of force. Associated with such a gyrating motion, a magnetic moment μ_B may be defined, with a magnitude

$$\mu_B = \frac{W_{\perp}}{B} \tag{6.32}$$

directed oppositely to the local magnetic field; $w_{\perp} = (1/2)mv_{\perp}^2$ is the kinetic energy in the perpendicular directions. The magnetic moment (6.32) is propor-

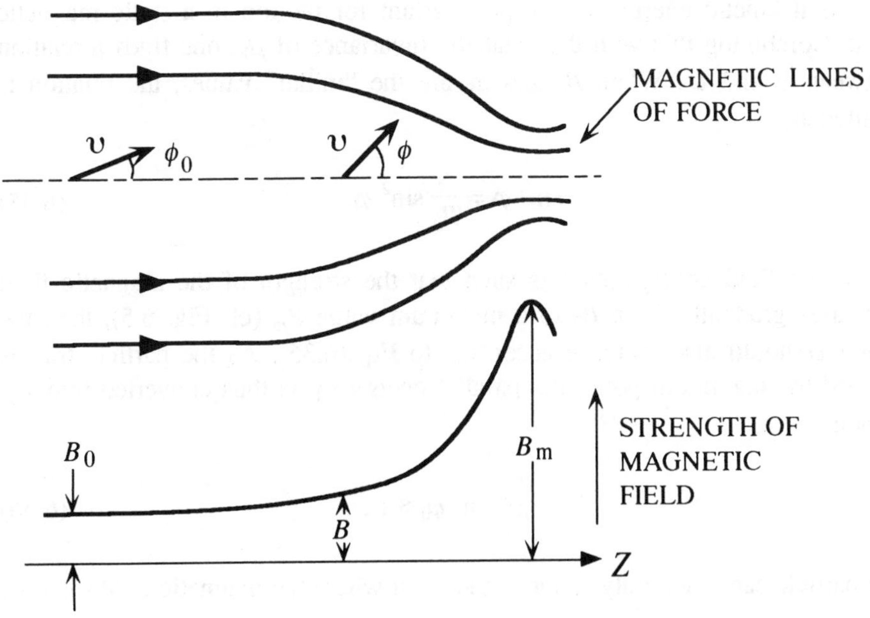

Figure 6.5 Magnetic mirror.

tional to the action integral associated with the periodic cyclotron motion of the charged particle; hence it it an adiabatic invariant. For an infinitesimally small rate of change in the magnetic field in space and time, the magnetic moment is kept invariant [e.g., Chandrasekhar 1960].

The concept of the *magnetic-mirror field* stems directly from the adiabatic invariance of the magnetic moment. Consider a charged particle moving with velocity **v** at an angle φ relative to the magnetic-field configuration as depicted in Fig. 6.5; we then have

$$\frac{v_{\perp}}{v_{\parallel}} = \tan\varphi \tag{6.33}$$

and the magnetic moment

$$\mu_B = \frac{W}{B}\tan\varphi \ . \tag{6.34}$$

The total kinetic energy w is kept invariant for motion in a static magnetic field. Combining this with the adiabatic invariance of μ_B, one finds a relation between B and φ. When B_0 and φ_0 are the "initial" values, the relation is written as

$$\sin^2 \varphi = \frac{B}{B_0} \sin^2 \varphi_0 . \tag{6.35}$$

If the field configuration is such that the strength of the magnetic field increases gradually from B_0 to a maximum value B_m (cf. Fig. 6.5), then the angle φ should also increase according to Eq. (6.35). As the particle travels toward the maximum point, the parallel energy $w_{\parallel}$ is thus converted into $w_{\perp}$. When

$$\frac{B_m}{B_0} \sin^2 \varphi_0 > 1 , \tag{6.36}$$

the particle can reach only as far as the point where the magnetic field satisfies

$$\frac{B}{B_0} \sin^2 \varphi_0 = 1 . \tag{6.37}$$

There $\varphi = \pi/2$, i.e., $v_{\parallel} = 0$; the particle is reflected at this point and starts to move away from the magnetic mirror.*

If, on the other hand, the *mirror ratio* B_m/B_0 or the initial angle φ_0 is not large enough to meet with (6.36), the particle is no longer reflected by such a field configuration; it will be lost from the system. For a given mirror ratio B_m/B_0, the maximum angle φ_L for the loss is thus calculated to be

$$\varphi_L = \sin^{-1} \sqrt{\frac{B_0}{B_m}} . \tag{6.38}$$

This angle determines the *loss cone* of the field configuration; those particles whose velocity vectors lie within the cone are immediately lost from the system.

When two magnetic mirrors are separated at a distance to produce the magnetic-field configuration of Fig. 6.6, then those charged particles with sufficiently large φ are reflected back and forth between the mirrors and are thus

*Fermi [1949] proposed a theory of cosmic-ray acceleration by collision against moving magnetic-field irregularities in interstellar space; in this theory he evokes such a mirror mechanism (called a "Type A" reflection) and a reflection following a curve of the line of force (a "Type B" reflection).

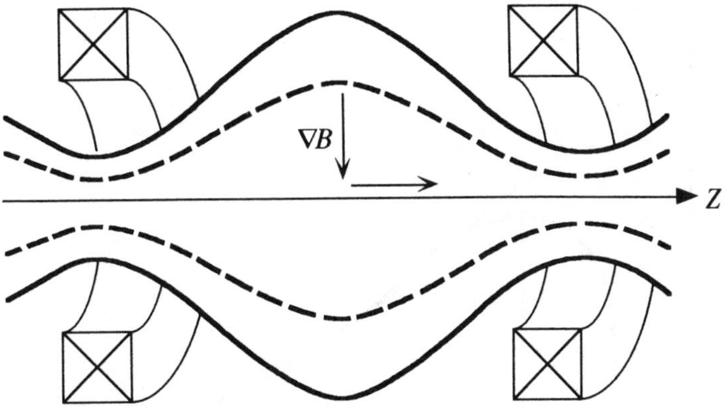

Figure 6.6 A magnetic bottle.

trapped. Such a scheme of plasma confinement is called a magnetic bottle; intensive effort has been focused on the possible use of such a mirror-confinement scheme for controlled thermonuclear fusion research.

For a plasma confined in such a magnetic-mirror device, the velocity distribution, therefore, deviates significantly from a Maxwellian. It now takes a form characterized as a *loss-cone distribution,* in which for a given value of $v_{\parallel}$ those particles with $v_\perp < \alpha v_{\parallel}$ are absent from the system. The parameters α and v_m (cf., Fig. 6.7) thus characterize the loss-cone distribution.

The plasma-wave instability associated with such a loss-cone distribution may be understood in terms of those instabilities arising from a doubly humped velocity distribution. Consider a one-dimensional distribution defined by

$$f(v_x) \equiv \int_{-\infty}^{\infty} dv_y \int_{-\infty}^{\infty} dv_{\parallel} f(v_\perp, v_{\parallel}) \,. \tag{6.39}$$

Its gradient at $v_x = v_0$ is then calculated as

$$f'(v_x)|_{v_x=v_0} = \int_{v_0}^{\infty} dv_\perp \int_{-\infty}^{\infty} dv_{\parallel} \frac{v_0}{\sqrt{v_\perp^2 - v_0^2}} \frac{\partial f}{\partial v_\perp} \,. \tag{6.40}$$

We thus find that $f'(v_0)$ can be positive if $v_0 < v_m$; the one-dimensional distribution (6.39) takes a doubly humped shape, as depicted in Fig. 6.8. The system is

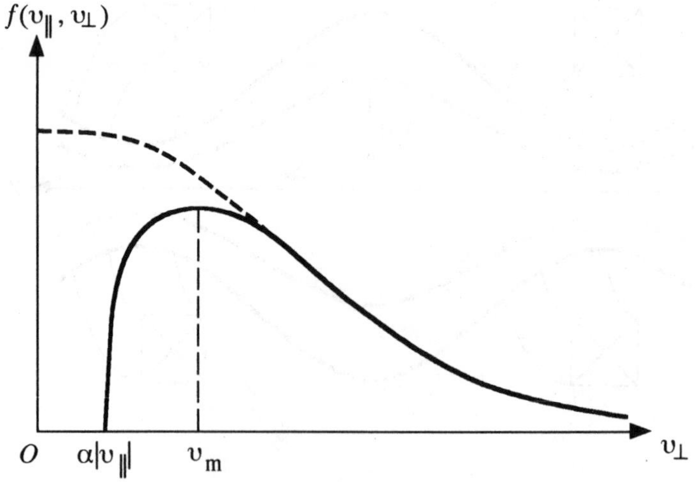

Figure 6.7 Loss-cone distribution.

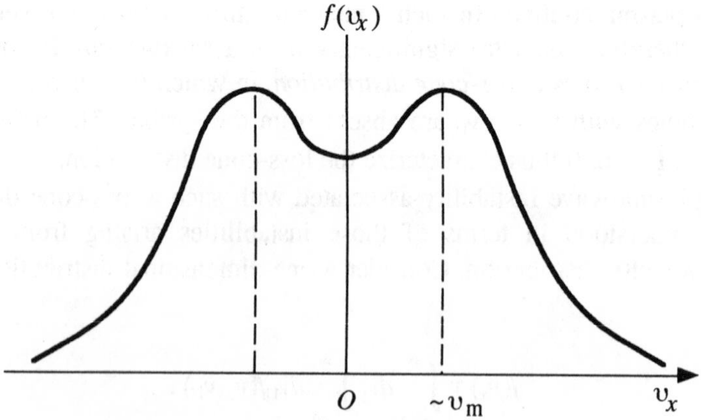

Figure 6.8 Doubly humped distribution from the loss-cone distribution of Fig. 6.7.

therefore unstable against excitations of those plasma waves in the regime approximately

$$\frac{\omega}{k_\perp} < v_{\mathrm{m}} . \tag{6.41}$$

A more accurate analysis of instability begins with the dielectric response function of Eq. (3.137). We are still with the notation of Section 6.3A, so that the quantities without a subscript refer to those associated with the ions. Neglecting the electron Landau damping in the cold-electron approximation, we find Eq. (6.25) from the real part of the dispersion relation. The condition for a coupling at $\omega_k = n\Omega$ yields the critical density (6.28). The growth rate of the loss-cone instability is then calculated from the imaginary part of the dispersion relation as

$$-\gamma_k = \pi^2 \frac{(n\Omega)^2 \omega_i^2}{k^2 |k_{||}|} \int_0^\infty dv_\perp \left[\frac{\partial f}{\partial v_\perp}\right]_{v_{||}=0} J_n\left(\frac{kv_\perp}{\Omega}\right)^2 . \tag{6.42}$$

Here we have assumed a symmetry in the loss-cone distribution such that $(\partial f/\partial v_{||})_{v_{||}=0} = 0$. Each function in the integrand of Eq. (6.42) behaves as in Fig. 6.9. Thus, the major contributions to the integral of Eq. (6.42) come from the vicinity of the first peaks of J_n^2. When the position p of this peak falls into the positive domain of $(\partial f/\partial v_{||})_{v_{||}=0}$, that is, when $q p$ in Fig. 6.9, the $-\gamma_k$ given by Eq. (6.42) will take on a positive value; an exponentially growing wave or an instability is indicated. Since $p \approx n$ and $q \approx k_\perp v_{\mathrm{m}}/\Omega$, the criterion for the instability is approximately

$$k_\perp v_{\mathrm{m}} > n\Omega , \tag{6.43}$$

a condition identical to (6.41).

A detailed investigation of Eq. (6.42) reveals that for fixed values of ω_e^2 (i.e., the density) and v_{m} (i.e., the magnetic-mirror ratio) a maximum growth rate takes place when the harmonic number n is small (i.e., $n = 1$) and $k_\perp$ is large (i.e., $k_\perp \approx k$). In these curcumstances, we estimate

$$-\gamma_k \approx \frac{\omega_e \omega_i^2}{\Omega^2} . \tag{6.44}$$

A remarkably large growth rate is thus indicated.

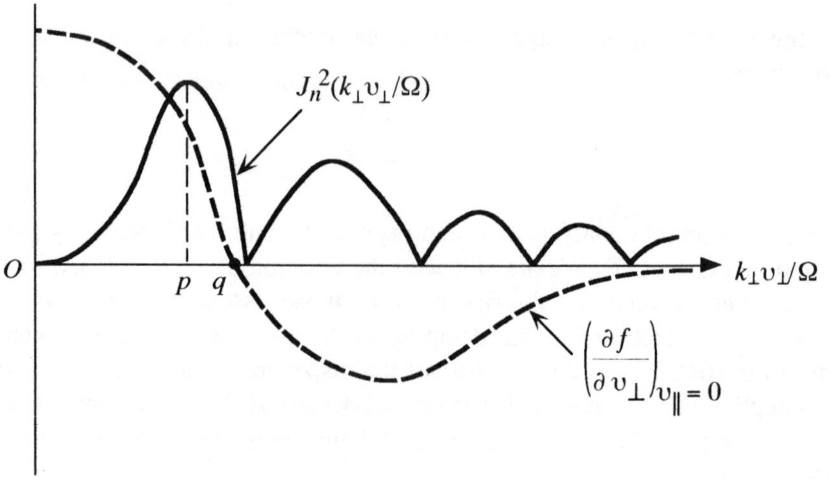

Figure 6.9 Behavior of J_n^2 and $(\partial f/\partial v_\perp)_{v_\parallel=0}$.

C. Electromagnetic Instabilities

Thus far we have been concerned with instabilities of longitudinal electrostatic oscillations. We can extend the arguments to the cases for instabilities in the transverse electromagnetic waves. Among various possibilities of instabilities, we here confine ourselves to the simplest cases of electromagnetic instabilities propagating along the magnetic field.

Since $k_\perp = 0$ and $|k_{\parallel}| = k$, it is useful to transform the dielectric tensor through the unitary matrix (4.84) as in Section 4.6A. The dielectric functions appropriate to the right- or left-circularly polarized waves may thereby be calculated as

$$\varepsilon_{r(l)}(k,\omega) = 1 + \sum_{\sigma} \frac{\omega_\sigma^2}{\omega^2} \int d\mathbf{v} \frac{v_\perp}{2} \left[(\omega - kv_{\parallel}) \frac{\partial f_\sigma}{\partial v_\perp} + kv_\perp \frac{\partial f_\sigma}{\partial v_{\parallel}} \right]$$

$$\times \frac{1}{\omega - kv_{\parallel} \pm \Omega_\sigma + i\eta} \tag{6.45}$$

where the velocity integration has been defined by Eq. (3.130).

Let us investigate the dielectric functions Eq. (6.45) for an electron-ion plasma in the limit of low frequencies such that $|\omega - kv_{\parallel}| << \Omega_i << |\Omega_e|$. We may then expand the denominator in Eq. (6.45) and carry out the partial integrations to obtain

$$\varepsilon_{T(l)}(k,\omega) \to 1 \pm \sum_\sigma \frac{\omega_\sigma^2}{\omega^2 \Omega_\sigma} \int d\mathbf{v}(kv_{\parallel} - \omega) f_\sigma$$

$$+ \sum_\sigma \frac{\omega_\sigma^2}{\omega^2 \Omega_\sigma^2} \int d\mathbf{v}(kv_{\parallel} - \omega)^2 f_\sigma + \cdots$$

$$- \sum_\sigma \frac{\omega_\sigma^2}{\omega^2 \Omega_\sigma^2} \int d\mathbf{v} \frac{k^2 v_\perp^2}{2} f_\sigma + \cdots .$$

Assuming no net flows of the particles,

$$\int d\mathbf{v} v_{\parallel} f_\sigma = 0,$$

and compensation of the net space charges,

$$\sum_\sigma Z_\sigma n_\sigma = 0 \, ,$$

we find that the leading terms of the expansion are expressed as

$$\varepsilon_{T(l)}(k,\omega) \to 1 + \frac{4\pi \rho_m c^2}{B^2} + \frac{4\pi k^2 c^2}{\omega^2 B^2}(p_{\parallel} - p_\perp) \, . \tag{6.46}$$

Here $\rho_m \equiv \sum_\sigma m_\sigma n_\sigma$ is the mass density of the plasma, and

$$p_{\parallel} = \sum_\sigma m_\sigma n_\sigma \int d\mathbf{v} v_{\parallel}^2 f_\sigma \, , \tag{6.47a}$$

$$p_\perp = \sum_\sigma m_\sigma n_\sigma \int d\mathbf{v} \frac{v_\perp^2}{2} f_\sigma \, , \tag{6.47b}$$

are the kinetic pressures of the plasma parallel and perpendicular to the magnetic field.

The dispersion relation for the electromagnetic wave has been given by Eq. (4.87). Substitution of (6.47) into this equation then yields

$$\omega = \pm \frac{kcc_{\mathrm{A}}}{\sqrt{c^2 + c_{\mathrm{A}}^2}} \sqrt{1 - \frac{p_{\|} - p_{\perp}}{B^2/4\pi}}$$

$$= \pm kc_{\mathrm{A}} \sqrt{1 - \frac{p_{\|} - p_{\perp}}{B^2/4\pi}} \,. \tag{6.48}$$

In the last expression of Eq. (6.48), we have assumed that $c >> c_{\mathrm{A}}$, the Alfvén velocity defined by Eq. (4.107). For a tenuous plasma such that $B^2/4\pi >> p_{\|}$ and $p_{\perp}$, or for a dense isotropic plasma with $p_{\|} = p \gtrsim B^2/4\pi$, Eq. (6.48) describes a propagation of a stable Alfvén wave. If, however, a dense anisotropic plasma is considered such that $p_{\|} > p_{\perp} + B^2/4\pi$, Eq. (6.48) becomes

$$\omega = \pm i \frac{k}{\sqrt{\rho_{\mathrm{m}}}} \sqrt{p_{\|} - p_{\perp} - \frac{B^2}{4\pi}} \,. \tag{6.49}$$

Onset of an instability is thus predicted. This instability is called the *fire-hose instability*.

The physical mechanism of such an instability and the origin of the nomenclature are easily understood: Suppose that the magnetic lines of force are bent slightly due to electomagnetic perturbations. Since the charged particles are tied to the magnetic lines of force, their velocity components parallel to the magnetic field exert centrifugal force upon the lines of force and thereby act to enhance the original perturbation. Observed here is a similarity between this instability and the well-known unstable motion of a fire hose when the water velocity inside becomes very high. Opposing this action of the parallel pressure which excites the instability, the perpendicular pressure and the tension of the magnetic lines of force act to stabilize the disturbances and restore the system to the state without perturbations. The stabilizing action of the perpendicular pressure arises in the following way. When a curvature of the lines of force is created, the intensity of the magnetic field also deviates from uniformity. Since the magnetic moment (6.32) is an adiabatic invariant, the density of perpendicular energy becomes greater where the magnetic field is stronger. Development of perturbations will thus be resisted by the local concentration of the perpendicular energy density.

6.4 LONGITUDINAL DIELECTRIC FUNCTIONS FOR INHOMOGENEOUS PLASMAS

We now turn to consideration of those instabilities characteristic of inhomogeneous plasmas in a magnetic field. We begin this section with a survey of the drift motions of charged particles produced by the external force fields and/or the inhomogeneities acting across the magnetic field [Rosenbluth & Longmire 1957]. The main objective is to derive those dielectric response functions for the inhomogeneous plasmas through which various drift-induced instabilities may be analyzed.

A. Drifts of Charged Particles in Magnetic Fields

A *diamagnetic drift* arises as a result of interplay between a spatial inhomogeneity and the finiteness of the Larmor radii. When a density gradient exists, the current contributions from the cyclotron orbitals in the high-density side exceed those coming from the low-density side; a net current appears, as Fig. 6.10a illustrates. Similarly, when a temperature gradient exists, the balance in the average velocities between the high- and low-temperature domains accounts for the appearance of a current flow, as shown in Fig. 6.10b. When a Maxwellian velocity distribution is assumed locally, the average drift velocity $\mathbf{v}_D$ of a charged particle associated with such a diamagnetic current is calculated (Problem 6.7) to be

$$\mathbf{v}_D = -\frac{cT}{ZeB^2}\left(\frac{\nabla n}{n}+\frac{\nabla T}{T}\right)\times \mathbf{B} \,. \tag{6.50}$$

For a weakly coupled plasma, which we here assume, the kinetic pressure P is given by that of an ideal gas

$$P = nT \tag{6.51}$$

(cf., Eq. 1.30). Equation (6.50) may then be simplified as

$$\mathbf{v}_D = -\frac{c}{nZeB^2}\nabla P \times \mathbf{B} \,. \tag{6.52}$$

It is therefore a *pressure-gradient drift* as well.

When an external force $\mathbf{F}$ is applied to a magnetized plasma, a drift arises in the direction perpendicular to both $\mathbf{F}$ and $\mathbf{B}$. In the direction parallel to the magnetic field, the particle performs a trivial motion with a constant acceleration. For components perpendicular to the magnetic field, the equation of motion is

$$m\frac{d\mathbf{v}_\perp}{dt} = \mathbf{F}_\perp + \frac{Ze}{c}\mathbf{v}_\perp \times \mathbf{B} \,. \tag{6.53}$$

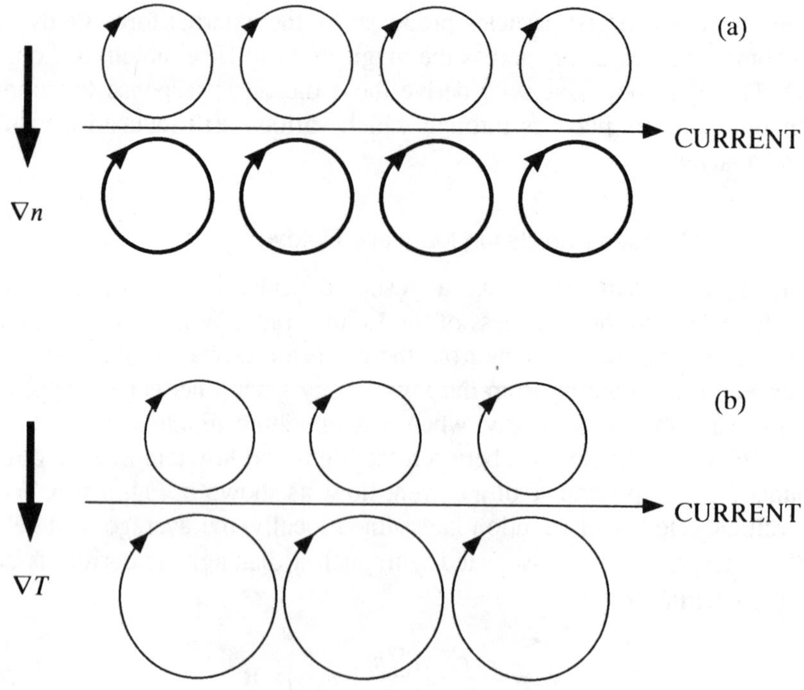

Figure 6.10 Diamagnetic currents. The line thickness indicates the number of particles contained in each orbital.

Setting $\mathbf{v} = \mathbf{v}_F + \mathbf{v}_1$ with

$$\mathbf{v}_F \equiv \frac{c}{ZeB^2}\mathbf{F}_\perp \times \mathbf{B}\,, \tag{6.54}$$

one finds that $\mathbf{v}_1$ obeys

$$m\frac{d\mathbf{v}_1}{dt} = \frac{Ze}{c}\mathbf{v}_1 \times \mathbf{B}\,,$$

the equation of motion for a simple cyclotron motion. Equation (6.54) therefore describes a uniform drift motion across the external field and the magnetic field.

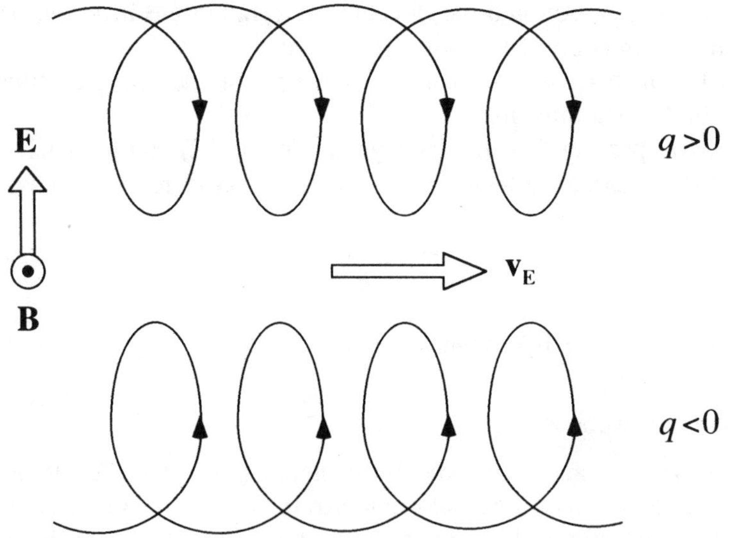

Figure 6.11 The $\mathbf{E}\times\mathbf{B}$ drift.

In particular, when the force arises from the electric field **E**, $\mathbf{F} = Ze\mathbf{E}$ is substituted in Eq. (6.54) to yield

$$\mathbf{v}_E = \frac{c}{B^2}\mathbf{E}\times\mathbf{B} , \tag{6.55}$$

the well-known $\mathbf{E}\times\mathbf{B}$ *drift*

$$\mathbf{v}_g = \frac{mc}{ZeB^2}\mathbf{g}\times\mathbf{B} , \tag{6.56}$$

the *gravitational drift.*

The physical origin of such a drift is easily understood. Consider for example the case with $\mathbf{E}\times\mathbf{B}$ drift illustrated in Fig. 6.11. A particle gyrates in the direction determined by the sign of Z at a radius proportional to $v_\perp$; it is accelerated also by the electric field. On the side where the kinetic energy of the particle is raised by the electric field, the particle orbit assumes a larger radius of curvature; on the other side, where the particle energy is lowered, the particle moves with a smaller radius of curvature. As a result, the drift motion shown in Fig. 6.11 appears. If the sign of the charge is changed, both the direction of acceleration and the direction of gyration are inverted; hence no change in the direction of the drift. The gravitational drift can be understood analogously; in this case, a change in the sign of the charge results in an inversion in the drift direction.

We add in passing that the diamagnetic drift (6.52) can be obtained from Eq. (6.54) if we take the effective force due to the pressure gradient as

$$\mathbf{F} = -\frac{1}{n}\nabla P\,, \tag{6.57}$$

a natural choice from a hydrodynamic consideration.

B. Dielectric Response

Instabilities may be analyzed through the dielectric response functions. In the inhomogeneous plasmas, the results obtained in Section 3.4D need to be modified so that the various drifts in the preceding subsection may be properly taken into account [Rosenbluth, Krall & Rostoker 1962; Rosenbluth 1965; Mikhailovsky 1983].

We assume the geometry of the plasma under consideration to be as follows: Inhomogeneities as well as a uniform force field $\mathbf{F} = F\hat{\mathbf{x}}$ exist in the x direction; a uniform magnetic field $\mathbf{B} = B\hat{\mathbf{z}}$ is applied. Drift motions of particles then appear in the y direction.

In the absence of collisions in a weakly coupled plasma (cf. Section 1.4C), the kinetic energy of a particle

$$W = \frac{1}{2}mv^2 - Fx \tag{6.58}$$

is a constant of motion. Since the system is translationally invariant in the y and z directions, the canonical momenta

$$p_y = mv_y + \frac{ZeB}{c}x\,,$$

$$p_z = mv_{\parallel}\,, \tag{6.59}$$

should also be conserved. Unperturbed distributions must then be expressible as functions of those three conserved quantities. The simplest choice of such a function accounting for the first-order effects in the inhomogeneity may be given by

$$f(x,\mathbf{v}) = \left(\frac{m}{2\pi T}\right)^{3/2} \left[1 + \alpha \left(x + \frac{v_y}{\Omega}\right)\right] \exp\left[-\left(\frac{mv^2}{2T} - \frac{Fx}{T}\right)\right], \tag{6.60}$$

where α is the parameter characterizing the spatial variation of the density. Equation (6.60) describes the first-order effect of the density gradient in the vicinity of $x = 0$, and would recover the Maxwellian (1.1) in the absence of the inhomogeneity and the force field. The distribution, however, ignores the possible spatial variations and/or anisotropy in the temperatures. We assume that essential features in the inhomogeneous plasmas are represented in the adoption of Eq. (6.60), with the merit of ensuing simplicity in the mathematics.

The density gradient at $x = 0$ may be calculated from Eq. (6.60) as

$$\frac{1}{n}\frac{dn}{dx} = \alpha + \frac{F}{T}. \tag{6.61}$$

We assume that the spatial variation of the density is small over a distance of the average Larmor radius, so that

$$\left|\frac{1}{n}\frac{dn}{dx}\right| << \frac{|\Omega|}{\langle v_\perp \rangle}. \tag{6.62}$$

It is further assumed that the space-charge neutrality is maintained not only at $x = 0$ but also in the vicinity of $x = 0$, to the first order in the inhomogeneity and the force field. For a two-component, electron-ion plasma, this condition is expressed as

$$\alpha_e + \frac{F_e}{T_e} = \alpha_i + \frac{F_i}{T_i}. \tag{6.63}$$

We distinguish between F_e and F_i because a force field may act differently on an electron and an ion. To the lowest order in α and F, the drift velocity is obtaimed as

$$\mathbf{v}_d = \int d\mathbf{v}\mathbf{v}f(x,\mathbf{v}) = \hat{\mathbf{y}}\alpha\frac{T}{m\Omega}$$

$$= \hat{\mathbf{y}}\left(\frac{T}{nm\Omega}\frac{dn}{dx} - \frac{F}{m\Omega}\right). \tag{6.64}$$

The first term on the right-hand side of Eq. (6.64) corresponds to the diamagnetic drift Eq. (6.50) with $\nabla T = \mathbf{0}$; the second term represents Eq. (6.54).

The Vlasov equation for the plasma under consideration may be written as

$$\frac{\partial f}{\partial t}+\mathbf{v}\cdot\frac{\partial f}{\partial \mathbf{r}}+\left[\frac{Ze}{m}\left(\mathbf{E}+\frac{\mathbf{v}}{c}\times\mathbf{B}\right)+\frac{1}{m}\mathbf{F}\right]\cdot\frac{\partial f}{\partial \mathbf{v}}=0 \tag{6.65}$$

where **E** contains the electric field produced by the space-charge distribution in the plasma. By virtue of Eq. (6.63), such a field vanishes to the first order in the inhomogeneities with an unperturbed state described by the distribution functions (6.60). It is then straightforward to show that Eq. (6.60) is a stationary solution to the Vlasov equation (6.65) in the absence of any additional perturbations. We may use Eq. (6.65) as the unperturbed distribution function, upon which the dielectric responses may be calculated.

Let us therefore apply an external electrostatic perturbation, to calculate the longitudinal (linear) response of the plasma. The electrostatic approximation to the dielectric response is known to be valid [Rosenbluth 1965] in those cases where the magnetic pressure $B^2/8\pi$ is much greater than the kinetic pressure, i.e., in the so-called low β plasmas.*

Since the system is inhomogeneous in the x direction, a Fourier decomposition of the physical variables in that direction is no longer a useful concept. Instead, we write the electrostatic potential in the plasma as

$$\Phi(\mathbf{r},t)=\Phi(x)\exp\left[i(k_y y+k_z z-\omega t)+\eta\, t\right]. \tag{6.66}$$

An appropriate functional form of $\Phi(x)$ should then be determined from the propagation characteristics of the field potential in the x direction. The distance over which such a propagation can have a significant effect may be measured by the average Larmor radius of the particles. If one passes to the limit of a very strong magnetic field, then such a distance would vanish. Perturbations may not effectively propagate in the x direction; we can approximate

$$\Phi(x)\approx\Phi(0) \tag{6.67}$$

in these circumstances. The situation is quite different in the y direction, however. Here the modes arising from the drift motions Eq. (6.64) can maintain communication between two points separated in the y direction. In fact, the presence of these drift modes is one of the most essential features in the inhomogeneous plasmas. The effects of spatial dispersion in the y direction must therefore be fully taken into account.

---

*In this terminology, β refers to the ratio of the plasma kinetic pressure to the magnetic pressure of the confinement field.

On the basis of the foregoing arguments, we shall henceforth adopt the approximation (6.67). General mathematical procedures going beyond this approximation are also available [Rosenbluth 1965; Mikhailovsky 1983]. Writing thus the distribution function in the vicinity of $x=0$ as a summation of the unperturbed part Eq. (6.60) and a perturbation

$$f(\mathbf{r},\mathbf{v};t)=f(\mathbf{x},\mathbf{v})+\delta f(\mathbf{v})\exp[i(k_y y+k_z z-\omega t)+\eta t]$$

and linearizing the Vlasov equation (6.65) with respect to the perturbations, we find

$$\left\{\frac{\partial}{\partial t}+\mathbf{v}\cdot\frac{\partial}{\partial\mathbf{r}}+\left[\Omega(\mathbf{v}\times\hat{\mathbf{z}})+\frac{F}{m}\hat{\mathbf{x}}\right]\cdot\frac{\partial}{\partial\mathbf{v}}\right\}\delta f(\mathbf{v})\exp[i(k_y y+k_z z-\omega t)+\eta t]$$

$$=i\Phi(0)\frac{Ze}{m}(k_y\mathbf{y}+k_z\mathbf{z})\cdot\frac{\partial f}{\partial\mathbf{v}}\exp[i(k_y y+k_z z-\omega t)+\eta t]\,. \qquad (6.68)$$

We solve this equation for $\delta f(\mathbf{v})$, to establish the relation between the potential Eq. (6.66) and the induced field in the plasma.

The solution may be obtained through the method of integration along the unperturbed trajectories in the phase space as described in Section 3.4B. The trajectories should be determined from a solution to the equations of motion with the force field $\mathbf{F}$ dictated by the left-hand side of Eq. (6.68). We thus find

$$\delta f(\mathbf{v})=-\Phi(0)\frac{Ze}{T}f(x,\mathbf{v})\left\{1+i(\omega-\omega^*)\int_0^\infty d\tau\exp[-i\varphi(\tau)-\eta\tau]\right\}. \qquad (6.69)$$

Here

$$\omega^*=\frac{\alpha T}{m\Omega}k_y=\mathbf{k}\cdot\mathbf{v_d} \qquad (6.70)$$

is the characteristic frequency of the drift mode,

$$\varphi(\tau)=k_y\left[\frac{v_x}{\Omega}(1-\cos\Omega\tau)+\frac{v_y-v_F}{\Omega}\sin\Omega\tau+v_F\tau\right]+k_z v_{\parallel}\tau-\omega\tau\,, \qquad (6.71)$$

$$v_F=-\frac{F}{m\Omega} \qquad (6.72)$$

is the magnitude of the drift (6.54) in the y direction, and τ was given by Eq. (3.118).

The induced potential field at $x = 0$ is calculated from Eq. (6.69) as

$$\Phi_{\text{ind}}(0) = \frac{4\pi Zen}{k^2}\left[\int d\mathbf{v}\,\delta f(\mathbf{v})\right]_{x=0}.$$

The dielectric response function is then obtained from the relation (cf. Eq. 5.2)

$$\varepsilon(k_y,k_z;\omega) = 1 - \frac{\Phi_{\text{ind}}(0)}{\Phi(0)}.$$

Extending to the cases with a multicomponent plasma, we thus find

$$\varepsilon(k_y,k_z;\omega) = 1 + \sum_\sigma \frac{k_\sigma^2}{k^2}$$

$$\times\left\{1 + (\omega - \omega_\sigma^*)\sum_n \frac{\Lambda_n(\beta_\sigma)}{\tilde{\omega}_\sigma - n\Omega_\sigma}\left[W\left(\frac{\tilde{\omega}_\sigma - n\Omega}{|k_z|\sqrt{T_\sigma/m_\sigma}}\right) - 1\right]\right\}, \quad (6.73)$$

where

$$\tilde{\omega}_\sigma = \omega - k_y v_{F\sigma}. \quad (6.74)$$

Specific choices of the force field and the inhomogeneity will determine the cases under consideration, and hence the dielectric function (6.73). Combinations of those choices together with the frequency domains of the collective modes will then offer various possibilities for investigating instabilities in the inhomogeneous plasmas.

6.5 RAYLEIGH-TAYLOR INSTABILITIES

The instability of a heavy liquid supported by a light liquid against the gravitational field is a classic problem in hydrodynamics. A similar problem can be considered for a plasma supported by a magnetic field against gravity or some other force field. Such instabilities are generally categorized as the *Rayleigh-Taylor instabilities*; in particular, those associated with gravity are called the *flute instabilities* or *interchange instabilities*. These are typical magnetohydrodynamic instabilities involving macroscopic mass motion of plasmas.

The dielectric response function calculated microscopically in the previous section can describe such a macroscopic instability as well. The analysis based on the dielectric response function not only reproduces the results of a macroscopic calculation but also clarifies additional features arising from micro-

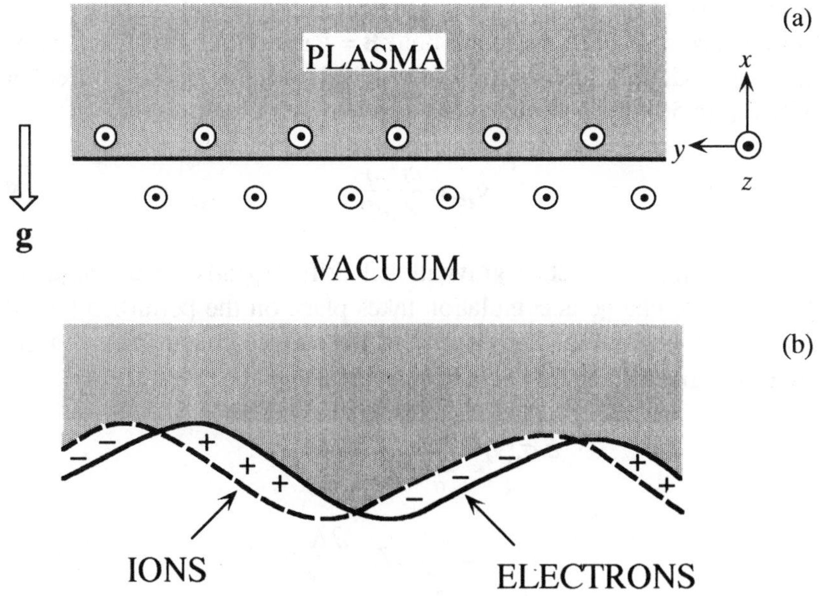

Figure 6.12 Gravitational instability.

scopic effects. In this section, following an orbit-theoretical introduction to a flute instability, we consider microscopic theories of the Rayleigh-Taylor instabilities.

A. Orbit-Theoretical Introduction

Consider the configuration of a plasma and a magnetic field shown in Fig. 6.12a. The plasma, occupying the region above the plane $x = 0$, is sustained by the magnetic field in the z direction against the gravitational field pointed downward in the x direction. If the magnetic field is strong enough, then the effect of the gravitational field is only to induce drift motion of the charged particles in the y direction. The plasma could conceivably float above the vacuum supported by the magnetic field. The question remains whether such a configuration is stable [Kruskal & Schwarzschild 1954; Rosenbluth & Longmire 1957].

To examine such a question of stability, we apply a small perturbation

$$\Delta x = \xi \sin ky \tag{6.75}$$

to the boundary plane $x = 0$ of the macroscopically neutral plasma consisting of electrons and ions. The magnetic field $\mathbf{B} = B\hat{\mathbf{z}}$ and the gravitation acceleration $\mathbf{g} = -g\hat{\mathbf{x}}$ induce a drift motion of charged particles in the y direction as given by Eq. (6.56); for a particle of σ species, the velocity is

$$v_{g\sigma} = \frac{cm_\sigma g}{Z_\sigma eB}. \tag{6.76}$$

Since the direction of such a gravitational drift depends on the sign of the charge, a surface-charge accumulation takes place on the perturbed boundary, as illustrated in Fig. 6.12b. The density of the surface charge $\delta\sigma_e(y)$ accumulated during a time interval δt is then calculated as

$$\begin{aligned}\delta\sigma_e(y) &= \sum_\sigma n_\sigma Z_\sigma \delta(\Delta x)_\sigma \\ &= \sum_\sigma n_\sigma Z_\sigma e \frac{\partial \Delta x}{\partial y} v_{g\sigma}\delta t.\end{aligned} \tag{6.77}$$

Hence

$$\sigma_e(y) = \frac{\rho_m cgk}{B}\left(\int^t dt\xi\right)\cos ky \tag{6.78}$$

with $\rho_m = \sum_\sigma n_\sigma m_\sigma$, the mass density.

Associated with the surface charge Eq. (6.78), electric field is produced; it is determined from the Poisson equation

$$\nabla\cdot(\varepsilon\mathbf{E}) = 4\pi\,\sigma_e(y)\delta(x). \tag{6.79}$$

For slowly varying perturbations in directions perpendicular to the magnetic field, we may use Eq. (4.109) for the dielectric constant. The solution to Eq. (6.79) determines the electric fields E_x and E_y, which in turn induce drift motion of the charged particles according to Eq. (6.55). Let us recall that the velocity of this drift is independent of the particle species; it therefore creates a mass motion in the plasma. The flow velocity is accordingly calculated as

$$v_x = \frac{4\pi\rho_m c^2}{\varepsilon_A B^2}gk\left(\int^t dt\xi\right)\sin ky\exp(-kx), \tag{6.80a}$$

$$v_y = \frac{4\pi \rho_m c^2}{\varepsilon_A B^2} gk \left(\int^t dt\xi \right) \cos ky \exp(-kx) . \tag{6.80b}$$

Note that Eqs. (6.80) satisfy $\nabla \cdot \mathbf{v} = 0$; this flow pattern does not induce any density irregularities inside the plasma. It does act to shift the boundary plane of the plasma, however; the speed of this motion is then given by $v_x|_{x=0}$ in Eq. (6.80a). Substituting Eqs. (4.109) and (6.75) in this evaluation and differentiating once again with respect to time, we find

$$\ddot{\xi} = gk\xi . \tag{6.81}$$

Equation (6.81) has the solutions $\exp(\pm\sqrt{gk}\,t)$; an instability with the growth rate of $\sqrt{gk}$ is thus indicated. The plasma cannot be supported by the magnetic field alone in a stable manner against the gravitational field.

It may then be instructive to consider what would happen if the plasma occupied the region $x < 0$ in Fig. 6.12a rather than the region x 0. Such a situation can be treated by simply inverting the direction of gravity or by replacing g by $-g$ in the foregoing analysis. The displacement ξ now exhibits an oscillatory behavior; the system is therefore stable, as we might have expected.

B. Gravitational Instability

Let us return to the dielectric response function Eq. (6.73) and apply it to the cases of the flute instability which we have just studied orbit-theoretically. In so doing we shall discover those additional effects arising from the finiteness of the Larmor radii of the ions which act to stabilize the flute instabilities [Rosenbluth, Krall & Rostoker 1962; Krall & Rosenbluth 1963].

The cases appropriate to the flute mode may be analyzed through the following specifications:* The plasma is macroscopically neutral; hence we set $n_e = n_i \equiv N$, and $(1/n_e)dn_e/dx = (1/n_i)dn_i/dx = (1/N)dN/dx$, assuming $Z_i = 1$. The latter condition has been expressed in Eq. (6.63). The magnetic field is strong so that $\beta_e = k_y^2 T_e / m_e \Omega_e^2 \to 0$; we thereby neglect the finite Larmor-radius effects of the electrons. The perturbation modes are of a flute type, constant along the magnetic lines of force; $k_z = 0$ and thus the W functions in the dielectric function vanish. Finally the low-frequency magnetohydrodynamic modes are to be considered; $\tilde{\omega} \ll \Omega_i$, hence only the $n = 0$ terms remain in Eq. (6.73).

---

*In order to avoid possible confusion with the harmonic number n, we employ the upper-case letter N in this chapter to denote the number density of the electrons or the ions (with $Z_i = 1$), which are mutually compensated.

Taking account of the foregoing specifications, we find that the dielectric function appropriate to the flute mode is

$$\varepsilon(k,\theta;\omega) = 1 + \frac{k_e^2}{k^2}\left[1 - \frac{\omega - (\alpha_e T_e/m_e\Omega_e)k}{\omega + (F_e/m_e\Omega_e)k}\right]$$

$$+ \frac{k_i^2}{k^2}\left[1 - \frac{\omega - (\alpha_e T_i/m_i\Omega_i)k}{\omega + (F_e/m_i\Omega_i)k}\right]. \tag{6.82}$$

The dispersion relation, $\varepsilon(k,0;\omega) = 0$, then becomes a quadratic equation for ω, with a solution

$$\omega = -\frac{F_i k}{m_i\Omega_i}\left[X + \frac{F_e m_i\Omega_i}{F_i m_e\Omega_e}\right]. \tag{6.83}$$

Here

$$X = \frac{1}{2}\left\{\mu + \nu(1-\Lambda_0) \pm \sqrt{[\mu + \nu(1-\Lambda_0)]^2 - 4\mu\,\nu}\right\}, \tag{6.84}$$

with

$$\mu = 1 - \frac{F_e m_i\Omega_i}{F_i m_e\Omega_e}, \tag{6.85a}$$

$$\nu = -\frac{1}{N}\frac{dN}{dx}\frac{T_i}{F_i[(k/k_i)^2 + (1-\Lambda_0)]} \tag{6.85b}$$

and Λ_0 is given by Eq. (4.13) for the ions. When the frequency takes on a complex value, that is, when the inside of the square root in Eq. (6.84) becomes negative, the system develops an instability of the flute type.

Let us apply the foregoing calculation to the analysis of the gravitational instability in Fig. 6.12 with

$$F_e = -m_e g\,,\; F_i = -m_i g\,. \tag{6.86}$$

Since $m_i >> m_e$, $\mu = 1$. As we shall see shortly, the most important cases of the magnetohydrodynamic instability take place in the long-wavelength regime, so that we may set

$$1 - \Lambda_0 \approx (kR_L)^2$$

where

$$R_L = \sqrt{\frac{T_i}{m_i \Omega_i^2}} \tag{6.87}$$

is an average Larmor radius for the ions. In light of the numerical estimate through Eq. (4.110), we can ordinarily assume that $(k_i R_L)^2 >> 1$. Hence the condition for instability is obtained from Eq. (6.84) as

$$k^2 < \frac{4\Omega_i^2}{g} \frac{1}{N} \frac{dN}{dx} \left(1 + \frac{T_i}{m_i g} \frac{1}{N} \frac{dN}{dx} \right)^{-2} . \tag{6.88}$$

The instability, when it takes place, is thus confined in the long-wavelength regime. For the existence of such a regime, the right-hand side of (6.88) should be positive, i.e.,

$$g \frac{dN}{dx} > 0 . \tag{6.89}$$

The gravitational field must be directed toward the direction of the density decrease (cf. Eq. 6.86); otherwise the system will be gravitationally stable. The criterion (6.88) demonstrates a stabilization effect due to the finiteness of the Larmor radii. As the thermal velocity $\sqrt{T_i/m_i}$ and hence the Larmor radii of the ions increase, the upper bound of the unstable k regime determined from the right-hand side of (6.88) decreases. Since the values of the wave number available to the system are limited from below by geometrical reasons, the wave-number regime for the instability will diminish and may eventually vanish with a further increase of the ion Larmor radii. We shall later return to a discussion of analogous effects in connection with the stability of a magnetic-mirror confinement system.

In the limit of long wavelengths and low temperatures, Eq. (6.83) is transformed as

$$\omega = \frac{g}{2\Omega_i} k \pm i \sqrt{\frac{g}{N} \frac{dN}{dx}} . \tag{6.90}$$

The growth rate in this equation differs from that obtained in Eq. (6.81). The reason is traced simply to the fact that the case treated in Section 6.5A was concerned with a discontinuous change of the plasma density at $x = 0$, while we here treat the case with a continuous density variation. The growth rate of Eq. (6.90) can be naturally recovered when the present case is analyzed through the orbital-theoretical method of Section 6.5A (Problem 6.10).

C. Confinement in Curved Magnetic Fields

The driving force for the gravitational instability discussed above was provided by the electric field arising from the charge separation at the density gradient. Such a charge separation in turn was brought about by the gravitational field, whose direction of acceleration is insensitive to the sign of the electric charge of a particle. A charge-insensitive force is not limited to gravity, however. One can develop a similar line of argument to derive a stability criterion whenever a charge-insensitive force field exists in a direction perpendicular to the magnetic field. Hence, for a plasma confined in a configuration of the magnetic field, if there is a force acting on charged particles, regardless of their signs, in the direction from the plasma to vacuum, then the system is found to be unstable with respect to an interchange mode of the Rayleigh-Taylor type. If the force is directed in the opposite direction, the system may be stable as far as such a Rayleigh-Taylor mode is concerned.

An important example for such a charge-insensitive field arises when the magnetic lines of force have finite radii of curvature. Since the radius vector $\mathbf{R}$ of the curvature is given by

$$\frac{\mathbf{R}}{R^2} = -(\hat{\mathfrak{b}} \cdot \nabla)\hat{\mathfrak{b}} \tag{6.91}$$

where $\hat{\mathfrak{b}} = \mathbf{B}/B$ is the unit vector directed along a magnetic line of force, a charged particle moving along the magnetic field experiences a centrifugal force

$$\mathbf{F}_R = \frac{mv_{\parallel}^2\mathbf{R}}{R^2} = \frac{2w_{\parallel}\mathbf{R}}{R^2}\,. \tag{6.92}$$

The curvature also produces a field gradient in the direction perpendicular to the magnetic lines of force. Within the range of validity for the adiabatic invariance of the magnetic moment Eq. (6.32), such a field gradient then induces an effective force field in the perpendicular direction given by

$$\mathbf{F}_G = -\mu_B \nabla_{\perp} B = -\frac{w_{\perp}}{B}\nabla_{\perp} B \tag{6.93}$$

where $\nabla_{\perp}$ is the differential operator in the direction perpendicular to the magnetic field. A vacuum field relation $\nabla \times \mathbf{B} = 0$, which one can reasonably assume for a low-β plasma-confinement system, gives

$$\frac{\nabla_{\perp} B}{B} = -\frac{\mathbf{R}}{R^2}\,. \tag{6.94}$$

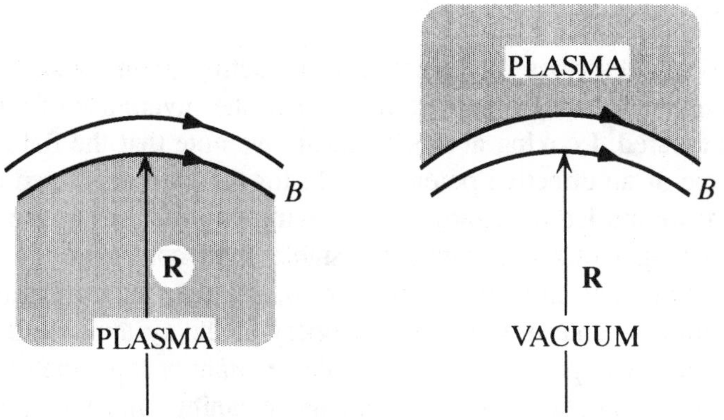

(a) (b)

Figure 6.13 Stability of magnetic confinement.

Collecting the two contributions Eqs. (6.92) and (6.93), we may apply the results of gravitational-instability analysis to the magnetic-confinement cases through a replacement

$$\mathbf{g} \rightarrow \frac{w_\perp + 2w_{\parallel}}{m}\frac{\mathbf{R}}{R^2} = -\left(v_{\parallel}^2 + \frac{v^2}{2}\right)\frac{\nabla_\perp B}{B}. \tag{6.95}$$

In light of the criterion on the gravitational instability, we then find that a system in which the magnetic field confines the plasma in a convex shape (as in Fig. 6.13a) is unstable because **R** in this case is directed from the plasma to vacuum; on the other hand, a system like Fig. 6.13b is stable.

In the dielectric function Eq. (6.82), we may now substitute

$$F_e = -\frac{2T_e}{R},\ F_i = -\frac{2T_i}{R}. \tag{6.96}$$

The direction of the radius-of-curvature vector is chosen in such a way as to make the system unstable when $dN/dx > 0$. In Eq. (6.85a), $\mu = 1 + T_e/T_i$. The growth rate of the instability in the limit of the long wavelengths is thus found to be

$$|\gamma| = \sqrt{\frac{1}{N}\frac{dN}{dx}\frac{2(T_e + T_i)}{m_i R}}. \tag{6.97}$$

D. Magnetic Well

Equation (6.95) describes the effective force acting upon a particle in directions perpendicular to the magnetic field when the invariance of the magnetic moment is assured. Looking at this equation, we note that the field strength B plays the role of an effective potential field for the particle. Based on such an observation, we are led to regard the following two features as the essence of designing a magnetohydrodynamically stable field configuration for plasma confinement: (1) The field is nowhere zero; (2) the field strength increases outward from the center toward the periphery of the system. Field configurations satisfying those properties are known as *minimum-B* geometries.

Property 1 is necessary to maintain the meaning, and thus the adiabatic invariance, of the magnetic moment throughout the system. Property 2 means that B increases outward from a point, or a closed curve, so that surfaces on which B is constant, the *magnetic isobars*, are closed and nested about this point or curve. A contour plot of B therefore gives the shape of a three-dimensional *magnetic well*. It should be noted that the isobars are not the same as the surfaces constructed by the magnetic lines of force, although they may coincide in a few very simple geometries.

The best-known example of the minimum-B geometry is the hybrid configuration of the mirror and cusp fields first constructed by Ioffe and his coworkers [Gott, Ioffe & Telkovsky 1962; Ioffe 1965]. The problem of equilibrium and stability of plasma in such a system has been theoretically investigated by Taylor [1963, 1965]. A simple mirror field like the one in Fig. 6.6 increases in strength outward along the z axis from the center plane but decreases radially. To obtain Property 2 of a minimum-B configuration, we may superpose a multipole field produced by an even number of straight rods laid parallel to the z axis (see Fig. 6.14a). Odd and even rods carry current in alternate directions to produce a series of cusped fields (see Fig. 6.14b); these fields obviously increase radially. A minimum-B geometry is thus realized when the rod current exceeds a certain critical value.

Increasing the current through this critical value therefore offers a simple and unambiguous experimental test for the stability properties of the hybrid configuration. Such experiments by Ioffe et al. have indeed shown an abrupt increase of the plasma lifetime by more than an order of magnitude when the strength of the cusp fields reaches a certain value, a clear demonstration of the stability achieved by the minimum-B principle.

E. Finite Larmor-Radius Stabilization

In order to study the stabilizing effects of the finite Larmor radii, let us consider the plasma confined in a magnetic bottle of Fig. 6.6. In such an axially

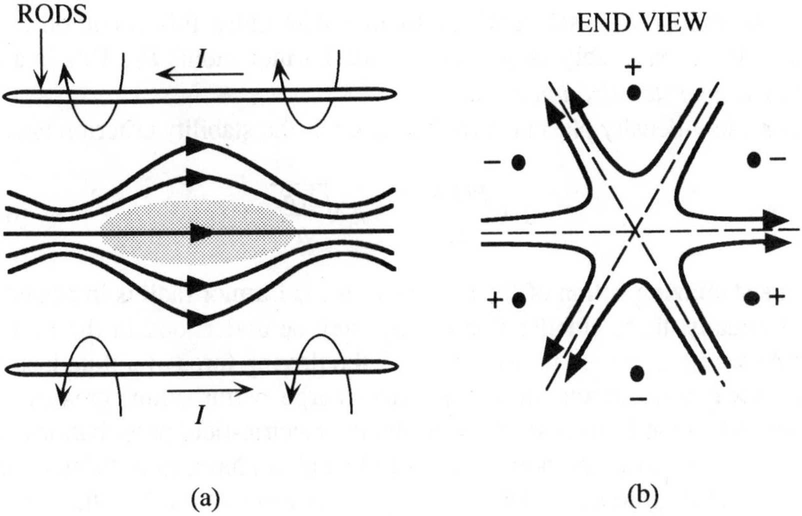

Figure 6.14 Mirror-cusp field.

symmetric system, the wave numbers k of the flute mode describe variations in the azimuthal direction. Geometrically, therefore, the available values of k are bounded from below by approximately an inverse of the radius of the confined plasma. We also recall that the most unstable flute mode occurs in the small k domain. Hence estimates

$$k \approx \frac{1}{N}\frac{dN}{dx} \equiv \frac{1}{r}, \tag{6.98}$$

$$\nu(1-\Lambda_0) \approx \frac{R}{2r}\frac{\varepsilon_A}{\varepsilon_A+1},$$

$$\nu \approx \frac{Rr}{2R_L^2}\frac{\varepsilon_A}{\varepsilon_A+1}$$

follow with $\varepsilon_A \equiv (k_i R_L)^2$.

For a plasma with $\varepsilon_A >> 1$ (cf., Eq. 4.109), the stability criterion from Eq. (6.84) reads

$$\frac{R_L}{r}\left(1+\frac{2\mu r}{R}\right) > 2\sqrt{\frac{2\mu r}{R}}. \tag{6.99}$$

When the ratio r/R is substantially smaller than unity, this condition may be satisfied by a reasonably large value of the Larmor radius R_L. This is a *finite Larmor-radius stabilization* effect.

For a low-density plasma such that $\varepsilon_A << 1$, the stability criterion becomes

$$1+\frac{R(k_i R_L)^2}{2\mu L} > k_i\sqrt{\frac{2rR}{\mu}} . \tag{6.100}$$

Again a stabilizing action of the finiteness of the Larmor radii is indicated.

Physically, these stabilization effects may be understood in the following way: As we have argued in Section 6.5C, the driving force of a flute instability is the $\mathbf{E}\times\mathbf{B}$ drift arising from the space-charge perturbations created in the plasma. When the Larmor radii are finite, the electric-field perturbations which an ion feels are different, both in magnitude and in phase, from those observed by the ion at its guiding center and therefore by an electron. (For the electrons, the Larmor-radius effects have been neglected.) In addition, the actual space-charge perturbations also differ from those obtained solely from consideration of the guiding-center drifts such as Eq. (6.78). The combined effects of these differences are such as to cause slight out-of-phase motion between the ions and the electrons leading to the stabilization.

6.6 DRIFT WAVES AND INSTABILITY

When a group of particles move with a drift velocity $\mathbf{v}_d$, a collective mode is defined with the dispersion Eq. (6.21). The frequency (6.70) may thus be looked upon as the collective mode associated with a diamagnetic current in an inhomogeneous plasma. The extra free energy contained in the inhomogeneities is the exciter for such a collective motion. Accordingly the collective nature of an inhomogeneous plasma is modified significantly from that of a homogeneous plasma. A most remarkable feature is the appearance of drift waves and the associated instability [Rudakov & Sagdeev 1959; Galeev, Oraevskii & Sagdeev 1963; Mikhailovsky & Rudakov 1963; Krall & Rosenbluth 1963].

A. Elementary Survey

Let us begin with consideration of the space-time perturbations in a plasma,

$$\Phi(\mathbf{r},t) = \Phi \exp\,[i(k_{yy} + k_z z - \omega t)]\ ,$$
$$n_e(\mathbf{r},t) = N + \delta n_e \exp\,[i(k_y y + k_z z - \omega t)]\ ,$$
$$n_i(\mathbf{r},t) = N + \delta n_i \exp\,[i(k_y y + k_z z - \omega t)]\ .$$

The equilibrium density N changes gradually in the x direction. We assume slow space-time variations so that the electrons can follow the potential perturbation adiabatically; hence

$$\frac{\delta n_e}{N} = \frac{e\Phi}{T_e}. \tag{6.101}$$

The behavior of the ions is governed by the continuity equation

$$\frac{\partial n_i}{\partial t} + \nabla \cdot (n_i \mathbf{v}) = 0 \tag{6.102}$$

and the equation of motion

$$\frac{\partial v}{\partial t} = -\frac{e}{m_i}\nabla\Phi + \Omega_i(\mathbf{v} \times \hat{\mathbf{z}}). \tag{6.103}$$

In the low-frequency domain $|\omega| << \Omega_i$, Eqs. (6.102) and (6.103) may be linearized and solved as

$$\frac{\delta n_i}{N} = \left(\frac{\omega_e^*}{\omega} + \frac{T_e k_z^2}{m_i \omega^2}\right)\frac{e\Phi}{T_e}. \tag{6.104}$$

Here ω_e^* is the drift frequency Eq. (6.70) for the electrons in which $F_e = 0$ and hence $\alpha = (1/N)(dN/dx)$. For long-wavelength excitations, we may assume

$$\delta n_e \approx \delta n_i, \tag{6.105}$$

because otherwise the electrostatic energy in the induced space charge would become extremely large. Combination of Eqs. (6.101), (6.104), and (6.105) yields the dispersion relation

$$\omega^2 + \omega_e^* \omega - s^2 k_z^2 = 0 \tag{6.106}$$

where Eq. (4.50b) has been used for s, the propagation velocity for the ion-acoustic wave. The solution as a function of k_z is depicted in Fig. 6.15. When $s^2 k_z^2 >> (\omega_e^*)^2$, the two branches of the solution turn into the ordinary ion-acoustic wave Eq. (4.50a). Toward the small k_z regime the upper branch departs from the acoustic mode and merges ω_e^* at $k_z = 0$. Here the dispersion relation describes a collective mode propagating in the y direction carried by the diamagnetic drift motion of the electrons; such a collective mode is called the *drift wave*.

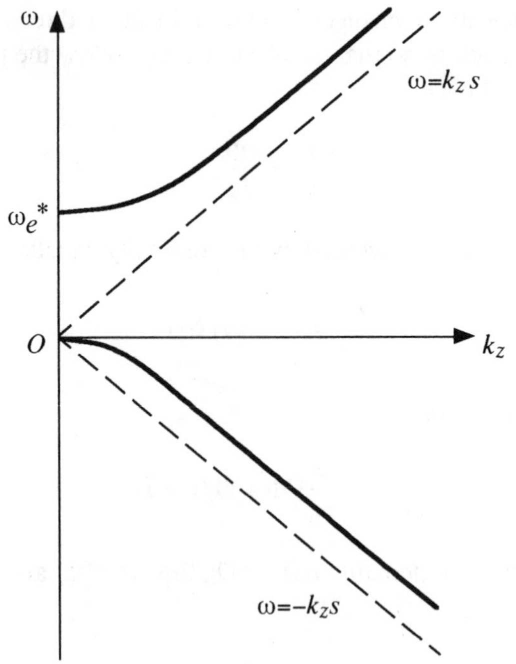

Figure 6.15 Drift waves.

B. Microscopic Calculation

A microscopic analysis of the drift wave may be carried out with the aid of the dielectric response function Eq. (6.73). Here we assume that the force field $F=0$; hence $\alpha_e=\alpha_i=(1/N)(dN/dx)$. The frequency domain to be considered is the same as the case of the ion-acoustic wave, specified by (4.47). The dielectric response function in these circumstances may thus be simplified as

$$\varepsilon(k_y,k_z;\omega)=1+\frac{k_e^2}{k^2}\left\{1+i\sqrt{\frac{\pi}{2}}\,\frac{\omega-\omega_e^*}{|k_z|\sqrt{T_e/m_e}}\exp\left[-\frac{\omega^2}{2k_z^2(T_e/m_e)}\right]\right\}$$

$$+\frac{k_i^2}{k^2}\left[1-\frac{\omega-\omega^*}{\omega}\left\{1+\frac{k_z^2(T_i/m_i)}{\omega^2}-i\sqrt{\frac{\pi}{2}}\right.\right.$$

$$\times \frac{\omega}{|k_z|\sqrt{T_i/m_i}} \exp\left[-\frac{\omega^2}{2k_z^2(T_i/m_i)}\right]\Lambda_0(\beta)\Bigg\}\Bigg] \tag{6.107}$$

where we have omitted the subscript *i* for some of the ionic quantities so that

$$\omega^* \equiv \omega_i^* = -\omega_e^* \frac{T_i}{T_e},$$

$$\beta \equiv \frac{k_y^2 T_i}{m_i \Omega_i^2} = k_y^2 R_L^2 .$$

The real part of the equation, $\varepsilon(k_y, k_z; \omega) = 0$, becomes

$$\frac{k^2}{k_i^2} + \frac{T_i}{T_e} + 1 - \frac{\omega - \omega^*}{\omega}\left[1 + \frac{k_z^2(T_i/m_i)}{\omega^2}\right]\Lambda_0(\beta) = 0 . \tag{6.108}$$

As we observed in Fig. 6.15, two characteristic frequencies are involved in the present problem: the frequency of a drift mode $|\omega_e^*|$ and the frequency of the acoustic mode $s|k_z|$. The relative magnitude between the two frequencies depends on the ratio $|k_z/k_y|$. For those wave vectors such that

$$|k_y|\sqrt{\frac{T_e}{T_i}} R_L \frac{1}{N}\left|\frac{dN}{dx}\right| >> |k_z| \tag{6.109}$$

the drift frequency represents the major effect; the solution to Eq. (6.108) then is

$$\omega = -\omega^* \frac{\Lambda_0(\beta)}{1 + (T_i/T_e) + (k_y/k_i)^2 - \Lambda_0(\beta)} . \tag{6.110}$$

Since $(R_L k_i)^2 >> 1$ in many practical cases, the term $(k_y/k_i)^2$ in the denominator is usually negligible. When $\beta << 1$, Eq. (6.110) gives $\omega = \omega_e^*$, a result obtained in the previous section.

From the imaginary part of Eq. (6.107), the growth rate $-\gamma$ of the drift wave may be calculated as

$$-\gamma = \sqrt{\frac{\pi}{2}} \frac{\omega^2}{\omega^* \Lambda_0(\beta)}\left\{\frac{T_i}{T_e} \frac{\omega - \omega_e^*}{|k_z|\sqrt{T_e/m_e}} + \frac{\omega - \omega^*}{|k_z|\sqrt{T_i/m_i}} \exp\left[-\frac{\omega^2}{2k_z^2(T_i/m_i)}\right]\right\} . \tag{6.111}$$

The second term on the right-hand side of Eq. (6.111) stems from the Landau damping due to ions; it is a negative-definite, decay contribution. This ion Landau-damping term may be negligible when k_z is small so that

$$\left|\frac{\omega}{k_z}\right| >> \sqrt{\frac{T_i}{m_i}} . \tag{6.112}$$

When $T_e > T_i$, a condition implied in Eq. (4.47), the constraint (6.109) is more restrictive than (6.112). For those wave vectors where the drift mode predominates over the acoustic mode, the ion Landau-damping term is thus negligible.

Keeping only the first term on the right-hand side of Eq. (6.111) and substituting (6.110) in it, we obtain

$$-\gamma = \sqrt{\frac{\pi}{2}} \frac{(\omega^*)^2 \Lambda_0(\beta) \left\{[1-\Lambda_0(\beta)](1+T_i/T_e)+(k_y/k_i)^2\right\}}{|k_z|\sqrt{T_e/m_e}\,[1-\Lambda_0(\beta)+(T_i/T_e)+(k_y/k_i)^2]^3} . \tag{6.113}$$

Note that Eq. (6.113) is a positive definite quantity, indicating a growing drift wave or an instability. As long as there is a density gradient across the magnetic field, the drift wave exists in the wave-vector regime of (6.109). A conclusion from Eq. (6.113), therefore, is that such a drift wave is *always* unstable. It is in this sense that the instability associated with the drift wave is sometimes called the *universal instability*.

As a function of k_y, the growth rate, Eq. (6.113), takes on a maximum value somewhere in the range

$$\frac{1}{R_L^2} << k_y^2 << k_i^2 . \tag{6.114}$$

This relation assumes $R_L^2 >> k_i^{-2}$; if this is not the case, (6.114) is replaced by $(k_y/k_i)^2 << 1$.

C. Stabilization

In actual plasma-confinement systems, there exist a number of factors which act to stabilize the drift waves. Hence the situations are not so catastrophic as the terminology "universal instability" might imply.

To see the origin of such a stabilization effect, let us return to (6.112), the condition to avoid the ion Landau damping. We note that the values of $|k_z|$ in actual plasmas are generally limited from below. The total length L of the plasma along the magnetic field, be it in an open-ended system or in a toroidal system, provides a measure of the lower limit for the available values of $|k_z|$.

If the geometrical condition of the system is such that, even with this smallest possible value of $|k_z|$, the condition (6.112) cannot be satisfied, then we may conclude that the ion Landau-damping term in Eq. (6.111) represents a major effect; the drift wave may be stabilized.

The critical longitudinal length L_{cr} of the plasma below which the drift waves are stabilized may thus be estimated from the equation $|\omega/k_z| = \sqrt{T_i/m_i}$ in which $|k_z| \approx L_{cr}$. The expression for the frequency ω has been obtained as Eq. (6.110) in which we may neglect $(k_y/k_i)^2$ by virtue of (6.114). Hence

$$L_{cr} \approx \frac{\pi}{|k_y|R_L} \frac{1 - \Lambda_0(\beta) + (T_i/T_e)}{\Lambda_0(\beta)} r \tag{6.115}$$

where $r \equiv N(dN/dx)^{-1}$ roughly measures a distance of plasma confinement across the magnetic field (cf., Eq. 6.98).

When $|k_y|$ is small so that $k_y^2 R_L^2 << 1$, Eq. (6.115) becomes $L_{cr} \approx \pi(T_i/T_e)(r/|k_y|R_L)$. An extremely large critical length is indicated here. Furthermore, the contribution to the growth rate stemming from Eq. (6.113), being proportional to k^4, becomes infinitesimally small in this regime of k_y. Taking account of these two effects, we conclude that the drift waves are ordinarily stable in the long-wavelength regime. When $|k_y|$ is large so that (6.114) is satisfied, Eq. (6.115) turns into

$$L_{cr} \approx \frac{(2\pi)^{3/2}}{2}\left(1 + \frac{T_i}{T_e}\right) r . \tag{6.116}$$

Still a substantially large critical length is indicated. Unless a fairly long and thin plasma is constructed, it is rather difficult for the drift waves to avoid the ion Landau damping.

The stabilizing action of the ion Landau damping can also be exploited when the magnetic field has a shear structure. We thus consider a field configuration

$$B_z = B \, , B_y = BSx \, ,$$

where S^{-1} is the shearing distance of the magnetic field. A drift wave with a wave vector $(0,k_y,0)$ at $x = 0$ may then be observed as having a slightly different wave vector $[0,k_y\sqrt{1 - S^2d^2},k_ySd]$ at $x = d$. The stability criterion $|\omega/k_z| < \sqrt{T_i/m_i}$ over the distance $d \approx r$ may then be expressed as

$$S > R_L \left(\frac{1}{N} \frac{dN}{dx} \right)^2 . \tag{6.117}$$

We thus find that a modest amount of shearing is sufficient to stabilize the drift waves.

D. Drift Cyclotron Instability

Thus far we have been concerned with the properties of drift waves in the low-frequency domain such that $|\omega| << \Omega_i$. As the frequency of the drift wave increases toward the vicinity of Ω_i, a coupling may take place between the drift wave and the fundamental Bernstein mode of the ions. Both the drift wave and the Bernstein mode propagate predominantly in directions perpendicular to the magnetic field, and may couple quite effectively. The extra free energy in the spatial inhomogeneity of the plasma is thereby fed into the ion Bernstein mode, resulting in the onset of an instability [Mikhailovsky & Timofeev 1963].

For the analysis of such an instability, we assume $F = 0$, $k_z = 0$ (i.e., $|k_y| = k$), and consider the dielectric response function Eq. (6.73) in the vicinity of $\omega \approx \Omega_i << |\Omega_e|$; hence

$$\varepsilon(k,0;\omega) = 1 + \frac{k_e^2}{k^2}\left[\frac{\omega_e^*}{\omega} + 1 - \Lambda_0(\beta_e)\right]$$

$$+ \frac{k_i^2}{k^2}\left\{1 - (\omega - \omega^*)\left[\frac{\Lambda_0(\beta)}{\omega} + \frac{\Lambda_1(\beta)}{\omega - \Omega_i}\right]\right\} . \tag{6.118}$$

Let us further assume that

$$\beta_e = k^2 \frac{T_e m_e}{T_i m_i} R_L^2 < 1 , \tag{6.119a}$$

$$\beta = k^2 R_L^2 >> 1 . \tag{6.119b}$$

By virtue of these assumptions, we may expand

$$\Lambda_0(\beta_e) \approx 1 - \beta_e ,$$

$$\Lambda_0(\beta) \approx \Lambda_1(\beta) \approx \frac{1}{\sqrt{2\pi}\, kR_L} \equiv \delta .$$

We expect that $\delta << 1$. The dispersion relation is then calculated from Eq. (6.118) as

$$1+\frac{k^2}{K^2}-\frac{\omega^*}{\omega}=\frac{\delta(\omega-\omega^*)}{\omega-\beta_i} \tag{6.120}$$

where

$$\frac{1}{K^2}=\frac{1}{k_i^2}+\frac{m_e}{m_i}R_L^2 . \tag{6.121}$$

Note that $(m_e/m_i)R_L^2k_i^2=(\omega_e/\Omega_e)^2$ ($<< 1$).

We now observe the following possibilities for decomposing Eq. (6.120): Unless $\omega \approx \Omega_i$, the right-hand side of Eq. (6.120) is negligibly small; we have a solution corresponding to a drift branch,

$$\omega_1=\frac{\omega^*}{1+(k/K)^2} . \tag{6.122a}$$

On the other hand, when $\omega^* << \Omega_i$, we may neglect ω^* in Eq. (6.120) in the vicinity of $\omega \approx \Omega_i$; hence we obtain a cyclotron branch

$$\omega_2=\Omega_i\left[1+\frac{\delta}{(k/K)^2}\right] . \tag{6.122b}$$

If these two branches intersect, then we expect complex frequency solutions in the vicinity of the intersections, indicating the onset of an instability.

Let us therefore seek the condition for such intersections to take place. The frequency ω_1 of the drift branch takes on a maximum value

$$\omega_{\max}=\frac{\alpha K R_L^2 \Omega_i}{2} \tag{6.123}$$

at $k=K$. The values of β_e and β at this wave number are

$$\beta_e \approx \frac{T_e\omega_e^2}{T_i\Omega^2},\ \beta \approx \frac{\omega_i^2}{\Omega_i^2} ,$$

so that the assumptions (6.119) may still be reasonably satisfied there. The condition for the interaction may then be expressed as $\omega_{\max} \ge \Omega_i$; we thus find

$$\alpha k_i R_L^2 \approx 2\sqrt{\frac{1+m_e}{m_i}}(k_i R_L)^2 \,. \tag{6.124}$$

Equation (6.120), being a quadratic equation for ω, can be solved analytically. We first write it as

$$\left(1+\frac{k^2}{K^2}-\delta\right)\omega^2-\left[\Omega_i\left(1+\frac{k^2}{K^2}\right)+\omega^*(1-\delta)\right]\omega+\omega^*\Omega_i=0 \,.$$

The condition for the instability then is

$$4\omega^*\Omega_i\left(1+\frac{k^2}{K^2}-\delta\right)>\left[\Omega_i\left(1+\frac{k^2}{K^2}\right)+\omega^*(1-\delta)\right]^2 \,.$$

In terms of the two characteristic frequencies Eqs. (6.122) and to the first order in δ, this condition can be expressed as

$$\frac{2k^2}{K^2}\omega_1(\omega_1+\omega_2)\delta>(\omega_1-\omega_2)^2\left(1+\frac{k^2}{K^2}-2\delta\right). \tag{6.125}$$

Because of the smallness of δ, it is necessary to satisfy the coupling condition $\omega_1 \approx \omega_2$ to observe an instability; only in the vicinity of $\omega_1 = \omega_2$ can the drift cyclotron instability take place.

Consequences of these drift-wave instabilities to magnetically confined plasmas have been extensively investigated [e.g., Horton 1984].

PROBLEMS

6.1 For the ion-acoustic wave instability of Section 6.2A, the boundary curve, $v_d = v_d(k)$, between the growing and damped waves [e.g., Ichimaru 1962] may be determined from the condition of marginal stability, $\gamma_k = 0$. Calculate the front edge of such a boundary curve from Eqs. (6.11) for $T_e >> T_i$.

6.2 The boundary curve obtained in Problem 6.1 exhibits a characteristic folding back toward the k axis when the temperature ratio, T_e/T_i, is sufficiently large. The critical wave number k_c, at which the instability first takes place with an increase of v_d, departs from zero and now takes on a finite value. Find the relation between the mass ratio m_e/m_i and the critical temperature ratio $(T_e/T_i)_c$ above which such a folding-back takes place.

6.3 With an appropriate change of variables, the dispersion relations (6.15) and (6.17) may be rewritten in the form

$$1 = \frac{1}{x^2} + \frac{\alpha}{(x-z)^2} .$$

Denoting by $y(x;z)$ the right-hand side of this equation, show that the curve $y(x;z)$ as a function of x is tangential to $y = 1$ when

$$z = \left(1 + \alpha^{1/3}\right)^{3/2} .$$

Use this result to derive Eq. (6.18) and explain the derivation.

6.4 Consider a case with Eq. (6.23) where the electrons as a whole drift with a velocity v_{d} relative to the ions along the magnetic field. Show that the condition for the energy of the drifting electrons channeled into the nth Bernstein mode is given by

$$v_{\mathrm{d}} > \left| \frac{n\Omega}{k_{\|}} \right| .$$

6.5 Instead of the high-frequency plasma oscillations (6.25) in the cold-electron limit of (6.24), we may consider the ion-acoustic waves of

$$\omega_k = \frac{\omega_i k}{\sqrt{k^2 + k_e^2}}$$

in the other limiting cases of the hot electrons

$$\frac{T_e}{m_e} >> \frac{T_i}{m_i} .$$

This sort of situation may arise when the major supply of the plasma energy comes from the electrons. Consider the coupling between such an ion-acoustic wave and the nth ionic Bernstein mode and show that the threshold density of the ions is given by

$$n_i > \frac{n^2 B^2}{4\pi m_i c^2} .$$

6.6 Show that the dielectric functions for the right- and left-circularly polarized waves in a plasma with drift v_d and anisotropic two-temperature ($T_\perp$ and $T_{||}$) distribution are given by

$$\varepsilon_{r(l)}(k,\omega) = 1 - \frac{\omega_p^2}{\omega^2}\left\{\frac{\tilde{\omega}}{\tilde{\omega} \pm \Omega}\left[1 - W\left(\frac{\tilde{\omega} \pm \Omega}{k\sqrt{T_{||}/m}}\right)\right] + \left(1 - \frac{T_\perp}{T_{||}}\right) W\left(\frac{\tilde{\omega} \pm \Omega}{k\sqrt{T_{||}/m}}\right)\right\}$$

where $\tilde{\omega}$ is the Doppler-shifted frequency, $\tilde{\omega} = \omega - kv_d$.

6.7 Consider an inhomogeneous plasma in the magnetic field where the guiding centers of the charged particles are distributed as

$$f_G(\mathbf{r}_\perp,\omega) = \left(\frac{m}{2\pi T(\mathbf{r}_\perp)}\right)^{3/2} \exp\left(-\frac{mv^2}{2T(\mathbf{r}_\perp)}\right).$$

Here the temperature varies in the directions perpendicular to the magnetic field; $\mathbf{r}_\perp$ represents the position vector in these directions. The cyclotron orbit of a charged particle in the phase space may be given by Eqs. (3.115) and (3.116). When the density of the guiding centers also varies as $n(\mathbf{r}_\perp)$, the flux of particle flow at the origin is calculated as

$$n\mathbf{v}_D = \int d\mathbf{v}\mathbf{v}n(-\mathbf{r}_\perp)f_G(-\mathbf{r}_\perp,\mathbf{v}) .$$

Carry out the cyclotron-orbital average $\langle n\mathbf{v}_D\rangle/n$ of the lowest-order contributions in the Taylor expansion of $n(-\mathbf{r}_\perp)$ and $f_G(-\mathbf{r},\mathbf{v})$ with respect to the spatial inhomogeneities, and prove Eq. (6.50).

6.8 Show that the distribution function (6.60) is a stationary solution of the Vlasov equation (6.65) in the absence of external perturbations.

6.9 Derive Eq. (6.69).

6.10 When the plasma density charges continuously in the x direction, rather than in the discontinuous way of Fig. 6.12(a), then the space-charge field $\delta\rho_e(y)$ will be created in the plasma as a result of perturbation Eq. (6.75). This field can be calculated quite analogously to the surface charge Eq. (6.77) and is now given by

$$\rho_e(y) = \frac{1}{N}\frac{dN}{dx}\frac{\rho_m c}{B} gk\left(\int^t d\tau \xi\right)\cos ky\,.$$

We may neglect the spatial dependence of $(1/N)(dN/dx)$ and determine the induced electric field from the Poisson equation. Thus, following the procedure of Section 6.5A, show that the growth rate of the gravitational instability in these circumstances is given by $\sqrt{(1/N)(dN/dx)g}$.

6.11 Prove relation (6.94).

CHAPTER **7**

Fluctuations

The macroscopic states of plasmas are usually specified in terms of a set of statistically averaged quantities. The number density, the kinetic temperature, and the current density are examples of such macroscopic quantities. Since a plasma consists of a large number of randomly moving particles, the instantaneous values of those quantities deviate from their mean values; the physical quantities are therefore said to fluctuate.

The study of fluctuations represents one of the central subjects in statistical physics. Fluctuations are, first of all, connected directly with the correlation functions. Since the interaction energies are calculated once the correlations are known, the thermodynamic properties of a many-particle system are formulated in terms of the fluctuations. Rates of the dynamic processes such as relaxation and transport are likewise controlled by fluctuations.

The spectral functions of fluctuations are in some cases referred to as structure factors. For a system in thermodynamic equilibrium, fluctuations remain small in magnitude; these are thermal fluctuations. When one of the plasma instabilities sets in, however, fluctuations are expected to grow enormously; the system goes over to a turbulent state. Regardless of the level, fluctuations play a vital role in the analyses of static and dynamic properties of plasma.

This chapter begins with description of fluctuations and correlations in plasmas, and then microscopically formulates space-time evolution of the fluctuations in the phase space; this part closely follows the kinetic theory developed in Chapter 2. Spectral repesentations for the fluctuations are presented with inclusion of treatment on strong Coulomb-coupling effects in the static and dynamic properties of the plasmas, along the ways elucidated in Chapter 3. Various elementary processes stemming from fluctuations are considered: These include collisional processes, density-fluctuation excitations, transport processes, and electromagnetic radiation.

7.1 EVOLUTION OF FLUCTUATIONS

The space-time evolution of microscopic fluctuations in Klimontovich distributions, as formulated in Section 2.1A, is analyzed for a classical multicomponent plasma; different particle species are distinguished by subscripts such as σ. Distinction between the spontaneous and induced (linear and nonlinear) fluctuations in the evolution processes is particularly stressed.

A. Description in Phase Space

The *Klimontovich distribution*

$$N_\sigma(X;t) \equiv \sum_i \delta[X - X_j(t)] \tag{7.1}$$

describes the microscopic density distribution for the plasma particles of the σ species in 6-dimensional phase space*

$$X \equiv (\mathbf{r},\mathbf{p}) \tag{7.2}$$

and $X_j(t)$ represents the trajectory of the jth particle in the phase space.

The distribution Eq. (7.1) satisfies the *Klimontovich equation*

$$\frac{\partial N_\sigma}{\partial t} + \mathbf{v}\cdot\frac{\partial N_\sigma}{\partial \mathbf{r}} + \dot{\mathbf{p}}\cdot\frac{\partial N_\sigma}{\partial \mathbf{p}} = 0 \tag{7.3}$$

where the force $\dot{\mathbf{p}}$ is given by

$$\dot{\mathbf{p}} = Z_\sigma e\left(\mathbf{E} + \frac{\mathbf{v}}{c}\times\mathbf{B}\right). \tag{7.4}$$

*In this and subsequent chapters, we shall specify the phase space in terms of the position and the kinetic momentum, rather than in terms of the position and the velocity as in Chapter 2, for a clearer connection to quantum and relativistic descriptions.

This is the *Lorentz force* acting on a particle with electric charge $Z_\sigma e$ at X in the microscopic (and external) electromagnetic fields, **E** and **B**. Mindful of the cases with relativistic plasmas, we write the kinetic momentum as

$$\mathbf{p} = \gamma_\sigma m_\sigma \mathbf{v} . \tag{7.5}$$

Here m_σ is the rest mass of a particle and γ_σ refers to the Lorentz factor

$$\begin{aligned} \gamma_\sigma &= \sqrt{1 + \frac{p^2}{m_\sigma^2 c^2}} \\ &= \left(1 - \frac{v^2}{c^2}\right)^{-1/2} . \end{aligned} \tag{7.6}$$

Equation (7.5) sets the *inverse mass tensor* via

$$\frac{1}{\mathbf{m}_\sigma} = \left(\frac{\partial \mathbf{v}}{\partial \mathbf{p}}\right)_\sigma . \tag{7.7}$$

The electromagnetic fields are determined from the solution to the set of Maxwell equations:

$$\nabla \times \mathbf{E} + \frac{1}{c}\frac{\partial \mathbf{B}}{\partial t} = 0 , \tag{7.8a}$$

$$\nabla \times \mathbf{B} - \frac{1}{c}\frac{\partial \mathbf{E}}{\partial t} = \frac{4\pi}{c}\mathbf{J} , \tag{7.8b}$$

$$\nabla \cdot \mathbf{E} = 4\pi \rho_e , \tag{7.8c}$$

$$\nabla \cdot \mathbf{B} = 0 , \tag{7.8d}$$

where the electric current density **J** and the electric charge density ρ_e are given by the sums between the microscopic and external contributions:

$$\mathbf{J} = \sum_\sigma Z_\sigma e \int d\mathbf{p} \mathbf{v} N_\sigma(X;t) + \mathbf{J}_{\text{ext}} , \tag{7.9a}$$

$$\rho_e = \sum_\sigma Z_\sigma e \int d\mathbf{p} N_\sigma(x;t) + \rho_{\text{ext}} . \tag{7.9b}$$

As we have argued in Section 2.1A, the Klimontovich equations are a rewriting of the microscopic equations of motion for the

$$N = \sum_{\sigma} N_\sigma$$

particles, and are thereby concerned with microscopic information in matching details. A correspondence to the macroscopic world in which we observe the plasma through a set of macroscopic variables may be provided by the averaging process in Gibbs' sense as follows: The angular brackets $\langle \cdots \rangle$ denote an ensemble average over replicas of the system with the same "macroscopic realizations," represented by e.g., thermodynamic, hydrodynamic, electromagnetic variables and the single-particle distribution functions.

In many cases, we may alternatively interpret $\langle A \rangle$ as the space-time average of a physical quantity A over scales substantially greater than those associated with the fluctuations. We thus consider a hierarchical structure to the characteristic scales in the space-time variations as discussed in Section 2.3A, and thereby assume the existence of space-time scales, l and τ, satisfying

$$l_1 >> l >> l_2 \text{ and } \tau_1 >> \tau >> \tau_2 \,, \tag{7.10}$$

where l_1, τ_1 and l_2, τ_2 are the characteristic scales associated with the single-particle and multiparticle distributions, respectively. The average $\langle \cdots \rangle$ is then carried out over the scales l and τ.

In either definition, the average of the Klimontovich distribution Eq. (7.1) yields the *single-particle distribution* $F_\sigma(X;t)$ with the normalization

$$\int d\mathbf{p} F_\sigma(X;t) = n_\sigma(\mathbf{r},t) \tag{7.11}$$

the number-density distribution in the hydrodynamics. A schematic drawing of $F_\sigma(X;t)$ is shown in Fig. 7.1 by a dashed line. In a spatially uniform plasma, $F_\sigma(X;t)$ gives the momentum distribution $F_\sigma(\mathbf{p})$.

We may thus split Eq. (7.1) into its average and the fluctuations:

$$N_\sigma(X;t) = F_\sigma(X;t) + \delta N_\sigma(X;t) \,. \tag{7.12}$$

The electromagnetic fields may likewise be written as the sums between the averages and the fluctuations:

$$\mathbf{E}(\mathbf{r},t) = \langle \mathbf{E}(\mathbf{r},t) \rangle + \delta \mathbf{E}(\mathbf{r},t) \,, \tag{7.13a}$$

$$\mathbf{B}(\mathbf{r},t) = \langle \mathbf{B}(\mathbf{r},t) \rangle + \delta \mathbf{B}(\mathbf{r},t) \,. \tag{7.13b}$$

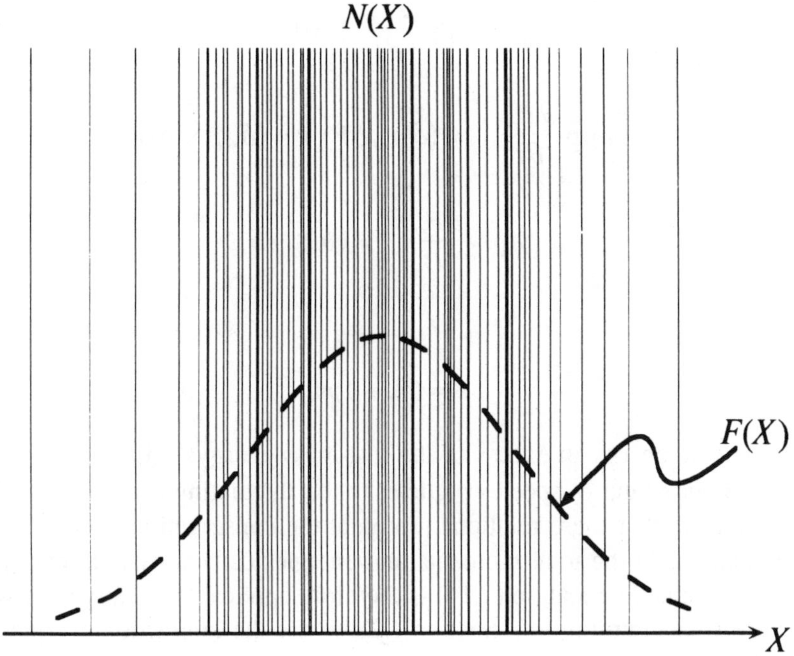

Figure 7.1 A schematic view of the Klimontovich distribution $N(X)$ and its average $F(X)$.

The Klimontovich equation (7.3) may be split into two equations: The averaged part then yields the *kinetic equation*

$$\frac{\partial F_\sigma}{\partial t} + \mathbf{v} \cdot \frac{\partial F_\sigma}{\partial \mathbf{r}} + Z_\sigma e \left[\langle \mathbf{E} \rangle + \frac{\mathbf{v}}{c} \times \langle \mathbf{B} \rangle \right] \cdot \frac{\partial F_\sigma}{\partial \mathbf{p}}$$

$$= -Z_\sigma e \frac{\partial}{\partial \mathbf{p}} \cdot \left\langle \delta N_\sigma \left(\delta \mathbf{E} + \frac{\mathbf{v}}{c} \times \delta \mathbf{B} \right) \right\rangle. \tag{7.14}$$

Clearly, the right-hand side represents the *collision term*, which we shall denote as $\partial F_\sigma / \partial t]_c$.

It is straightforward to see that the collision term satisfies the following moment relations:

$$\int d\mathbf{p} \frac{\partial F_\sigma}{\partial t}\bigg]_c = 0 \,, \tag{7.15a}$$

$$\int d\mathbf{p}\mathbf{p} \frac{\partial F_\sigma}{\partial t}\bigg]_c = \langle \delta \rho_{e\sigma} \delta \mathbf{E} \rangle + \frac{1}{c} \langle \delta \mathbf{J}_\sigma \times \delta \mathbf{B} \rangle \,, \tag{7.15b}$$

$$\int d\mathbf{p}\varepsilon \frac{\partial F_\sigma}{\partial t}\bigg]_c = \langle \delta \mathbf{J}_\sigma \cdot \delta \mathbf{E} \rangle \,. \tag{7.15c}$$

Here

$$\varepsilon = (\gamma_\sigma - 1) m_\sigma c^2 \tag{7.16}$$

is the kinetic energy in excess of the rest-mass energy; $\delta \mathbf{J}_\sigma$ and $\delta \rho_{e\sigma}$ are the contributions of the σ-species particles to the fluctuating parts of Eqs. (7.3).

It is clear that these relations represent the conservation laws of the particle numbers and the total momenta and energies carried by both particles and fields. The kinetic equation (7.14) is therefore applicable equally to those plasmas in strong coupling (cf., Section 1.2), where the interaction energies represent the dominant effects in the system.

B. Spontaneous and Induced Fluctuations

The fluctuation part of Eq. (7.3) can be written as

$$\frac{\partial \delta N_\sigma}{\partial t} + \mathcal{L}\, \delta N_\sigma = \mathcal{S} \tag{7.17}$$

where

$$\mathcal{L} = Z_\sigma e \left[\langle \mathbf{E} \rangle + \frac{\mathbf{v}}{c} \times \langle \mathbf{B} \rangle \right] \cdot \frac{\partial}{\partial \mathbf{p}} \,, \tag{7.18}$$

$$\mathcal{S}(X;t) = -Z_\sigma e \left(\delta \mathbf{E} + \frac{\mathbf{v}}{c} \times \delta \mathbf{B} \right) \cdot \frac{\partial F_\sigma}{\partial \mathbf{p}}$$

$$- Z_\sigma e \left(\delta \mathbf{E} + \frac{\mathbf{v}}{c} \times \delta \mathbf{B} \right) \cdot \frac{\partial \delta N_\sigma}{\partial \mathbf{p}} + Z_\sigma e \left\langle \left(\delta \mathbf{E} + \frac{\mathbf{v}}{c} \times \delta \mathbf{B} \right) \cdot \frac{\partial \delta N_\sigma}{\partial \mathbf{p}} \right\rangle . \tag{7.19}$$

The last term of Eq. (7.19) is there only to make $\langle \mathcal{S}(X;t) \rangle = 0$ as it should be.

A formal solution to Eq. (7.17) is obtained in the following way [Ichimaru 1977a]: We first note that Eq. (7.17) may be looked upon as an inhomogeneous differential equation for δN_σ with the source term $\mathcal{S}(X;t)$. Its general solution is then given by a summation between a general solution to the homogeneous equation

$$\frac{\partial\, \delta N_\sigma^{(0)}}{\partial t} + \mathcal{L}\, \delta N_\sigma^{(0)} = 0 \tag{7.20}$$

and a special solution $\delta N_\sigma^{(S)}$ to Eq. (7.17).

Let us treat $\delta N_\sigma^{(S)}$ first. We remark that $\mathcal{S}(X;t)$ plays the part of the source term by which to generate the fluctuations $\delta N_\sigma^{(S)}$ through Eq. (7.17). We note also that each term in Eq. (7.19) depends on the electromagnetic fluctuations, $\delta\mathbf{E}$ and $\delta\mathbf{B}$; in the present formalism the microscopic electromagnetic interactions between the discrete particles have been embodied only in those terms containing the fluctuating electromagnetic fields. Consequently, $\delta N_\sigma^{(S)}$ should correspond to that part of the fluctuations *induced* by the electromagnetic interactions between the particles.

The foregoing argument is useful also for the interpretation of that part of the fluctuations $\delta N_\sigma^{(0)}$ described by Eq. (7.20). Since Eq. (7.20) has been obtained by setting $\mathcal{S}(X;t) = 0$ in Eq. (7.17), $\delta N_\sigma^{(0)}$ should correspond to those *spontaneous* fluctuations in a fictitious, noninteracting (ideal-gas) system, arising from the random thermal motion of the discrete particles; such fluctuations exist even without the electromagnetic interactions.

The induced fluctuations $\delta N_\sigma^{(S)}$ can be formulated by integrating Eq. (7.17) along the unperturbed trajectories, that is, along the orbitals of particles in the averaged fields without fluctuations, as we considered in Section 3.4B. Let a solution to the dynamic equations

$$\frac{d\mathbf{r}}{dt} = \mathbf{v}\,, \quad \frac{d\mathbf{p}}{dt} = Z_\sigma e\left[\langle\mathbf{E}\rangle + \frac{\mathbf{v}}{c}\times\langle\mathbf{B}\rangle\right]$$

be expressed by the correspondence:

$$(\mathbf{r},\mathbf{p},t) \leftrightarrow (\mathbf{r}',\mathbf{p}',t')\,. \tag{7.21}$$

Then Eq. (7.17) is integrated along the unperturbed trajectory Eq. (7.21), to give

$$\delta N_\sigma^{(S)}(X;t) = \int_{-\infty}^{t} dt'\, \mathcal{S}(X';t') \tag{7.22}$$

with the "initial" conditions

$$\lim_{t \to -\infty} \delta N_\sigma^{(S)}(X;t) = 0 . \tag{7.23}$$

The boundary conditions reflect the principle of causality (cf., Section 3.1) in that $\delta N_\sigma^{(S)}(X;t)$ represents the consequence of fluctuations built up through the adiabatic turn-on of the microscopic electromagnetic interactions; it vanishes at $t \to -\infty$.

The spontaneous part of the fluctuations can likewise be expressed in terms of the unperturbed trajectories $X_j^{(0)}(t)$ as

$$\delta N_\sigma^{(0)}(X;t) = \sum_j \delta[X - X_j^{(0)}(t)] - \left\langle \sum_j \delta[X - X_j^{(0)}(t)] \right\rangle . \tag{7.24}$$

A set of initial conditions $X_j^{(0)}(t_0)$ at t_0 has been assumed in Eq. (7.24). We shall arbitrarily take $t_0 = 0$ and denote $X_j^{(0)}(0) = X_j^{(0)}$.

C. Fluctuations in "Uniform" Systems

In Section 7.1A, we introduced the "space-time average over scales substantially greater than those associated with fluctuations." Let us now observe the system in such scales and extend it over the entire space and time by adopting the periodic bounary conditions. The system so extended can then be looked upon as "uniform" in space and time as far as the fluctuating quantities are concerned.

Let such a fluctuating quantity be $\delta A(X;t)$. In a uniform system, it is useful to work with the Fourier transformation (see the footnote on p. 20), so that

$$\delta A(X;t) = \sum_{\mathbf{k}} \int_{-\infty}^{\infty} \frac{d\omega}{2\pi} \delta A(\mathbf{k},\omega;\mathbf{p}) \exp[i(\mathbf{k}\cdot\mathbf{r} - \omega t)] . \tag{7.25}$$

By definition, $\delta A(\mathbf{k}=0,\omega;\mathbf{p}) = 0$, and $\delta A(\mathbf{k},\omega;\mathbf{p})$ does not contain a singular term proportional to $\delta(\omega)$.

In terms of the Fourier components, the solution to Eq. (7.17) is expressed as

$$\delta N_\sigma(\mathbf{k},\omega;\mathbf{p}) = \delta N_\sigma^{(0)}(\mathbf{k},\omega;\mathbf{p}) \tag{7.26a}$$

$$-Z_\sigma e \int_0^\infty d\tau \left[\frac{\partial F_\sigma(\mathbf{p}')}{\partial \mathbf{p}'} \right] \cdot \mathbf{T}(\mathbf{k},\omega;\mathbf{p}') \cdot \delta \mathbf{E}(\mathbf{k},\omega) \exp[-i\varphi(\tau) - \eta t] \tag{7.26b}$$

$$-Z_\sigma e \sum_{\mathbf{q}} \int_{-\infty}^{\infty} \frac{d\omega'}{2\pi} \int_0^{\infty} d\tau \left[\frac{\partial}{\partial \mathbf{p}'} \delta N_\sigma(\mathbf{k}-\mathbf{q},\omega-\omega';\mathbf{p}') \right] \cdot \mathbf{T}(\mathbf{q},\omega';\mathbf{p}')$$

$$\cdot\, \delta\mathbf{E}(\mathbf{q},\omega') \exp\left[-i\varphi(\tau)-\eta\,\tau\right] . \tag{7.26c}$$

Here $\tau = t - t'$, $\varphi(\tau) = \mathbf{k}\cdot(\mathbf{r}-\mathbf{r}') - \omega\tau$ characterizes the phase evolution along the unperturbed trajectory, and η is a positive infinitesimal to insure the retarded boundary conditions. The magnetic-field fluctuations $\delta\mathbf{B}(\mathbf{k},\omega)$ are accounted for in Eqs. (7.26) via the projection tensor $\mathbf{T}(\mathbf{k},\omega;\mathbf{p})$ defined in Eq. (3.106).

The term (7.26a) obviously represents the spontaneous fluctuations arising from discreteness of the individual particles. Physically, it corresponds to the term (3.164a) in the equation of motion for the phase-space density fluctuations in the electron gas. In terms of the unperturbed trajectories $X_j^{(0)}(t)$ in Eq. (7.24), we may express

$$\delta N_\sigma^{(0)}(\mathbf{k},\omega;\mathbf{p}) = \frac{1}{V}\sum_{j=1}^{\infty} \int_{-\infty}^{\infty} dt\,\delta\left[\mathbf{p}-\mathbf{p}_j^{(0)}(t)\right] \exp\left\{-i[\mathbf{k}\cdot\mathbf{r}_j^{(0)} + \varphi(t)]\right\} \tag{7.27}$$

and choose $V \approx l^3$ in light of the specification for the "uniform" plasma with the assumption (7.10).

The second term (7.26b) describes that part of density fluctuation induced linearly by the microscopic electromagnetic-field fluctuations $\delta\mathbf{E}(\mathbf{k},\omega)$. It has a physical content the same as the term (3.164b) and gives rise to a mean-field contribution.

Finally, the third term (7.26c) represents that part of density fluctuation induced through nonlinear coupling between $\delta N_\sigma(\mathbf{k}-\mathbf{q},\omega-\omega';\mathbf{v}')$ and $\delta\mathbf{E}(\mathbf{q},\omega')$. It leads to the local-field corrections in the dielectric formulation of Section 3.2, to which we shall return in Section 7.3.

7.2 SPECTRAL FUNCTIONS

We begin this section with a review on the fundamental relation between fluctuations and correlations. Electromagnetic fluctuations in weakly coupled plasmas are then described; these calculations lead to an important notion concerning the superposition of dressed test-particles. Spectral representations of collision terms and of energy density in electrostatic fluctuations are finally presented.

A. Correlation Functions and Spectral Representations

The Fourier component of a fluctuating quantity $\delta A(\mathbf{r},t)$ is calculated as

$$\delta A(\mathbf{k},\omega)=\frac{1}{V}\int d\mathbf{r}\int_{-\infty}^{\infty} dt\,\delta A(\mathbf{r},t)\exp\left[-i(\mathbf{k}\cdot\mathbf{r}-\omega t)\right]. \tag{7.28}$$

The *correlation function* between such fluctuating quantities δA and δB is defined as

$$C_{AB}(\mathbf{r},t)=\frac{1}{2}\langle\delta A(\mathbf{r}'+\mathbf{r},t'+t)\delta B(\mathbf{r}',t')+\delta B(\mathbf{r}',t')\delta A(\mathbf{r}'+\mathbf{r},t'+t)\rangle$$

$$=\frac{1}{V}\sum_{\mathbf{k}}\int_{-\infty}^{\infty} d\omega\langle AB^*(\mathbf{k},\omega)\rangle\exp\left[i(\mathbf{k}\cdot\mathbf{r}-\omega t)\right]. \tag{7.29}$$

The spectral function $\langle AB^*(\mathbf{k},\omega)\rangle$ in Eq. (7.29) is therefore given by the Fourier transformation of the correlation function as

$$\langle AB^*(\mathbf{k},\omega)\rangle=\frac{1}{2\pi}\int_V d\mathbf{r}\int_{-\infty}^{\infty} d\omega\, C_{AB}(\mathbf{r},t)\exp\left[-i(\mathbf{k}\cdot\mathbf{r}-\omega t)\right]. \tag{7.30}$$

A useful relation for an explicit evaluation of the spectral function is*

$$\langle\delta A(\mathbf{k},\omega)\delta B(-\mathbf{k},\omega')\rangle=\frac{4\pi^2}{V}\langle AB^*(\mathbf{k},\omega)\rangle\delta(\omega+\omega') \tag{7.31}$$

(Problem 7.2) [e.g., Landau & Lifshitz 1969].

Let us apply Eq. (7.31) to the calculation of the spectral tensor for the spontaneous current fluctuations

$$\delta\mathbf{J}_\sigma^{(0)}(\mathbf{k},\omega)=Z_\sigma e\int d\mathbf{p}\mathbf{v}\,\delta N_\sigma^{(0)}(\mathbf{k},\omega;\mathbf{p}) \tag{7.32}$$

associated with Eq. (7.27) for the particles of σ species. We recall that those represent the spontaneous fluctuations in a fictitious, noninteracting (ideal-gas) system. The positions $\mathbf{r}_j^{(0)}$ and $\mathbf{r}_{j'}^{(0)}$ of two different ($j\neq j'$) particles at an arbitrarily chosen time $t=0$, for instance, are totally *uncorrelated.* Hence

$$\langle\mathbf{J}\mathbf{J}^*(\mathbf{k},\omega)\rangle_{\sigma\sigma'}^{(0)}=\delta_{\sigma\sigma'}(Z_\sigma e)^2 n_\sigma\int d\mathbf{p}\mathbf{v}\mathbf{v}f_\sigma(\mathbf{p})\delta(\omega-\mathbf{k}\cdot\mathbf{v}) \tag{7.33}$$

*Here and hereafter in this section, we assume a classical system, where $A(\mathbf{r},t)$ and $B(\mathbf{r}',t')$ commute.

when there are no externally applied force fields. Here n_σ and $f_\sigma(\mathbf{p})$ refer to the average number density and momentum distribution for particles of the σ species; the latter is normalized to unity after integration over the momentum space, in the sense that

$$\int d\mathbf{p} f_\sigma(\mathbf{p}) = 1 . \tag{7.34}$$

When a uniform magnetic field, $\mathbf{B} = B\hat{z}$, is externally applied to the plasma in the z direction, the spectral tensor takes the form

$$\langle \mathbf{J}\mathbf{J}^*(\mathbf{k},\omega)\rangle^{(0)}_{\sigma\sigma'} = \delta_{\sigma\sigma'}(Z_\sigma e)^2 n_\sigma \sum_{\nu=-\infty}^{\infty} \int d\mathbf{p}\Pi_\sigma(p_\perp,p_{\parallel};\nu)$$
$$\times f_\sigma(p_\perp,p_\parallel)\delta(\nu\Omega_\sigma + k_\parallel v_\parallel - \omega) . \tag{7.35}$$

Here

$$\Omega_\sigma = \frac{Z_\sigma e}{\gamma_\sigma m_\sigma c} , \tag{7.36}$$

the functions $\Pi_\sigma(p_\perp,p_\parallel;\nu)$, and the $\mathbf{p}$ integration are those defined in Eqs. (3.129) and (3.130).

The spontaneous density fluctuations are related with the spontaneous current fluctuations Eq. (7.32) via

$$\delta\rho^{(0)}_\sigma(\mathbf{k},\omega) \equiv \int d\mathbf{p}\delta N^{(0)}_\sigma(\mathbf{k},\omega;\mathbf{p}) = \frac{\mathbf{k}\cdot\delta\mathbf{J}^{(0)}_\sigma(\mathbf{k},\omega)}{Z_\sigma e\omega} . \tag{7.37}$$

Their spectral functions are thus calculated from Eq. (7.33) or (7.35) as

$$S^{(0)}_{\sigma\sigma'}(\mathbf{k},\omega) \equiv \langle\rho\rho^*(\mathbf{k},\omega)\rangle^{(0)}_{\sigma\sigma'} = \frac{\mathbf{k}\mathbf{k}:\langle\mathbf{J}\mathbf{J}^*(\mathbf{k},\omega)\rangle^{(0)}_{\sigma\sigma'}}{(Z_\sigma e\omega)^2} , \tag{7.38}$$

where : implies a double contraction between two tensors.

B. Electromagnetic Fluctuations in the RPA

The *random-phase approximation* (RPA) is applicable to the cases of weak fluctuations whereby the equation of motion may be linearized with respect to the fluctuations [Bohm & Pines 1953; see also Section 1.4]. The nonlinear term (7.26c) is thus negligible in this approximation; Eq. (7.26) becomes

$$\delta N_\sigma(\mathbf{k},\omega;\mathbf{p}) = \delta N_\sigma^{(0)}(\mathbf{k},\omega;\mathbf{p})$$

$$-Z_\sigma e \int_0^\infty d\tau \left[\frac{\partial F_\sigma(\mathbf{p}')}{\partial \mathbf{p}'}\right] \cdot \mathbf{T}(\mathbf{k},\omega;\mathbf{p}') \cdot \delta \mathbf{E}(\mathbf{k},\omega) \exp\left[-i\varphi(\tau) - \eta t\right], \quad (7.39)$$

an equation linear in the fluctuations.

The current-density fluctuations $\delta \mathbf{J}(\mathbf{k},\omega)$ are calculated through a moment integral of $\delta N_\sigma(\mathbf{k},\omega;\mathbf{p})$ in a way analogous to Eq. (7.32). Defining the fluctuations in the spontaneous current density via

$$\delta \mathbf{J}^{(0)}(\mathbf{k},\omega) = \sum_\sigma \delta \mathbf{J}_\sigma^{(0)}(\mathbf{k},\omega), \quad (7.40)$$

we find that Eq. (7.39) is transformed as

$$\delta \mathbf{J}^{(0)}(\mathbf{k},\omega) = \frac{i\omega}{4\pi}\left[\boldsymbol{\varepsilon}(\mathbf{k},\omega) - \left(\frac{ck}{\omega}\right)^2 \mathbf{I}_{\mathrm{T}}\right] \cdot \delta \mathbf{E}(\mathbf{k},\omega). \quad (7.41)$$

Here $\boldsymbol{\varepsilon}(\mathbf{k},\omega)$ and $\mathbf{I}_{\mathrm{T}}$ are the (RPA) dielectric tensor Eq. (3.128) or (3.132) and the transverse projection tensor Eq. (3.89) introduced in Chapter 3.

Note a formal similarity between Eqs. (7.41) and (3.87). The spontaneous current density $\delta \mathbf{J}^{(0)}(\mathbf{k},\omega)$ in Eq. (7.41) plays the part of the external current density in Eq. (3.87). The major difference in physical content between the two equations lies in the fact that $\delta \mathbf{J}^{(0)}(\mathbf{k},\omega)$ in Eq. (7.41) is a stochastic variable arising from the randomness in the motion of individual particles; the statistical nature is therefore carried over in the electric field fluctuations $\delta \mathbf{E}(\mathbf{k},\omega)$.

For the calculation of the spectral tensor associated with the electric-field fluctuations, one inverts Eq. (7.41) with the aid of the inverse dispersion tensor $\mathbf{Z}(\mathbf{k},\omega)$ defined by

$$\mathbf{Z}(\mathbf{k},\omega) \cdot \left[\boldsymbol{\varepsilon}(\mathbf{k},\omega) - \left(\frac{ck}{\omega}\right)^2 \mathbf{I}_{\mathrm{T}}\right] = \mathbf{I}. \quad (7.42)$$

The spectral tensor is thus obtained as

$$\langle \mathbf{E}\mathbf{E}^*(\mathbf{k},\omega)\rangle = \sum_\sigma \frac{16\pi^2}{\omega^2} \mathbf{Z}(\mathbf{k},\omega) \cdot \langle \mathbf{J}\mathbf{J}^*(\mathbf{k},\omega)\rangle_{\sigma\sigma'}^{(0)} \cdot \mathbf{Z}^\dagger(\mathbf{k},\omega) \quad (7.43)$$

where $\mathbf{Z}^\dagger(\mathbf{k},\omega)$ is the Hermitian-conjugate tensor to $\mathbf{Z}(\mathbf{k},\omega)$. Equation (7.43) describes the spectral distribution of the electromagnetic-field fluctuations in the RPA.

C. Superposition of the Dressed Test Particles

The result in Eq. (7.43) contains an important physical concept: Let us note that the spectral functions of the electromagnetic fluctuations calculated in the preceding subsection are applicable even for those plasmas in a nonequilibrium stationary state as long as the RPA is valid. The fluctuations are then expressed by superposing the fields of dynamically screened or "dressed" test particles without further consideration of the correlations between them. The theoretical scheme for calculating the fluctuation spectrum through such a superposition principle has been extensively investigated [Nozières & Pines 1958a, b; Thompson & Hubbard 1960; Hubbard, J. 1961a; Rostoker 1964a, b; Ichimaru 1965].

The idea of the superposition principle can be explained in the following way: We pick a particle in the σ species and regard it as a test charge. Suppose that such a particle is located initially at $\mathbf{r}_j^{(0)}$ and has momentum $\mathbf{p}_j^{(0)}$; in the absence of collisions and an external magnetic field, the Fourier components of its current-density field are

$$\delta \mathbf{J}_{\sigma}^{(0)}(\mathbf{k},\omega) = \frac{2\pi Z_{\sigma} e}{V} \sum_j \mathbf{v}_j^{(0)} \delta\left(\omega - \mathbf{k} \cdot \mathbf{v}_j^{(0)}\right) \exp\left(-i\mathbf{k} \cdot \mathbf{r}_j^{(0)}\right). \tag{7.44}$$

The field of the dressed particle can be constructed by taking additional account of the screening cloud induced by the test charge. The total dressed-particle fields Eq. (7.39), therefore, consist of the bare-particle field (7.26a) plus the induced screening cloud (7.26b).

The spectral tensor can now be calculated by superposing the fields produced by those individual dressed particles. In these processes the correlations between two different dressed particles may be neglected; the major effects in the correlations have already been taken into account in the construction of each dressed particle. Introducing the single-particle momentum distribution $f_{\sigma}(\mathbf{p})$ to carry out the required average over $\mathbf{p}_j^{(0)}$ as in Eq. (7.33), we obtain Eq. (7.43).

In terms of the Bogoliubov hierarchy (7.10) on the characteristic time scales (cf., Section 2.3B), we may describe the superposition principle in the following way. Let us again fix on a charged particle in the plasma, which we regard as a moving test charge. It will act to polarize the medium and so carry a screening cloud with it. This action corresponds to the establishment of a pair correlation; the characteristic time associated with such a process has been denoted by τ_2. In the course of its motion, however, the test charge collides with other field particles; its energy and momentum change abruptly. The polarization cloud originally associated with the test charge will no longer represent a screening cloud appropriate to the new situation; the polarization cloud must adjust itself to these revised circumstances. If the mean free time τ_1 between such colllisions is much larger than τ_2 in such an event, the adjustment will take place quickly so that, for most of the time between two successive short-range collisions, the test charge can be considered as carrying a well-established cloud. It is clear that the superposition calculation of Eq. (7.43) amounts to assuming that each particle, not almost always, but always carries such an equilibrated screening cloud. We thus argue that the superposition principle represents a good approximation as long as $\tau_1 >> \tau_2$. For a stable plasma in weak coupling, such a condition is well satisfied.

It is instructive to consider yet another argument, which exhibits quantum-theoretically the validity of the superposition principle. For this purpose, we note that a fictitious neutral counterpart may be constructed to a many-particle system consisting of charged particles, by adiabatically turning off the long-range part of the Coulomb interaction between the particles [Nozières & Pines 1958a, b; Ichimaru 1965]. The turning off can be achieved by subtracting the average self-consistent field of each particle, and thus the screened short-range Coulomb forces remain in the resulting fictitious system.

D. Collision Terms in the RPA

In terms of the spectral functions of fluctuations, the collision term of the kinetic equation, represented by the right-hand side of Eq. (7.14), is expressed as

$$\left.\frac{\partial F_\sigma(\mathbf{p})}{\partial t}\right]_c = -Z_\sigma e \frac{\partial}{\partial \mathbf{p}} \cdot \sum_{\mathbf{k}} \int d\omega \, \mathbf{T}(\mathbf{k},\omega;\mathbf{p}') \cdot \langle N_\sigma \mathbf{E}^*(\mathbf{k},\omega;\mathbf{p})\rangle \qquad (7.45)$$

where the Fourier analyses are made with the periodic boundary conditions in a unit volume. Assuming the RPA evolution Eq. (7.39) of the fluctuations in the absence of the external magnetic field, we write

$$\langle \delta N_\sigma(\mathbf{k},\omega;\mathbf{p})\delta\mathbf{E}(-\mathbf{k},\omega')\rangle = \frac{4\pi}{i\omega'}\langle \delta N_\sigma^{(0)}(\mathbf{k},\omega;\mathbf{p})\delta\mathbf{J}_0(-\mathbf{k},\omega')\rangle \cdot \mathbf{Z}^\dagger(\mathbf{k},-\omega')$$

$$- Z_\sigma e\left[\frac{\partial F_\sigma(\mathbf{p})}{\partial \mathbf{p}}\right] \cdot \frac{\mathbf{T}(\mathbf{k},\omega;\mathbf{p}')}{i(\mathbf{k}\cdot\mathbf{v}-\omega-i\eta)} \cdot \langle \delta\mathbf{E}(\mathbf{k},\omega)\delta\mathbf{E}(-\mathbf{k},\omega')\rangle \, .$$

With the aid of Eqs. (7.33) and (7.43), Eq. (7.45) is calculated as

$$\left.\frac{\partial F_\sigma(\mathbf{p})}{\partial t}\right]_c = -4\pi (Z_\sigma e)^2 \sum_{\mathbf{k}} \mathbf{k}\cdot\frac{\partial}{\partial \mathbf{p}}\left\{\frac{F_\sigma(\mathbf{p})}{(\mathbf{k}\cdot\mathbf{v})^2}\,\mathrm{Im}[\mathbf{v}\cdot\mathbf{Z}(\mathbf{k},\mathbf{k}\cdot\mathbf{v})\cdot\mathbf{v}]\right.$$

$$\left. -\left[\frac{2\pi Z_\sigma e}{(\mathbf{k}\cdot\mathbf{v})^2}\right]^2\left[\mathbf{k}\cdot\frac{\partial F_\sigma(\mathbf{p})}{\partial \mathbf{p}}\right]\int d\mathbf{p}' F_\sigma(\mathbf{p}')\delta(\mathbf{k}\cdot\mathbf{v}-\mathbf{k}\cdot\mathbf{v}')|\,\mathbf{v}\cdot\mathbf{Z}(\mathbf{k},\mathbf{k}\cdot\mathbf{v})\cdot\mathbf{v}'\,|^2\right\}. \tag{7.46}$$

This is a generalization of the *BGL collision term* [Balescu 1960; Guernsey 1960; Lenard 1960; see also Section 2.4] to cases with inclusion of the electromagnetic interactions.

The BGL collision term Eq. (2.68) is recovered from Eq. (7.65) when one adopts the longitudinal approximation

$$\mathbf{Z}(\mathbf{k},\omega) \to \frac{\mathbf{I}_\mathrm{L}}{\varepsilon(\mathbf{k},\omega)}. \tag{7.47}$$

Here $\varepsilon(\mathbf{k},\omega)$ and $\mathbf{I}_\mathrm{L}$ are the *dielectric response function* Eq. (3.136) and the longitudinal projection tensor Eq. (3.93). Since

$$\mathrm{Im}\,\varepsilon(\mathbf{k},\omega) = -\sum_\sigma \left(\frac{2\pi Z_\sigma e}{k}\right)^2 \int d\mathbf{p}\,\delta(\omega-\mathbf{k}\cdot\mathbf{v})\mathbf{k}\cdot\frac{\partial F_\sigma(\mathbf{p})}{\partial \mathbf{p}}, \tag{7.48}$$

Eq. (7.46) in the longitudinal approximation takes the form

$$\left.\frac{\partial F_\sigma(\mathbf{p})}{\partial t}\right]_c = -16\pi^3 \sum_{\sigma'}\sum_{\mathbf{k}'} (Z_\sigma Z_{\sigma'} e^2)^2 \frac{\mathbf{k}}{k}\cdot\frac{\partial}{\partial \mathbf{p}}\int d\mathbf{p}'\,\frac{\mathbf{k}}{k^2}$$

$$\cdot\left[\left(\frac{\partial}{\partial \mathbf{p}'}-\frac{\partial}{\partial \mathbf{p}}\right)F_\sigma(\mathbf{p})F_{\sigma'}(\mathbf{p}')\right]\frac{\delta(\mathbf{k}\cdot\mathbf{v}-\mathbf{k}'\cdot\mathbf{v}')}{|\,\varepsilon(\mathbf{k},\mathbf{k}\cdot\mathbf{v})\,|^2}, \tag{7.49}$$

which is identical to Eq. (2.68). Properties of these BGL collision terms have been investigated in Section 2.4B.

E. Energy Spectra of Fluctuations

Electromagnetic fluctuations carry energies with them; an energy spectrum associated with such electromagnetic fluctuations is thus defined.

The energy density $\mathcal{E}(\mathbf{r},t)$ of the electrostatic-field fluctuations in a dispersive medium is determined from the conservation properties in the Maxwell equations as [Landau & Lifshitz 1960]

$$\frac{\partial \mathcal{E}}{\partial t} = \frac{1}{4\pi}\mathbf{E}\cdot\frac{\partial \mathbf{D}}{\partial t} \tag{7.50}$$

where $\mathbf{D}(\mathbf{r},t)$ is the electric displacement.

The energy spectrum may be calculated from a statistical analysis of Eq. (7.50). Energy contained in a *collective mode* can then be obtained from such a calculation. More general treatments including the energy in the electromagnetic fluctuations become important in connection with the radiation spectra in the plasma; we shall treat such a problem in Section 7.5.

The fluctuating field quantities are expanded as in Eq. (7.25); one then sets

$$\delta\mathbf{D}(\mathbf{k},\omega) = \varepsilon(\mathbf{k},\omega)\delta\mathbf{E}(\mathbf{k},\omega)\,. \tag{7.51}$$

Integrating the statistical average of Eq. (7.50) with respect to time, one finds

$$\begin{aligned}\langle\mathcal{E}\rangle &= \frac{1}{4\pi}\int_{-\infty}^{t} dt' \int_{\text{unit volume}} d\mathbf{r}\left\langle \mathbf{E}(\mathbf{r},t')\cdot\frac{\partial \mathbf{D}(\mathbf{r},t')}{\partial t'}\right\rangle \\ &= \frac{1}{32\pi^3}\sum_{\mathbf{k}}\int_{-\infty}^{\infty} d\omega \int_{-\infty}^{\infty} d\omega' \,\frac{\partial[\omega'\varepsilon(\mathbf{k},\omega')]}{\partial\omega'} \\ &\quad \times\langle\delta\mathbf{E}(\mathbf{k},\omega)\cdot\delta\mathbf{E}(-\mathbf{k},-\omega')\rangle \exp[-i(\omega-\omega')t]\,. \end{aligned} \tag{7.52}$$

Here the spectral function of the electric-field fluctuations is defined in accord with Eq. (7.31) as

$$\begin{aligned}\langle\delta\mathbf{E}(\mathbf{k},\omega)\cdot\delta\mathbf{E}(-\mathbf{k},-\omega')\rangle &= (2\pi)^2\mathrm{Tr}\langle\mathbf{E}\mathbf{E}^*(\mathbf{k},\omega)\rangle\delta(\omega-\omega')\,, \\ \langle| E^2 |(\mathbf{k},\omega)\rangle_{\text{coll}} &\simeq \langle| E^2 |(\mathbf{k})\rangle_{\text{coll}}\,\delta(\omega-\omega_{\mathbf{k}})\,. \end{aligned} \tag{7.53}$$

Equation (7.52) describes the total energy density resulting from the summation of various field fluctuations over the entire $\mathbf{k}$ and ω spaces. This expression thus enables one to decompose the energy density into its spectral components

$$\mathcal{E}(\mathbf{k},\omega) = \frac{1}{8\pi}\frac{\partial[\omega\varepsilon(\mathbf{k},\omega)]}{\partial\omega}\langle| E^2 |(\mathbf{k},\omega)\rangle\,. \tag{7.54}$$

The total energy contained in the $\mathbf{k}$ mode of the fluctuations is then given by

$$\mathcal{E}_{\mathbf{k}} = \frac{1}{8\pi}\int_{-\infty}^{\infty} d\omega\,\frac{\partial[\omega\varepsilon(\mathbf{k},\omega)]}{\partial\omega}\langle| E^2 |(\mathbf{k},\omega)\rangle\,. \tag{7.55}$$

These are the general formulas relating the electrostatic fluctuations to their energy spectra.

A particularly important case of application for these formulas occurs in the fluctuations associated with a relatively undamped collective mode. As we have noted in Section 3.2D, the dispersion relation to the collective modes in the density fluctuation excitations is given by the solution to Eq. (3.54). This equation determines a complex frequency: $\omega = \omega_k - i\gamma_k$. When $|\gamma_k/\omega_k| << 1$, Eq. (7.54) exhibits a peak structure representing the collective mode at $\omega = \omega_k$, so that

$$\langle | E^2 |(\mathbf{k},\omega)\rangle_{\text{coll}} \simeq \langle | E^2 |(\mathbf{k})\rangle_{\text{coll}} \delta(\omega - \omega_k) . \tag{7.56}$$

The energy density of the electrostatic fluctuations in the collective mode with the wave vector **k** is thus given by

$$\mathcal{E}_{\mathbf{k}} = \frac{\omega_k}{8\pi} \left(\frac{\partial \varepsilon_1}{\partial \omega} \right)_{\omega = \omega_k} \langle | E^2 |(\mathbf{k})\rangle_{\text{coll}} \tag{7.57}$$

where the dielectric response function is split into its real and imaginary parts as in Eq. (3.55). In an RPA plasma without an external magnetic field, one finds

$$\mathcal{E}_{\mathbf{k}} = - \left. \frac{\omega_k f(\mathbf{p})}{\mathbf{k} \cdot \partial f(\mathbf{p}) / \partial \mathbf{p}} \right|_{\omega_k = \mathbf{k} \cdot \mathbf{v}} \tag{7.58}$$

(cf., Problem 7.4).

7.3 DENSITY FLUCTUATION EXCITATIONS

Density-density correlations and fluctuations are described in the *dielectric formulation*, which expresses the static and dynamic structure factors in terms of the dielectric response functions. Strong-coupling effects beyond the RPA are represented via the local field corrections.

A. Structure Factors

We begin a mathematical formulation of the density fluctuation excitations in a uniform and stationary OCP by expressing the spatial Fourier components of the density fluctuations as

$$\rho_{\mathbf{k}}(t) = \int d\mathbf{r} \left\{ \sum_{j=1}^{N} \delta[\mathbf{r} - \mathbf{r}_j(t)] - n \right\} \exp(-i\mathbf{k} \cdot \mathbf{r})$$

$$= \sum_{j=1}^{N} \exp[-i\mathbf{k} \cdot \mathbf{r}_j(t)] - N\delta(\mathbf{k},0)\,, \qquad (7.59)$$

where $\delta(\mathbf{k},\mathbf{k}')$ is the three-dimensional Kronecker's delta, and n is the average number density. Written as a sum of extremely numerous terms with randomly varying phases, these Fourier components represent a set of random variables. Such random variables may be analyzed in terms of their spectral functions defined through the statistical averages.

A simple and nontrivial function describing a spectral distribution of the density fluctuations may be obtained from the mean square values of Eq. (7.59). The *static structure factor* is thus defined as

$$S(\mathbf{k}) = \frac{1}{N}\left\langle |\rho_{\mathbf{k}}(t)|^2 \right\rangle = \frac{1}{N}\left\langle \rho_{\mathbf{k}}(t)\rho_{-\mathbf{k}}(t) \right\rangle \qquad (7.60)$$

where the angular brackets denote a statistical-ensemble average. Because the system is stationary, Eq. (7.60) is a function of the wave vector only; the time argument may be suppressed there. The static structure factor represents a power spectrum of the density fluctuations in the **k** space.

For a system consisting of statistically uncorrelated particles, one finds

$$S_0(\mathbf{k}) = 1 - \delta(\mathbf{k},0)\,. \qquad (7.61)$$

This can be proved by rewriting Eq. (7.60) with the aid of Eq. (7.59) as

$$S(\mathbf{k}) = 1 - N\delta(\mathbf{k},0) + \frac{1}{N}\left\langle \sum_{j \neq j'}^{N} \exp[-i\mathbf{k} \cdot (\mathbf{r}_j - \mathbf{r}_{j'})] \right\rangle. \qquad (7.62)$$

When the positions of two different particles are uncorrelated, the last term in Eq. (7.62) yields $(N-1)\delta(\mathbf{k},0)$; hence Eq. (7.61).

The static structure factor is closely related to the pair correlation function. The *radial distribution function*, defined as

$$g(\mathbf{r}) = 1 + \frac{1}{N(N-1)}\left\langle \sum_{i \neq j}^{N} \delta[\mathbf{r} - \mathbf{r}_i(t) + \mathbf{r}_j(t)] \right\rangle \qquad (7.63)$$

(cf., Eq. 2.30b) is connected with the static structure factor via the Fourier transformation as

$$g(\mathbf{r}) = 1 + \frac{1}{n}\int \frac{d\mathbf{k}}{(2\pi)^3}[S(\mathbf{k}) - 1]\exp(i\mathbf{k}\cdot\mathbf{r}) . \tag{7.64}$$

The *interaction-energy density* is then calculated as

$$\begin{aligned} U_{\text{int}} &= 2\pi(Ze)^2 n \int \frac{d\mathbf{k}}{(2\pi)^3}\frac{1}{k^2}[S(\mathbf{k}) - 1] \\ &= \frac{(Zen)^2}{2}\int d\mathbf{r}\,\frac{1}{r}[g(\mathbf{r}) - 1] . \end{aligned} \tag{7.65}$$

The static structure factor introduced in the foregoing paragraphs does not exhaust all the information contained in the random variables, $\rho_{\mathbf{k}}(t)$. In particular, it does not describe the time-dependent behavior or the dynamical structure of those random variables. Such information may be retrieved in the *dynamic structure factor* defined as

$$S(\mathbf{k},\omega) = \frac{1}{2\pi V}\int_{-\infty}^{\infty} dt \langle \rho_{\mathbf{k}}(t' + t)\rho_{-\mathbf{k}}(t')\rangle \exp(i\omega t) . \tag{7.66}$$

Since the system under consideration is stationary, Eq. (7.66) is independent of t'. It follows from Eq. (7.66) that the dynamic structure factor is real and positive definite; this function satisfies a sum rule,

$$\frac{1}{n}\int_{-\infty}^{\infty} d\omega\, S(\mathbf{k},\omega) = S(\mathbf{k}) . \tag{7.67}$$

Physically, the dynamic structure factor represents the power spectrum of the density fluctuations in the frequency and wave vector space. One way of seeing this is to note that Eq. (7.66) is a version of the *Wiener-Khinchine theorem* in the theory of random noise. One likewise notes the relation

$$\int \frac{d\mathbf{k}}{(2\pi^3)}\int_{-\infty}^{\infty} d\omega\, S(\mathbf{k},\omega) = \left\langle \left\{ \sum_{j=1}^{N} \delta[\mathbf{r} - \mathbf{r}_j(t)] - n \right\}^2 \right\rangle , \tag{7.68}$$

an analogue of *Parseval's theorem*.

B. Dielectric Formulations

In the dielectric formulation, presented in Section 3.2, the states of plasmas have been analyzed through consideration of those effective interactions which renormalize the strong coupling effects arising from interparticle correlations.

The ideas of renormalized potentials and responses to external disturbances may be generalized to descriptions of the dynamic as well as static correlations in quantum as well as classical plasmas. The scheme to be considered in this subsection follows the density-response formalism or the dielectric formulation [Pines & Nozières 1966; Ichimaru, Iyetomi & Tanaka 1987]. For simplicity we begin with consideration of the electron OCP at finite temperatures (cf., Section 3.5); the system is characterized by the electric charge $-e$, mass m, number density n and temperature T.

A weak external potential field $\Phi_{\mathrm{ext}}(\mathbf{k},\omega)$ which varies spatially and temporally with the wave vector $\mathbf{k}$ and the frequency ω is applied to the plasma. The density fluctuation $\delta n(\mathbf{k},\omega)$ induced in the system is then expressed through a linear-response relation as [e.g., Ichimaru 1982; see Section 3.5B]

$$\delta n(\mathbf{k},\omega) = \chi_0(\mathbf{k},\omega)\left\{\Phi_{\mathrm{ext}}(\mathbf{k},\omega) + v(k)[1 - G(\mathbf{k},\omega)]\delta n(\mathbf{k},\omega)\right\}, \quad (7.69)$$

$$\chi_0(\mathbf{k},\omega) = 2\int \frac{d\mathbf{p}}{(2\pi\hbar)^3}\frac{f_0(\mathbf{p}) - f_0(\mathbf{p}+\hbar\mathbf{k})}{\hbar\omega + \varepsilon_{\mathbf{p}} - \varepsilon_{\mathbf{p}+\hbar\mathbf{k}} + i\eta}. \quad (7.70a)$$

Here $v(k) = 4\pi e^2/k^2$ and $\chi_0(\mathbf{k},\omega)$ is the retarded free-particle polarizability [Lindhard 1954] given by Eq. (3.169).

The function $G(\mathbf{k},\omega)$ introduced in Eq. (7.69) is the *dynamic local-field correction*. Physically it stems from the nonlinear coupling term (7.26c) of the fluctuations. If one sets $G(\mathbf{k},\omega) = 0$ in Eq. (7.69), then the RPA relations are recovered. The dynamic local-field correction thus describes the strong Coulomb-coupling effects between the fluctuations beyond RPA. Once $G(\mathbf{k},\omega)$ is given, the dielectric response function is defined (cf., Section 3.2A) and calculated as

$$\varepsilon(\mathbf{k},\omega) = 1 - \frac{v(k)\chi_0(\mathbf{k},\omega)}{1 + v(k)G(\mathbf{k},\omega)\chi_0(\mathbf{k},\omega)}. \quad (7.70b)$$

The dynamic structure factor is then evaluated as in Eq. (3.64) with the aid of the *fluctuation-dissipation theorem* [Callen & Welton 1951; Kubo 1957; see Eq. 3.64].

In some cases of dense-plasma problems, one adopts an approximation whereby the dynamic local-field correction is replaced by its static value,

$$G(\mathbf{k}) \equiv G(\mathbf{k},\omega = 0)\,, \tag{7.71}$$

ignoring the frequency dependence in the local field correction. Such will be called the *static local-field correction* approximation.

The static structure factor is calculated from Eq. (3.64) in accord with Eq. (7.67). In the classical limit ($\hbar \to 0$), one then finds (cf., Eq. 3.66)

$$S(\mathbf{k}) = \frac{k^2}{k^2 + k_{\mathrm{D}}^2[1 - G(\mathbf{k})]}\,. \tag{7.72}$$

This equation has been shown equivalent to the *Ornstein-Zernike relation* [e.g., Hansen & McDonald 1986; Ichimaru, Iyetomi & Tanaka 1987] in the theory of classical liquids. The static local-field correction approximation thus leads to a correct description of the static correlations and thermodynamic functions in classical systems.

Once the quantum effects start to play a role, the static local-field correction approximation ceases to provide a rigorous theoretical description for those static properties. We remark, however, that a number of theoretical schemes [e.g., Ichimaru, Iyetomi & Tanaka 1987; see also Section 3.5E] based on such a static approximation have succeeded in predicting or reproducing the values of the correlation energy close to those obtained in the GFMC method [Ceperley & Alder 1980] for degenerate electron liquids at metallic densities.

The extension of this density-response formalism to a multicomponent system (electric charge $Z_\sigma e$, mass m_σ, number density n_σ, and temperature T, where σ refers to the particle species) is straightforward [Vashishta, Bhattacharyya & Singwi 1974; Ichimaru, Mitake, Tanaka & Yan 1985]. We apply to the system a weak (fictitious) external potential field $\Phi_\sigma^{(e)}(\mathbf{k},\omega)$ which couples only to the density field $n_\sigma(\mathbf{k},\omega)$ of the σ-species particles. The induced density fluctuations $\delta n_\sigma(\mathbf{k},\omega)$ are then expressed as

$$\delta n_\sigma(\mathbf{k},\omega) = \sum_\tau \chi_{\sigma\tau}(\mathbf{k},\omega)\Phi_\tau^{(e)}(\mathbf{k},\omega)\,, \tag{7.73}$$

which define the density-density response functions $\chi_{\sigma\tau}(\mathbf{k},\omega)$ between the σ and τ species of particles. The dielectric response function is thus given by

$$\frac{1}{\varepsilon(\mathbf{k},\omega)} = 1 + v(k)\sum_{\sigma,\tau} Z_\sigma Z_\tau \chi_{\sigma\tau}(\mathbf{k},\omega)\,. \tag{7.74}$$

The linear-response relation (7.69) is rewritten in this case as

$$\delta n_\sigma(\mathbf{k},\omega) = \chi_\sigma^{(0)}(\mathbf{k},\omega)$$
$$\times \left\{ \Phi_\sigma^{(e)}(\mathbf{k},\omega) + v(k) \sum_\tau Z_\sigma Z_\tau [1 - G_{\sigma\tau}(\mathbf{k},\omega)] \delta n_\tau(\mathbf{k},\omega) \right\}, \tag{7.75}$$

where $\chi_\sigma^{(0)}(\mathbf{k},\omega)$ is the free-particle polarizability, Eq. (7.70a), of the σ-species particles and $G_{\sigma\tau}(\mathbf{k},\omega)$ refer to the dynamic local-field corrections between the σ and τ species of particles. Comparison between Eqs. (7.73) and (7.75) yields the expressions for $\chi_{\sigma\tau}(\mathbf{k},\omega)$. For simplicity in notation, we write the results for a two-component plasma (TCP with $\sigma,\tau = 1,2$), suppressing the frequency and wave-vector arguments:

$$\chi_{11} = \chi_1^{(0)} \left\{ 1 - Z_2^2 v \chi_2^{(0)} [1 - G_{22}] \right\} / D \tag{7.76a}$$

$$\chi_{22} = \chi_2^{(0)} \left\{ 1 - Z_1^2 v \chi_1^{(0)} [1 - G_{11}] \right\} / D \tag{7.76b}$$

$$\chi_{12} = Z_1 Z_2 v \chi_1^{(0)} \chi_2^{(0)} [1 - G_{12}] / D \tag{7.76c}$$

$$\chi_{21} = Z_1 Z_2 v \chi_1^{(0)} \chi_2^{(0)} [1 - G_{21}] / D \tag{7.76d}$$

where

$$D = \left\{ 1 - Z_1^2 v \chi_1^{(0)} [1 - G_{11}] \right\} \left\{ 1 - Z_2^{2v} \chi_2^{(0)} [1 - G_{22}] \right\}$$
$$- Z_1^2 Z_2^2 v^2 \chi_1^{(0)} \chi_2^{(2)} [1 - G_{12}][1 - G_{21}] . \tag{7.77}$$

The dynamic structure factors $S_{\sigma\tau}(\mathbf{k},\omega)$, the static structure factors $S_{\sigma\tau}(k)$, and the radial distribution functions $g_{\sigma\tau}(r)$ are given by

$$S_{\sigma\tau}(\mathbf{k},\omega) = -\frac{\hbar}{2\pi} \coth\left(\frac{\hbar\omega}{2T} \right) \mathrm{Im}\chi_{\sigma\tau}(\mathbf{k},\omega) , \tag{7.78}$$

$$S_{\sigma\tau}(\mathbf{k}) = \frac{1}{\sqrt{n_\sigma n_\tau}} \int_{-\infty}^{\infty} d\omega \, S_{\sigma\tau}(\mathbf{k},\omega) , \tag{7.79}$$

$$g_{\sigma\tau}(\mathbf{r}) = 1 + \frac{1}{\sqrt{n_\sigma n_\tau}} \int \frac{d\mathbf{k}}{(2\pi)^3} [S_{\sigma\tau}(\mathbf{k}) - \delta_{\sigma\tau}] \exp(i\mathbf{k} \cdot \mathbf{r}) , \tag{7.80}$$

where $\delta_{\sigma\tau}$ represents Kronecker's delta.

C. RPA Structure Factors in Nonequilibrium Plasmas

In Section 7.2C, we noted that a superposition principle of the dressed test particles is applicable to the RPA calculation of fluctuations in weakly coupled plasmas; the method is valid even under nonequilibrium circumstances.

Following the superposition principle, we thus evaluate the dynamic structure factor in the OCP, by picking each particle in a plasma and regarding it as a test charge. Suppose that such a particle is located initially at $\mathbf{r}_j$ and has a momentum $\mathbf{p}_j$. In the absence of collisions, the Fourier components of its density field are

$$\rho_j^{(0)}(\mathbf{k},\omega) = 2\pi \exp(-i\mathbf{k}\cdot\mathbf{r}_j)\delta(\omega - \mathbf{k}\cdot\mathbf{v}_j)\,. \tag{7.81}$$

The field of the dressed particle can be constructed by taking additional account of the induced screening cloud; hence

$$\rho_j^{(S)}(\mathbf{k},\omega) = \frac{\rho_j^{(0)}(\mathbf{k},\omega)}{\varepsilon(\mathbf{k},\omega)}\,. \tag{7.82}$$

A dressed particle, therefore, consists of a bare particle field Eq. (7.81) and its induced screening cloud. The total density fields resulting from superposition of those dressed particles are

$$\begin{aligned}\rho^{(S)}(\mathbf{k},\omega) &= \sum_{j=1}^{N} \rho_j^{(S)}(\mathbf{k},\omega)\\ &= \frac{2\pi}{\varepsilon(\mathbf{k},\omega)}\sum_{j=1}^{N} \exp(-i\mathbf{k}\cdot\mathbf{r}_j)\delta(\omega - \mathbf{k}\cdot\mathbf{v}_j)\,.\end{aligned} \tag{7.83}$$

We may now use the expression (7.83) for the calculation of the dynamic structure factor in accord with Eq. (7.32). In these processes, correlations between two different dressed particles may be neglected; the major effects of Coulomb-induced correlations have already been taken into account in the structure of each dressed particle Eq. (7.82). We thus have

$$\begin{aligned}S(\mathbf{k},\omega)\delta(\omega-\omega') &= \frac{1}{(2\pi)^2 V}\left\langle \rho^{(S)}(\mathbf{k},\omega)\rho^{(S)}(-\mathbf{k},-\omega')\right\rangle\\ &= n\frac{\langle\delta(\omega - \mathbf{k}\cdot\mathbf{v})\rangle}{|\,\varepsilon(\mathbf{k},\omega)\,|^2}\,\delta(\omega-\omega')\,.\end{aligned}$$

Introducing the single-particle momentum distribution $F(\mathbf{p})$ to carry out the required average, we finally obtain

$$S(\mathbf{k},\omega)=\frac{S^{(0)}(\mathbf{k},\omega)}{|\,\varepsilon(\mathbf{k},\omega)\,|^2}, \tag{7.84}$$

where

$$S^{(0)}(\mathbf{k},\omega)=\int d\mathbf{p}F(\mathbf{p})\delta(\omega-\mathbf{k}\cdot\mathbf{v}) \tag{7.85}$$

is the dynamic structure factor for the noninteracting particles. For a Maxwellian plasma with the distribution Eq. (1.1), Eq. (7.85) can also be expressed as

$$S^{(0)}(\mathbf{k},\omega)=\frac{n}{k}\sqrt{\frac{m}{2\pi T}}\exp\left(-\frac{m\omega^2}{2Tk^2}\right) \tag{7.86}$$

in the RPA.

The dynamic structure factor Eq. (7.84) clearly demonstrates both the collective and individual-particle aspects of fluctuations in the plasma (cf., Section 1.4B). As we have just observed, the spectral function $S^{(0)}(\mathbf{k},\omega)$ stems from the superposition of self-motion of the individual particles, and thereby represents the fluctuations associated with motion of the individual particles (cf., Fig. 1.5).

The collective mode, characterized by the dispersion relation Eq. (3.54), greatly affects the shape and strength of the spectral function. Generally, it gives rise to a sharp peak in the frequency spectrum, as illustrated in Fig. 1.4. Integration across such a resonance pole should then yield a measure of the total energy contained in fluctuations associated with the given collective mode, as we have treated in Section 7.2E.

Such an integration may be carried out through the method described in Section 3.2D. We first note in Eq. (7.84) that the dynamic structure factor is generally written in a form proportional to $|\,\varepsilon(\mathbf{k},\omega)\,|^{-2}$, which we rewrite

$$\frac{1}{|\,\varepsilon(\mathbf{k},\omega)\,|^2}=-\frac{1}{\mathrm{Im}\varepsilon(\mathbf{k},\omega)}\mathrm{Im}\frac{1}{\varepsilon(\mathbf{k},\omega)}.$$

In the vicinity of a weakly damped collective mode $\omega=\omega_{\mathbf{k}}$, we may use an approximate expression

$$-\mathrm{Im}\frac{1}{\varepsilon(\mathbf{k},\omega)}\simeq\pi\left(\frac{\partial\varepsilon_1}{\partial\omega_{\mathbf{k}}}\right)^{-1}\delta(\omega-\omega_{\mathbf{k}}) \tag{7.87}$$

where

$$\frac{\partial \varepsilon_1}{\partial \omega_{\mathbf{k}}} = \left[\frac{\partial \varepsilon_1(\mathbf{k},\omega)}{\partial \omega} \right]_{\omega=\omega_{\mathbf{k}}} . \tag{7.88}$$

Since the decay rate γ_k of the collective mode is given by

$$\gamma_k = \left(\frac{\partial \varepsilon_1}{\partial \omega_{\mathbf{k}}} \right)^{-1} \mathrm{Im}\,\varepsilon(\mathbf{k},\omega_{\mathbf{k}}) , \tag{7.89}$$

we obtain

$$\frac{1}{|\varepsilon(\mathbf{k},\omega)|^2} = \frac{\pi}{\gamma_k} \left(\frac{\partial \varepsilon_1}{\partial \omega_{\mathbf{k}}} \right)^{-2} \delta(\omega - \omega_{\mathbf{k}}) . \tag{7.90}$$

That part of the dynamic structure factor Eq. (7.84) which results from the collective mode is thus expressed as

$$S_{\mathrm{coll}}(\mathbf{k},\omega) = \frac{\pi}{\gamma_k} \left(\frac{\partial \varepsilon_1}{\partial \omega_{\mathbf{k}}} \right)^{-2} S^{(0)}(\mathbf{k},\omega_{\mathbf{k}})\,\delta(\omega - \omega_{\mathbf{k}}) . \tag{7.91}$$

Integration of this spectrum across ω_k then yields the static structure factor associated with this particular collective mode:

$$S_{\mathrm{coll}}(\mathbf{k}) = \frac{\pi}{n\gamma_k} \left(\frac{\partial \varepsilon_1}{\partial \omega_{\mathbf{k}}} \right)^{-2} S^{(0)}(\mathbf{k},\omega_{\mathbf{k}}) . \tag{7.92}$$

The mean square value of density fluctuations in a given collective mode is therefore proportional to $1/\gamma_k$; the fluctuations would increase as the decay rate decreases.

In Eq. (7.92), we thus observe the possibility that the level of the fluctuations in the collective mode grows enormously as the decay rate approaches zero. Such may be looked upon as the critical fluctuations associated with the onset of a plasma-wave instability, quite analogous to the critical opalescence in the vicinity of the critical point for liquid-gas phase transition. A concrete example of such a *plasma critical fluctuation* at the onset of the ion-acoustic wave instability (cf., Section 6.2A) has been considered [Ichimaru, Pines, & Rostoker 1962].

For a Maxwellian plasma, we calculate the contribution of the two plasma-wave solutions $\omega = \pm\omega_p$ from Eq. (7.92) as

$$S_{\mathrm{pl}}(\mathbf{k}) = \frac{k^2}{k_{\mathrm{D}}^2} \qquad (k^2 << k_{\mathrm{D}}^2)\,. \tag{7.93}$$

The density fluctuations associated with the plasma oscillation thus exhaust the entire strength of the RPA structure factor, Eq. (7.72) with $G(\mathbf{k}) = 0$, in the limit of long wavelengths.

Before closing this subsection, let us briefly describe an application of the superposition principle to a TCP. We denote by $\rho_1^{(0)}(\mathbf{k},\omega)$ and $\rho_2^{(0)}(\mathbf{k},\omega)$ the density fields associated with the bare electrons and ions, respectively. The fields of the dressed electrons and ions are then calculated in the RPA as

$$\rho_1^{(S)}(\mathbf{k},\omega) = \left[1 - Z_1^2 v(k)\frac{\chi_1^{(0)}(\mathbf{k},\omega)}{\varepsilon(\mathbf{k},\omega)}\right]\rho_1^{(0)}(\mathbf{k},\omega)$$

$$+ Z_1 Z_2 v(k)\frac{\chi_2^{(0)}(\mathbf{k},\omega)}{\varepsilon(\mathbf{k},\omega)}\rho_1^{(0)}(\mathbf{k},\omega)\,, \tag{7.94a}$$

$$\rho_2^{(S)}(\mathbf{k},\omega) = Z_1 Z_2 v(k)\frac{\chi_1^{(0)}(\mathbf{k},\omega)}{\varepsilon(\mathbf{k},\omega)}\rho_2^{(0)}(\mathbf{k},\omega)$$

$$+\left[1 - Z_2^2 v(k)\frac{\chi_2^{(0)}(\mathbf{k},\omega)}{\varepsilon(\mathbf{k},\omega)}\right]\rho_2^{(0)}(\mathbf{k},\omega)\,, \tag{7.94b}$$

where $\chi_1^{(0)}(\mathbf{k},\omega)$ and $\chi_2^{(0)}(\mathbf{k},\omega)$ are the free-particle polarizabilities of the electrons and ions. The first terms on the right-hand sides of Eqs. (7.94) consist of the electron density fluctuations, while the second terms consist of the ion density fluctuations. Such a mixing of the individual electron and ion coordinates to form a correlated screening cloud for a test charge is a feature characteristic of the correlations in a TCP. Because of such a mixing, the electron density fluctuations contained in the dressed ions Eq. (7.94b), for example, may have to be regarded as moving with the ion velocity.

The dynamic structure factor of the electrons may now be calculated by superposing the first terms of Eqs. (7.94). Neglecting the correlations between the dressed particles, we thus find

$$S_{11}(\mathbf{k},\omega) = S_1^{(0)}(\mathbf{k},\omega)\left[1 - Z_1^2 v(k)\frac{\chi_1^{(0)}(\mathbf{k},\omega)}{\varepsilon(\mathbf{k},\omega)}\right]^2$$

$$+ S_2^{(0)}(\mathbf{k},\omega)\left[Z_1 Z_2 v(k)\frac{\chi_1^{(0)}(\mathbf{k},\omega)}{\varepsilon(\mathbf{k},\omega)}\right]^2, \tag{7.95}$$

where $S^{(0)}_{1(2)}(\mathbf{k},\omega)$ is the dynamic form factor of a noninteracting electron (ion) system defined analogously to Eq. (7.94). Again we remark that the last term exhibits the influence of the ion momentum distribution on the frequency spectrum of the electron density fluctuations.

D. Static Local-Field Corrections

In the polarization potential approach to the theory of condensed plasmas, as exemplified in Eq. (7.69), the effective interaction potentials arising from fluctuating fields play an important part. Such a potential, describing effective interaction between particles, differs from the bare potential of binary interaction owing to many-body correlations in the system. The difference between the effective and bare potentials is referred to as the local field correction. This function, therefore, measures the extent to which the many-body correlational effects are involved in the description of plasma properties.

In this subsection, we formulate the static local-field corrections produced by Coulomb correlations in a strongly coupled, classical OCP. Since the local field corrections are directly related to the static structure factors via Eq. (7.72), those are the essential quantities in such theoretical evaluation of the thermodynamic quantities as internal energies and compressibilities (cf., Sections 3.1C, 3.2E, and 7.3B).

A number of approximation schemes have been advanced for the calculation of the local field corrections in a strongly coupled plasma. Notable among them are those proposed by Singwi and his collaborators [Singwi et al. 1968] and by Ichimaru and Totsuji [Ichimaru 1970; Totsuji & Ichimaru 1973, 1974], called the STLS scheme and the CA scheme, respectively; the latter employs the *convolution approximation* (CA) to the treatment of the triple correlation functions [e.g., Hansen & McDonald 1986; Ichimaru, Iyetomi & Tanaka 1987].

The static local-field correction $G(k)$ can be expressed in terms of the pair and triple correlation functions without resort to a perturbation calculation. A rigorous formulation of the local field corrections in a classical OCP (with $Z = 1$) may be obtained through examination of the low-frequency behavior of the Klimontovich distribution,

$$\rho_{\mathbf{k}}(\mathbf{p}) = \sum_{j=1}^{N} \delta(\mathbf{p} - \mathbf{p}_j) \exp(-i\mathbf{k} \cdot \mathbf{r}_j) ,$$

as follows [Tago, Utsumi & Ichimaru 1981]:

We first note that the distribution obeys the equation of motion,

$$\left(\frac{\partial}{\partial t}+i\frac{\mathbf{k}\cdot\mathbf{p}}{m}\right)\rho_{\mathbf{k}}(\mathbf{p})=i\frac{v(k)}{V}\mathbf{k}\cdot\frac{\partial}{\partial\mathbf{p}}\rho_0(\mathbf{p})\rho_{\mathbf{k}}$$
$$+i\sum_{\mathbf{q}}{}'\frac{v(q)}{V}\mathbf{q}\cdot\frac{\partial}{\partial\mathbf{p}}\rho_{\mathbf{k}-q}(\mathbf{p})\,\rho_{\mathbf{q}} \tag{7.96}$$

where

$$\rho_{\mathbf{k}}=\sum_{j=1}^{N}\exp\left(-i\mathbf{k}\cdot\mathbf{r}_j\right).$$

Multiplying Eq. (7.96) by $(-i)\mathbf{k}\cdot\mathbf{p}/m$ and integrating the result over $\mathbf{p}$, we find

$$\frac{\partial^2}{\partial t^2}\rho_{\mathbf{k}}-\frac{v(k)}{V}\int d\mathbf{p}\frac{\mathbf{k}\cdot\mathbf{p}}{m}\mathbf{k}\cdot\frac{\partial}{\partial\mathbf{p}}\rho_0(\mathbf{p})\rho_{\mathbf{k}}+\int d\mathbf{p}\left(\frac{\mathbf{k}\cdot\mathbf{p}}{m}\right)^2\rho_{\mathbf{k}}(\mathbf{p})$$
$$=-\frac{\omega_p^2}{N}\sum_{\mathbf{q}}{}'\frac{\mathbf{k}\cdot\mathbf{q}}{k^2}\rho_{\mathbf{k}-q}\rho_{\mathbf{q}}, \tag{7.97}$$

with the aid of a continuity equation,

$$\frac{\partial}{\partial t}\rho_{\mathbf{k}}=-i\int d\mathbf{p}\frac{\mathbf{k}\cdot\mathbf{p}}{m}\rho_{\mathbf{k}}(\mathbf{p}).$$

It should be remarked that in Eq. (7.97) the product $\rho_{\mathbf{k}-q}\rho_{\mathbf{q}}$ is given by

$$\rho_{\mathbf{k}-q}\rho_{\mathbf{q}}=\sum_{j\neq l}^{N}\exp\left[-i(\mathbf{k}-\mathbf{q})\cdot\mathbf{r}_j\right]\exp\left(-i\mathbf{q}\cdot\mathbf{r}_l\right) \tag{7.98}$$

and does not contain the self-interaction terms, $j=l$, in the summation. Observe also a similarity between Eqs. (7.97) and (1.46).

The right-hand side of Eq. (7.97) represents those nonlinear mode-coupling terms beyond the RPA, which lead to the local-field correction effects in Eq. (7.91). In the low-frequency limit, it may thus be expanded as

$$\omega_p^2G(k)\rho_{\mathbf{k}}-\frac{1}{\tau(k)}\frac{\partial}{\partial t}\rho_{\mathbf{k}}$$

where $1/\tau(k)$ refers to the relaxation rate of the density fluctuation excitations. Hence

$$G(k)\rho_{\mathbf{k}} - \frac{1}{\omega_p^2 \tau(k)} \frac{\partial}{\partial t}\rho_{\mathbf{k}} = -\frac{1}{2N} \sum_{\mathbf{q}}{}' K(\mathbf{k},\mathbf{q})\rho_{\mathbf{k}-\mathbf{q}}\rho_{\mathbf{q}} \tag{7.99}$$

with

$$K(\mathbf{k},\mathbf{q}) = \frac{\mathbf{k}\cdot\mathbf{q}}{q^2} + \frac{\mathbf{k}\cdot(\mathbf{k}-q)}{|\mathbf{k}-q|^2} \tag{7.100}$$

representing a symmetrized Coulomb interaction.

A statistical average of the product between Eq. (7.99) and $\rho_{-\mathbf{k}}$ yields

$$G(k) = -\frac{1}{2N} \sum_{\mathbf{q}}{}' K(\mathbf{k},\mathbf{q}) \frac{\left\langle \rho_{\mathbf{k}-\mathbf{q}}\rho_{\mathbf{q}}\rho_{-\mathbf{k}} \right\rangle}{NS(k)} \tag{7.101}$$

since $\langle \rho_{-\mathbf{k}}(\partial/\partial t)\rho_{\mathbf{k}} \rangle = 0$ for a stationary system. The statistical average in Eq. (7.101) can be expressed in terms of the pair and triple correlation functions, to yield

$$G(k) = -\frac{1}{NS(k)} \sum_{\mathbf{q}}{}' K(\mathbf{k},\mathbf{q}) \left[S(|\mathbf{k}-q|) - 1 + \frac{1}{2} h_3(\mathbf{k}-q,\mathbf{q}) \right]. \tag{7.102}$$

Here $h_3(\mathbf{k}_1,\mathbf{k}_2)$ is the Fourier transform of the triple correlation function $h_3(\mathbf{r}_{12},\mathbf{r}_{23},\mathbf{r}_{31})$ calculated as

$$h_3(\mathbf{k}_1,\mathbf{k}_2) = n^2 \int d\mathbf{r}_{13} \int d\mathbf{r}_{23} h_3(\mathbf{r}_{12},\mathbf{r}_{23},\mathbf{r}_{31}) \exp\left[-i(\mathbf{k}_1\cdot\mathbf{r}_{13} + \mathbf{k}_2\cdot\mathbf{r}_{23})\right] \tag{7.103}$$

with $\mathbf{r}_{12} = \mathbf{r}_1 - \mathbf{r}_2$, etc. The triple correlation function in turn is expressed in terms of the two- and three-particle radial distribution functions, $g(r)$ and $g_3(r_{12},r_{23},r_{31})$ defined in Eqs. (2.30b) and (2.30c), as

$$h_3(\mathbf{r}_{12},\mathbf{r}_{23},\mathbf{r}_{31}) = g_3(\mathbf{r}_{12},\mathbf{r}_{23},\mathbf{r}_{31}) - g(\mathbf{r}_{12}) - g(\mathbf{r}_{23}) - g(\mathbf{r}_{31}) + 2\,. \tag{7.104}$$

Equation (7.102) gives an exact expression for the local field correction in terms of the pair and triple correlation functions. With a little algebra it is in fact possible to show that a combination of Eq. (7.102) with the fluctuation-dissipation theorem Eq. (7.72) amounts to the Born-Green equation (2.32) between the two- and three-particle radial distribution functions (Problem 7.8).

To proceed further, one adopts an approximation to the triple correlation function in Eq. (7.102); such a treatment leads to an approximate expression for $G(k)$. Truncation of the hierarchical structure of the correlation is thereby completed. Combination of such a result with Eq. (7.72) makes a set of self-consistent equations to determine $S(k)$ and $G(k)$.

An ansatz based on the CA scheme [Ichimaru 1970] reads

$$h_3(\mathbf{k}_1,\mathbf{k}_2) = S(\mathbf{k}_1)S(\mathbf{k}_2)S(|\,\mathbf{k}_1+\mathbf{k}_2\,|) - S(\mathbf{k}_1) - S(\mathbf{k}_2) - S(|\,\mathbf{k}_1+\mathbf{k}_2\,|) + 2\,. \quad (7.105)$$

Substitution of this expression in Eq. (7.103) yields

$$G^{\mathrm{CA}}(k) = -\frac{1}{N}\sum_{\mathbf{q}}{}' K(\mathbf{k},\mathbf{q})\frac{1+S(q)}{2}[S(|\,\mathbf{k}-\mathbf{q}\,|)-1]\,. \quad (7.106)$$

This is the local field correction in the CA scheme; numerical consequences of this scheme have been investigated [Totsuji & Ichimaru 1973, 1974].

Equation (7.106) takes a form nonlinear with respect to the structure factor $S(k)$, which is an unknown function for the iterative solution to the resultant self-consistent equations. Such a nonlinear structure in the local field corrections may impair the numerical accuracy of the solution when the integrations required are carried through. Besides, it is instructive to explore physical contents in various expressions for the local field correction.

For such a purpose, it is useful to express the local field corrections in a general form as

$$G(k) = -\frac{1}{N}\sum_{\mathbf{q}}{}' K(\mathbf{k},\mathbf{q})R(q)[S(|\,\mathbf{k}-\mathbf{q}\,|)-1]\,. \quad (7.107)$$

The function $R(q)$ introduced here may be interpreted as a screening factor to the generalized Coulomb interaction $K(\mathbf{k},\mathbf{q})$ in the local field correction.

In the STLS scheme (Section 3.5E), $R(q)$ is taken to be unity, so that

$$G^{\mathrm{STLS}}(k) = -\frac{1}{N}\sum_{\mathbf{q}}{}' \frac{\mathbf{k}\cdot\mathbf{q}}{q^2}[S(|\,\mathbf{k}-q\,|)-1] \quad (7.108)$$

(cf., Problem 3.23). The STLS local-field correction thus ignores the screening effect and is thereby linearized in the structure factor. The STLS scheme has been used extensively in analysis of strongly coupled classical and quantum plasmas [Singwi et al. 1968; Singwi & Tosi 1981].

An approximation scheme which retains the features of the CA scheme Eq. (7.107) as well as attains such a linearization is offered by the *modified convolution approximation* (MCA) [Tago, Utsumi & Ichimaru 1981]. Its screening function in Eq. (7.107) takes the form

$$R^{\mathrm{MCA}}(q) = \frac{1}{2}[S^{\mathrm{IS}}(q)-1] \quad (7.109)$$

where

$$S^{\mathrm{IS}}(q) = \frac{k^2}{k^2 + k_{\mathrm{IS}}^2} \tag{7.110}$$

with the parameter k_{IS} determined from a self-consistency condition

$$\int_0^\infty dk[S^{\mathrm{IS}}(k) - 1] = \int_0^\infty dk[S(k) - 1] \tag{7.111a}$$

or equivalently, in light of Eq. (7.65),

$$k_{\mathrm{IS}} = -\frac{2}{a\Gamma}\frac{U_{\mathrm{int}}}{nT}. \tag{7.111b}$$

The MCA scheme has also been used extensively in analysis of various strongly coupled plasmas including multicomponent systems [Yan & Ichimaru 1987; Tanaka & Ichimaru 1989; Tanaka, Yan & Ichimaru 1990].

Finally, we remark that the *hypernetted-chain (HNC) approximation* scheme in the theory of classical liquids [e.g., Hansen & McDonald 1986; Ichimaru, Iyetomi & Tanaka 1987] can be expressed in terms of the HNC local-field correction

$$G^{\mathrm{HNC}}(k) = -\frac{1}{N}\sum_{\mathbf{q}}{}' \frac{\mathbf{k}\cdot\mathbf{q}}{q^2}[S(|\mathbf{k}-\mathbf{q}|) - 1] \times\left\{1 - [G^{\mathrm{HNC}}(q) - 1][S(q) - 1]\right\} \tag{7.112}$$

[Choquard 1978]. It is observed here that the STLS scheme may be recovered if $G^{\mathrm{HNC}}(q)$ is set equal to unity on the right-hand side of Eq. (7.112), while the CA scheme is obtained if $G^{\mathrm{HNC}}(q) = 0$ there. These schemes therefore represent two extreme cases of approximation oppositely directed from the point of view of the HNC approximation.

E. Fully Convergent Collision Terms

The dynamic and transport properties of plasmas are described in terms of the kinetic equations (7.14) which govern the evolution of the single-particle distribution functions in phase space. The right-hand side of Eq. (7.14) represents the collision term of the kinetic equation, $\partial F_\sigma/\partial t]_c$. By its construction the collision term conserves the particle number, the momentum, and the internal energy.

In formulating the collision term by fluctuations, the electrostatic Coulomb interactions only are assumed between the charged particles, so that the Fourier components of the electromagnetic fluctuations are given by

$$\delta \mathbf{B}(\mathbf{k},\omega)=0 \tag{7.113a}$$

$$\delta \mathbf{E}(\mathbf{k},\omega)=-i\mathbf{k}\Phi(\mathbf{k},\omega) \tag{7.113b}$$

with $\Phi(\mathbf{k},\omega)$ denoting the potential fluctuations. The collision term is then rewritten for the OCP as

$$\left.\frac{\partial F(\mathbf{p})}{\partial t}\right]_c = -iZe\int\frac{d\mathbf{k}}{(2\pi)^3}\int_{-\infty}^{\infty} d\omega\,\mathbf{k}\cdot\frac{\partial}{\partial \mathbf{p}}\left\langle N\Phi^*(\mathbf{k},\omega;\mathbf{p})\right\rangle \tag{7.114}$$

in the spectral notation of Eq. (7.30).

The microscopic density fluctuations $\delta N(\mathbf{k},\omega;\mathbf{p})$ may be separated into two parts, spontaneous and induced fluctuations, as

$$\delta N(\mathbf{k},\omega;\mathbf{p}) = \delta N^{(0)}(\mathbf{k},\omega;\mathbf{p}) + \delta N^{(S)}(\mathbf{k},\omega;\mathbf{p}) \tag{7.115}$$

(cf., Section 7.1B). In the static local-field correction approximation, the induced part may be expressed as

$$\delta N^{(S)}(\mathbf{k},\omega;\mathbf{p}) = -Ze\mathbf{k}\cdot\frac{\partial F(\mathbf{p})}{\partial\,\mathbf{p}}\,\frac{\Phi(\mathbf{k},\omega)}{\omega-\mathbf{k}\cdot\mathbf{v}+i\eta}\,[1-G(k)]\,. \tag{7.116}$$

In this expression the strong Coulomb-coupling effects between density fluctuations have been taken into account through the static local-field correction $G(k)$.

The fluctuating potential fields are connected with the density fluctuations via the Poisson equation. With the aid of Eqs. (7.115) and (7.116), one thus finds

$$\Phi(\mathbf{k},\omega)=\frac{V(k)}{Ze}\,\frac{\delta n^{(0)}(\mathbf{k},\omega)}{\tilde{\varepsilon}\,(\mathbf{k},\omega)}\,. \tag{7.117}$$

Here

$$V(k)=Z^2 v(k)=\frac{4\pi(Ze)^2}{k^2}\,, \tag{7.118}$$

$$\tilde{\varepsilon}(\mathbf{k},\omega) = 1 - V(k)[1 - G(k)]\chi_0(\mathbf{k},\omega)\,, \tag{7.119}$$

$$\chi_0(\mathbf{k},\omega) = -\int d\mathbf{p}\,\frac{1}{\omega - \mathbf{k}\cdot\mathbf{v} + i\eta}\,\mathbf{k}\cdot\frac{\partial F(\mathbf{p})}{\partial \mathbf{p}}\,. \tag{7.120}$$

The collision term Eq. (7.114) is thus calculated as

$$\left.\frac{\partial F(\mathbf{p})}{\partial t}\right]_c = \pi\int\frac{d\mathbf{k}}{(2\pi)^3}\mathbf{k}\cdot\frac{\partial}{\partial \mathbf{p}}\int d\mathbf{p}'\frac{V(k)^2[1 - G(k)]}{|\tilde{\varepsilon}(\mathbf{k},\omega)|^2}\delta(\mathbf{k}\cdot\mathbf{v} - \mathbf{k}\cdot\mathbf{v}')$$

$$\times\,\mathbf{k}\cdot\left(\frac{\partial}{\partial \mathbf{p}} - \frac{\partial}{\partial \mathbf{p}'}\right)F(\mathbf{p})F(\mathbf{p}')\,. \tag{7.121}$$

This collision term differs from the BGL term Eq. (7.49) in that the binary interaction is renormalized by a factor, $1 - G(k)$; such a renormalization affects one of the $V(k)$s in the numerator and the screening function (7.119) in the denominator. This renormalization factor thus accounts for those strong correlation effects beyond the Born approximation between the scattering particles. Higher-order scattering processes are included through this factor since a partial summation of the fluctuation terms to all the orders has been carried out in Eq. (7.116) [Ichimaru 1977a; Wallenborn & Baus 1978].

Generally for a classical plasma, $G(k)$ approaches unity in the limit of large k (Problem 7.9); the factor $1 - G(k)$ thus vanishes in this limit. Physically this stems from the strong Coulomb repulsion at short distances, which acts to create a Coulomb hole in the radial distribution function. The **k** integration in Eq. (7.121) is now convergent in the large-k domain because $1 - G(k)$ vanishes there. Such a feature is absent in the BGL collision term, where one usually introduces a cutoff to the upper bound of the **k** integration around the inverse of the Landau parameter, Eq. (2.71), to achieve convergence. The function $G(k)$ takes care of such short-range effects in a natural way for Eq. (7.149).

As in the case of the BGL collision term, the **k** integration in Eq. (7.121) converges in the small-k domain owing to the dielectric screening. The collision term (7.121) is thus convergent both in the large- and small-k limits of integration; it automatically unifies the treatments on the long- and short-range Coulomb collisions. A number of investigators [Hubbard 1961b; Kihara & Aono 1963; Gould & DeWitt 1967] proposed such a unification scheme to Coulomb-collision theory in the limit of weak coupling by introducing an artificial splitting of the integration into two separate parts. The collision term Eq. (7.121) leads naturally to convergent expressions for transport coefficients, and is applicable for plasmas in intermediate coupling as well as in weak coupling.

In the strong coupling regime ($\Gamma > 1$), the approximation (7.116) for the induced fluctuations is not justified. The strong and dynamic interparticle correlations will produce dissipative imaginary parts other than those in the free-particlelike contributions $(\omega - \mathbf{k}\cdot\mathbf{v} + i\eta)^{-1}$ on the right-hand side of Eq. (7.116). A straightforward extension of Eq. (7.116) is to replace $G(k)$ by the dynamic local-field correction $G(\mathbf{k},\omega)$, through which one can take account of the relaxation effects arising from the interparticle correlations:

$$\delta N^{(S)}(\mathbf{k},\omega;\mathbf{p}) = -Ze\mathbf{k}\cdot\frac{\partial F(\mathbf{p})}{\partial \mathbf{p}}\frac{\Phi(\mathbf{k},\omega)}{\omega - \mathbf{k}\cdot\mathbf{v} + i\eta}[1 - G(\mathbf{k},\omega)]\,, \tag{7.122}$$

a relation consistent with Eq. (7.69).

The collision term with the dynamic local-field correction is expressed as [Ichimaru & Tanaka 1986]

$$\left.\frac{\partial F}{\partial t}\right]_c = C_{\mathrm{R}} + C_{\mathrm{I}} \tag{7.123}$$

with

$$C_{\mathrm{R}} = \pi\int\frac{d\mathbf{k}}{(2\pi)^3}\mathbf{k}\cdot\frac{\partial}{\partial\mathbf{p}}\int d\mathbf{p}'\frac{V(k)^2[1 - \mathrm{Re}G(\mathbf{k},\mathbf{k}\cdot\mathbf{v})]}{|\tilde{\varepsilon}(\mathbf{k},\omega)|^2}$$

$$\times\,\delta(\mathbf{k}\cdot\mathbf{v} - \mathbf{k}\cdot\mathbf{v}')\mathbf{k}\cdot\left(\frac{\partial}{\partial\mathbf{p}} - \frac{\partial}{\partial\mathbf{p}'}\right)F(\mathbf{p})F(\mathbf{p}') \tag{7.124a}$$

and

$$C_{\mathrm{I}} = \int\frac{d\mathbf{k}}{(2\pi)^3}\mathbf{k}\cdot\frac{\partial}{\partial\mathbf{p}}\int d\mathbf{p}'V(k)^2\frac{\mathcal{P}}{\mathbf{k}\cdot\mathbf{v} - \mathbf{k}\cdot\mathbf{v}'}$$

$$\times\left[\frac{\mathrm{Im}G(\mathbf{k},\mathbf{k}\cdot\mathbf{v}')}{|\tilde{\varepsilon}(\mathbf{k},\mathbf{k}\cdot\mathbf{v}')|^2}\mathbf{k}\cdot\frac{\partial}{\partial\mathbf{p}} + \frac{\mathrm{Im}G(\mathbf{k},\mathbf{k}\cdot\mathbf{v})}{|\tilde{\varepsilon}(\mathbf{k},\mathbf{k}\cdot\mathbf{v})|^2}\mathbf{k}\cdot\frac{\partial}{\partial\mathbf{p}'}\right]F(\mathbf{p})F(\mathbf{p}')\,, \tag{7.124b}$$

where $\mathcal{P}$ stands for the principal part, and the screening function now takes the form

$$\tilde{\varepsilon}(\mathbf{k},\omega) = 1 - V(k)[1 - G(\mathbf{k},\omega)]\chi_0(\mathbf{k},\omega)\,. \tag{7.125}$$

The convergence of these collision terms in the large-k domain is still maintained with such a frequency-dependent local field correction.

F. Quantum-Mechanical Collision Terms

In dense plasmas, where the Coulomb coupling between the ions is strong (i.e., $\Gamma \geq 1$), the degeneracy parameter Θ for the electrons decreases; hence the quantum-mechanical effects become significant. The quantum-mechanical transport equations for the electrons are obtained through analyses analogous to the classical cases described in the previous section.

We consider the kinetic equation for electrons in random potential fields $\Phi_t(\mathbf{k},\omega)$ [Ichimaru & Tanaka 1985]. Let $c_\mathbf{p}^+$ and $c_\mathbf{p}$ be the creation and annihilation operators for an electron in a plane-wave state with momentum $\mathbf{p}$. The spin indexes are suppressed for simplicity, and we shall take account of spin dependence only through its degree of freedom 2. The Hamiltonian for the electron system is written as

$$H = 2\sum_\mathbf{p} \frac{p^2}{2m} c_\mathbf{p}^\dagger c_\mathbf{p} - \frac{e}{V}\sum_\mathbf{k}\sum_\mathbf{p}\int_{-\infty}^{\infty}\frac{d\omega}{2\pi}\rho(\mathbf{p},-\mathbf{k})\Phi_t(\mathbf{k},\omega)\exp(-i\omega t)\,, \tag{7.126}$$

where

$$\rho(\mathbf{p},\mathbf{k}) = 2c_\mathbf{p}^\dagger c_{\mathbf{p}+\hbar\mathbf{k}} \tag{7.127}$$

is the electron-hole operator. The Heisenberg equation of motion (cf., Eq. 3.163) for this operator reads

$$i\frac{\partial}{\partial t}\rho(\mathbf{p},\mathbf{q}) = (\Delta_\mathbf{p}^\mathbf{q}\varepsilon_\mathbf{p})\rho(\mathbf{p},\mathbf{q})$$

$$+\frac{e}{2V}\sum_\mathbf{k}\int_{-\infty}^{\infty}\frac{d\omega}{2\pi}\Delta_\mathbf{p}^\mathbf{k}\,\rho(\mathbf{p},\mathbf{q}-k),\Phi_t(\mathbf{k},\omega)\ \exp(-i\omega t) \tag{7.128}$$

Here

$$\varepsilon_\mathbf{p} = \frac{p^2}{2m}\,, \tag{7.129}$$

$$A,B = AB + BA\,, \tag{7.130}$$

and $\Delta_\mathbf{p}^\mathbf{k}$ is a difference operator in the momentum space, so that in general

$$\Delta_\mathbf{p}^\mathbf{k} f(\mathbf{p}) = \frac{1}{\hbar}\,[f(\mathbf{p}+\hbar\mathbf{k}) - f(\mathbf{p})]\,. \tag{7.131}$$

Equation (7.128) may be transformed into a transport equation through its statistical average denoted by $\langle\ \rangle$. For the potential field, we assume

$$\Phi_t(\mathbf{k},\omega) = \langle\Phi_t(\mathbf{k},\omega)\rangle + \Phi(\mathbf{k},\omega)$$

$$= 2\pi i\frac{\mathbf{k}}{k^2}\cdot \mathbf{E}\delta_{\mathbf{k},0}\delta(\omega) + \Phi(\mathbf{k},\omega)\,, \qquad (7.132)$$

so that a uniform dc electric field $\mathbf{E}$ is applied to the system; $\Phi(\mathbf{k},\omega)$ then represents the fluctuating internal potential fields produced by the electrons and ions. We also note that the Wigner distribution $F_e(\mathbf{r},\mathbf{p})$ is given by

$$F_e(\mathbf{r},\mathbf{p}) = \frac{1}{V}\sum_{\mathbf{q}}\langle\rho\,(\mathbf{p},\mathbf{q})\rangle \exp\,(i\mathbf{q}\cdot\mathbf{r})\,. \qquad (7.133)$$

We are interested in analysis of the transport processes in the presence of the dc electric field $\mathbf{E}$ or the the weak spatial gradients in temperature or flow velocity. The Wigner distribution in these circumstances can be calculated by consideration of contributions only from the vicinity of $\mathbf{q}=0$ in Eq. (7.128): we thus find

$$\frac{\partial}{\partial t}F_e(\mathbf{r},\mathbf{p}) = -\frac{\partial \varepsilon_{\mathbf{p}}}{\partial \mathbf{p}}\cdot\frac{\partial F_e(\mathbf{r},\mathbf{p})}{\partial \mathbf{r}} + e\mathbf{E}\cdot\frac{\partial F_e(\mathbf{r},\mathbf{p})}{\partial \mathbf{p}}$$

$$-i\frac{e}{2V}\sum_{\mathbf{k}}\int_{-\infty}^{\infty}\frac{d\omega}{2\pi}\,\Delta_{\mathbf{p}}^{\mathbf{k}}\langle\,\rho\,(\mathbf{p},-\mathbf{k}),\Phi(\mathbf{k},\omega)\,\rangle \exp\,(-i\omega t)\,. \qquad (7.134)$$

Introducing the Fourier components of the electron density fluctuations in the phase space through

$$\delta N(\mathbf{k},\omega;\mathbf{p}) = \int_{-\infty}^{\infty} dt\langle\rho\,(\mathbf{p},\mathbf{k})\rangle \exp\,(i\omega t)\,, \qquad (7.135)$$

we obtain the transport equation

$$\frac{\partial}{\partial t}F_e(\mathbf{r},\mathbf{p}) + \frac{\mathbf{p}}{m}\cdot\frac{\partial F_e(\mathbf{r},\mathbf{p})}{\partial \mathbf{r}} - e\mathbf{E}\cdot\frac{\partial F_e(\mathbf{r},\mathbf{p})}{\partial \mathbf{p}}$$

$$= ie\int\frac{d\mathbf{k}}{(2\pi)^3}\int_{-\infty}^{\infty} d\omega\,\Delta_{\mathbf{p}}^{\mathbf{k}}\langle N\Phi^*(\mathbf{k},\omega;\mathbf{p})\rangle \qquad (7.136)$$

where $\langle N\Phi^*(\mathbf{k},\omega;\mathbf{p})\rangle$ is a spectral function defined via

$$\langle\delta N(\mathbf{k},\omega;\mathbf{p})\Phi(\mathbf{k},\omega')\rangle = \frac{(2\pi)^2}{V}\langle N\Phi^*(\mathbf{k},\omega;\mathbf{p})\rangle\delta(\omega+\omega')\,. \qquad (7.137)$$

The right-hand side of Eq. (7.136) can be expressed in terms of the spectral function $\langle | \Phi^2 |(\mathbf{k},\omega)\rangle$ of the potential fluctuations. To do so, we note the quantum version of the induced fluctuations in the static local-field correction approximation Eq. (7.116):

$$\delta N^{(S)}(\mathbf{k},\omega;\mathbf{p}) = e\frac{\Delta_{\mathbf{p}}^{\mathbf{k}}F_e(\mathbf{r},\mathbf{p})}{\omega - \omega_{\mathbf{pk}} + i\eta}\Phi(\mathbf{k},\omega)[1 - G(\mathbf{k})] , \qquad (7.138)$$

where

$$F_e(\mathbf{r},\mathbf{p}) = \langle \rho(\mathbf{p},0)\rangle , \qquad (7.139a)$$

$$\omega_{\mathbf{pk}} = \Delta_{\mathbf{p}}^{\mathbf{k}}\varepsilon_{\mathbf{p}} = \frac{\mathbf{k}\cdot\mathbf{p}}{m} + \frac{\hbar k^2}{2m} . \qquad (7.139b)$$

The $\mathbf{r}$ dependence in Eq. (7.139a) accounts for the weak spatial dependence in a hierarchical scale of l_1 in Eq. (7.10) arising from the temperature and flow-velocity gradients. Equation (3.136) thus leads to

$$\frac{\partial}{\partial t}F_e(\mathbf{r},\mathbf{p}) + \frac{\mathbf{p}}{m}\cdot\frac{\partial F_e(\mathbf{r},\mathbf{p})}{\partial \mathbf{r}} - e\mathbf{E}\cdot\frac{\partial F_e(\mathbf{r},\mathbf{p})}{\partial \mathbf{p}}$$

$$= ie^2\int\frac{d\mathbf{k}}{(2\pi)^3}\int_{-\infty}^{\infty}d\omega\,\Delta_{\mathbf{p}}^{\mathbf{k}}\left[\frac{\Delta_{\mathbf{p}}^{\mathbf{k}}F_e(\mathbf{r},\mathbf{p})}{\omega - \omega_{\mathbf{pk}} + i\eta}\right]\langle | \Phi^2 |(\mathbf{k},\omega)\rangle[1 - G(\mathbf{k})] . \qquad (7.140)$$

This gives a *quantum-mechanical collision term* for the electrons.

7.4 TRANSPORT PROCESSES

The transport processes induced by fluctuations in plasmas are described by the hydrodynamic equations obtained through moment integrations of the kinetic equations in momentum space. Starting with such hydrodynamic moment equations, this section treats the transport coefficients in weakly and strongly coupled plasmas, such as the electric and thermal resistivities and viscosity.

A. Hydrodynamic Moment Equations

Consider the kinetic equation (7.14). Its integration over the momentum space yields the *equation of continuity*,

$$\frac{\partial n_\sigma}{\partial t} + \frac{\partial}{\partial \mathbf{r}}\cdot(n_\sigma \mathbf{u}_\sigma) = 0 \qquad (7.141)$$

where

$$n_\sigma(\mathbf{r},t) = \int d\mathbf{p} F_\sigma(X;t)\,, \tag{7.142}$$

$$\mathbf{u}_\sigma(\mathbf{r},t) = \frac{1}{n_\sigma(\mathbf{r},t)} \int d\mathbf{p}\mathbf{v} F_\sigma(X;t)\,, \tag{7.143}$$

are the number density and flow velocity, respectively, of the σ species.

In a multicomponent plasma, the mass density and the plasma-flow velocity are defined as

$$\rho_\mathrm{m}(\mathbf{r},t) = \sum_\sigma m_\sigma n_\sigma(\mathbf{r},t)\,, \tag{7.144}$$

$$\mathbf{u}(\mathbf{r},t) = \frac{1}{\rho_\mathrm{m}(\mathbf{r},t)} \sum_\sigma \int d\mathbf{p}\mathbf{p} F_\sigma(X;t)\,. \tag{7.145}$$

The *equation of motion* or the momentum equation is obtained by integrating the product of Eq. (7.14) and momentum $\mathbf{p}$ over the momentum space:

$$m_\sigma n_\sigma \left(\frac{\partial}{\partial t} + \mathbf{u}_\sigma \cdot \frac{\partial}{\partial \mathbf{r}} \right) \mathbf{u}_\sigma + \frac{\partial}{\partial \mathbf{r}} \cdot \mathbf{P}_\sigma - Z_\sigma e n_\sigma \mathbf{E} - \frac{1}{c} \mathbf{J}_\sigma \times \mathbf{B}$$

$$= Z_\sigma e \langle \delta n_\sigma \delta \mathbf{E} \rangle + \frac{1}{c} \langle \delta \mathbf{J}_\sigma \times \delta \mathbf{B} \rangle \tag{7.146}$$

where

$$\mathbf{P}_\sigma(\mathbf{r},t) = \int d\mathbf{p} m_\sigma \mathbf{v}\mathbf{v} F_\sigma(X;t) - m_\sigma n_\sigma(\mathbf{r};t) \mathbf{u}_\sigma(\mathbf{r};t) \mathbf{u}_\sigma(\mathbf{r};t)\,, \tag{7.147}$$

$$\delta \mathbf{J}_\sigma(\mathbf{r},t) = Z_\sigma e \int d\mathbf{p}\mathbf{v}\, \delta N_\sigma(X;t) \tag{7.148}$$

refer to the ideal-gas pressure tensor and the current-density fluctuations for the σ species. Equation (7.146) is a plasma version of the *Navier-Stokes equation.*

The *energy equation* is likewise obtained by integrating the product between Eq. (7.14) and kinetic energy $p^2/2m_\sigma$ over the momentum space:

$$\frac{\partial W_\sigma}{\partial t} + \frac{\partial}{\partial \mathbf{r}} \cdot \mathbf{Q}_\sigma - \mathbf{J}_\sigma \cdot \mathbf{E} = \langle \delta \mathbf{J}_\sigma \cdot \delta \mathbf{E} \rangle\,. \tag{7.149}$$

Here

$$W_\sigma(\mathbf{r},t) = \int d\mathbf{p}\, \frac{m_\sigma}{2} v^2 F_\sigma(X;t)\,, \tag{7.150}$$

$$\mathbf{Q}_\sigma(\mathbf{r},t) = \int d\mathbf{p}\, \frac{m_\sigma}{2} v^2 \mathbf{v} F_\sigma(X;t) \tag{7.151}$$

are the kinetic-energy density and the energy flow. With the aid of Eqs. (7.141) and (7.146), Eq. (7.149) is transformed as

$$\left(\frac{\partial}{\partial t} + \mathbf{u}_\sigma \cdot \frac{\partial}{\partial \mathbf{r}}\right) U_\sigma + (\mathbf{P}_\sigma + U_\sigma \mathbf{I}) : \frac{\partial \mathbf{u}_\sigma}{\partial \mathbf{r}} + \frac{\partial \mathbf{q}_{T\sigma}}{\partial \mathbf{r}}$$
$$= \langle \delta \mathbf{J}_\sigma \cdot \delta \mathbf{E} \rangle - \mathbf{u}_\sigma \cdot \left[Z_\sigma e \langle \delta n_\sigma \delta \mathbf{E} \rangle + \frac{1}{c} \langle \delta \mathbf{J}_\sigma \times \delta \mathbf{B} \rangle \right], \tag{7.152}$$

where

$$U_\sigma(\mathbf{r},t) = \int d\mathbf{p}\, \frac{m_\sigma}{2} w^2 F_\sigma(X;t) \tag{7.153}$$

$$\mathbf{Q}_{T\sigma}(\mathbf{r},t) = \int d\mathbf{p}\, \frac{m_\sigma}{2} w^2 \mathbf{w} F_\sigma(X;t) \tag{7.154}$$

are the partial ideal-gas internal energy density and heat flow, respectively, with

$$\mathbf{w} = \mathbf{v} - \mathbf{u}_\sigma(\mathbf{r},t) \tag{7.155}$$

denoting the *peculiar velocity.*

These hydrodynamic moment equations form the basis of formulating the transport processes for the conserving quantities, such as particle number, momentum, and energy.

B. Electric and Thermal Resistivities

We consider a fully ionized TCP consisting of the electrons (mass $m_e = m$, electric charge $-e$, number density $n_e = n$) and the ions (mass $m_i = M$, electric charge Ze, number density $n_i = n/Z$). The electric resistivity ρ_E and the thermal resistivity ρ_T due to the electronic transport can be calculated from a solution to Eq. (7.140) under specific boundary conditions [Ichimaru & Tanaka 1985].

For this purpose we set the Wigner distribution $F_e(\mathbf{r},\mathbf{p})$ as a summation between an unperturbed distribution $F_0(\mathbf{p})$ and a perturbation $F_1(\mathbf{p})$ arising from the presence of a dc electric field or a temperature gradient:

$$F_e(\mathbf{r},\mathbf{p}) = F_0(\mathbf{p}) + F_1(\mathbf{p}) \,, \tag{7.156}$$

where

$$F_0(\mathbf{p}) \equiv 2f_0(k) = \frac{2}{\exp\left[(p^2/2mT) - \alpha\right] + 1} \tag{7.157}$$

is the Fermi distribution (cf., Eq. 3.140) with $\mathbf{p} = \hbar\mathbf{k}$. In Eq. (7.157), α is the normalized (ideal-gas) chemical potential defined and calculated as in Eqs. (3.144) and (3.145).

For the calculation of the electric resistivity stemming from scattering of the electronic current by the random potential fields produced by the ions, we set $\partial/\partial t = 0$, $\partial/\partial \mathbf{r} = 0$, and $F_e(\mathbf{r},\mathbf{p}) = F_0(\mathbf{p})$ on the left-hand side of Eq. (7.140). Correspondingly, substituted in the right-hand side is a displaced Fermi distribution

$$F_e(\mathbf{r},\mathbf{p}) = F_0(\mathbf{p} - m\mathbf{u}) \,; \tag{7.158}$$

the electric current density is then given by $\mathbf{J} = -en\mathbf{u}$. Since $m << M$, we may approximately express

$$\langle |\Phi|^2(\mathbf{k},\omega)\rangle = \frac{Zn}{e^2}\left|\frac{v(k)}{\tilde{\varepsilon}(k)}\right|^2 S_{ii}(k)\delta(\omega) \,. \tag{7.159}$$

Here $v(k) = 4\pi e^2/k^2$, $S_{ii}(k)$ is the static structure factor of the ions in the TCP, and

$$\tilde{\varepsilon}(k) = 1 - v(k)[1 - G_{ee}(k)]\chi_0(k,0) \tag{7.160}$$

is the static screening function of the electrons; the local field correction on the right-hand side of Eq. (7.140) should be that between electrons and ions, $G_{ei}(k)$. The electric resistivity is thus given by the ratio E/J in the limit of both E and J approaching zero.

The result of such a calculation yields

$$\rho_E = \frac{Zm^3}{12\pi^3\hbar^3e^2n}\int_0^\infty dk k^3 f_0\left(\frac{k}{2}\right)\left|\frac{v(k)}{\tilde{\varepsilon}(k)}\right|^2 [1 - G_{ei}(k)]S_{ii}(k) \,. \tag{7.161}$$

The factor $1 - G_{ei}(k)$ in this formula accounts for those short-range events where the electrons and ions, correlated strongly by Coulombic attraction, scatter each other repeatedly; hence it is an effect beyond the Born approximation and acts to enhance the resistivity.

In the limit of $T \to 0$, the Fermi function $f_0(k)$ becomes a unit step function, so that Eq. (7.161) reduces to the *Ziman formula* [Ziman 1961],

$$\rho_E = \frac{Zm^3}{12\pi^3\hbar^3 e^2 n}\int_0^{2k_F} dk k^3 \left|\frac{v(k)}{\tilde{\varepsilon}(k)}\right|^2 [1 - G_{ei}(k)] S_{ii}(k) , \tag{7.162}$$

if the strong electron-ion coupling is ignored, i.e., $G_{ei}(k) = 0$.

For the calculation of the thermal resistivity, where T and α are regarded as functions of $\mathbf{r}$, we substitute the Fermi distribution Eq. (7.157), independent of t, into the left-hand side of Eq. (7.140). Since the kinetic pressure

$$P = \int d\mathbf{p} \frac{p^2}{3m} F_0(\mathbf{p})$$

should be kept uniform (i.e., independent of $\mathbf{r}$), T and α are connected via

$$\frac{\partial \alpha}{\partial \mathbf{r}} = -\frac{5}{3}\frac{I_{3/2}(\alpha)}{I_{1/2}(\alpha)}\frac{\partial \ln T}{\partial \mathbf{r}} , \tag{7.163}$$

where $I_\nu(\alpha)$ are the Fermi integrals, Eq. (3.143).

In the right-hand side of Eq. (7.140), we substitute Eq. (7.156) with

$$F_1(\mathbf{p}) = \frac{3m^2}{2\pi(\hbar k_F)^7 \Sigma} \mathbf{q}_T \cdot \mathbf{p}\left[\left(\frac{p}{\hbar k_F}\right)^2 - \lambda\right]\frac{(2\pi\hbar)^3}{2}\frac{\partial F_0(\mathbf{p})}{\partial \alpha} , \tag{7.164}$$

where

$$\Sigma = \frac{7}{4I_{1/2}(\alpha)}\Theta^{9/2}\left\{I_{5/2}(\alpha)I_{1/2}(\alpha) - \frac{25}{21}[I_{3/2}(\alpha)]^2\right\} , \tag{7.165}$$

$$\lambda = \frac{5}{3}\Theta\frac{I_{3/2}(\alpha)}{I_{1/2}(\alpha)} . \tag{7.166}$$

Since

$$\int d\mathbf{p} F_1(\mathbf{p}) = \int d\mathbf{p}\mathbf{p} F_1(\mathbf{p}) = 0 , \tag{7.167}$$

$$\int d\mathbf{p}\varepsilon_{\mathbf{p}}\frac{\mathbf{p}}{m}F_1(\mathbf{p})=\mathbf{q}_T \tag{7.168}$$

$\mathbf{q}_T$ represents the heat flux transported by the electrons.

Setting thus

$$\rho_T\mathbf{q}_T=-\frac{\partial T}{\partial \mathbf{r}} \tag{7.169}$$

we solve Eq. (7.140) for ρ_T as

$$\rho_T=\frac{Z\sqrt{m}\Theta^{9/2}}{16\sqrt{2}\pi\Sigma^2T^{5/2}}\int_0^{\infty}dkk^3\left|\frac{v(k)}{\tilde{\varepsilon}(k)}\right|^2[1-G_{ei}(k)]S_{ii}(k)$$
$$\times\int_{k/2k_F}^{\infty}dxx(x^2-\lambda)^2\frac{\partial f_0(k_Fx)}{\partial\alpha}. \tag{7.170}$$

This calculation corresponds to Eq. (7.161) for the electric resistivity.

In the limit of $T\to 0$, Eq. (7.170) reduces to

$$\rho_T=\frac{Zm^2}{4\pi^5\hbar^3nT}\int_0^{2k_F}dkk^3\left|\frac{v(k)}{\tilde{\varepsilon}(k)}\right|^2[1-G_{ei}(k)]S_{ii}(k)\,. \tag{7.171}$$

The ratio between Eqs. (7.162) and (7.171) yields the usual *Wiedemann-Frantz relation* [e.g., Ziman 1972] at $\Theta << 1$, i.e.,

$$\frac{\rho_E}{\rho_T}=\frac{\pi^2}{3}\frac{T}{e^2}\,. \tag{7.172}$$

For a hydrogenic ($Z=1$) plasma, Eqs. (7.161) and (7.170) may be rewritten as

$$\rho_E=4\sqrt{\frac{2\pi}{3}}\frac{\Gamma^{3/2}}{\omega_p}L_E\,, \tag{7.173}$$

$$\rho_T=\frac{52\sqrt{6\pi}}{75}\frac{T\Gamma^{3/2}}{e^2\,\omega_p}L_T\,. \tag{7.174}$$

We shall call L_E and L_T the *generalized Coulomb logarithms*, which are defined and calculated according to

$$L_E(\Gamma,\Theta) = \frac{3\sqrt{\pi}\,\Theta^{3/2}}{4}\int_0^\infty \frac{dk}{k} f_0\left(\frac{k}{2}\right)\frac{[1-G_{ei}(k)]}{|\tilde{\varepsilon}(k)|^2}S_{ii}(k)\,, \tag{7.175}$$

$$L_T(\Gamma,\Theta) = \frac{75\sqrt{\pi}\,\Theta^{9/2}}{104\Sigma^2}\int_0^\infty \frac{dk}{k}\frac{[1-G_{ei}(k)]}{|\tilde{\varepsilon}(k)|^2}S_{ii}(k)$$

$$\times\int_{k/2k_F}^\infty dx\,x(x^2-\lambda)^2\frac{\partial f_0(k_F x)}{\partial\alpha}\,. \tag{7.176}$$

In the classical limit $\Theta >> 1$, both L_E and L_T approach the same expression:

$$L_0 = \int_0^\infty \frac{dk}{k}\frac{S_{ii}(k)}{|\tilde{\varepsilon}(k)|^2}\exp\left(-\frac{\hbar^2k^2}{8mT}\right). \tag{7.177}$$

Assuming the weak-coupling cases $\Gamma << 1$ with the RPA, where all the local field corrections may be ignored, we may substitute the Debye-Hückel formulas

$$S_{ii}(k) = \frac{k^2+k_D^2}{k^2+2k_D^2}\,, \tag{7.178}$$

$$\tilde{\varepsilon}(k) = 1 + \frac{k_D^2}{k^2} \tag{7.179}$$

in Eq. (7.177), to find

$$L_0 = \exp(\zeta)\mathrm{Ei}(\zeta) - \frac{1}{2}\exp\left(\frac{\zeta}{2}\right)\mathrm{Ei}\left(\frac{\zeta}{2}\right). \tag{7.180}$$

Here

$$\zeta = \frac{\hbar k_D^2}{4mT} = \frac{3}{2}\left(\frac{4}{9\pi}\right)^{2/3}\frac{\Gamma}{\Theta} \tag{7.181}$$

and

$$\mathrm{Ei}(x) = \int_x^\infty \frac{dt}{t}\exp(-t) \tag{7.182}$$

is the exponential integral. Since $\zeta << 1$, Eq. (7.180) may be expanded as

$$L_0 = -\frac{1}{2}\ln\zeta - \frac{1}{2}(\gamma + \ln 2) - \frac{3}{4}\zeta\left(\ln\zeta - 1 + \gamma + \frac{1}{3}\ln 2\right)$$
$$+ O(\zeta^2, \zeta^2 \ln\zeta), \qquad (7.183)$$

where $\gamma = 0.57721\ldots$ is Euler's constant. The leading terms in this expansion were obtained by Kivelson and DuBois [1964] with the aid of a quantum-mechanical version of the BGL collision term.

For strongly coupled hydrogenic plasmas, it has been argued that strong correlations between electrons and ions near *metal-insulator boundaries* need to be most carefully taken into consideration [Tanaka et al. 1990]. The condition for pressure ionization in hydrogen may be given approximately as

$$E_{\rm F} \approx {\rm Ry}, \qquad (7.184)$$

where Ry, defined by Eq. (3.186), is the ionization energy of a hydrogen atom in the ground state. Analogously, the condition for thermal ionization is estimated approximately as

$$T \approx {\rm Ry}. \qquad (7.185)$$

Hydrogen is in an ionized, metallic state when $E_{\rm F} > {\rm Ry}$ or $T > {\rm Ry}$.

The strong coupling effects in dense hydrogen plasmas have been analyzed through an integral-equation approach which adopts the HNC approximation (cf., Eq. 7.112) for the classical ion-ion correlation and the MCA (cf., Eqs. 7.107 and 7.109–7.111) for the quantum-mechanical electron-electron and electron-ion correlations [Tanaka et al. 1990]. The resultant HNC-MCA equations have been solved self-consistently for the structure factors, the local field corrections, the excess internal energies, and the electric and thermal resistivities at 48 combinations between Γ and Θ. In Table 7.1, the calculated values for the latter three quantities are tabulated at 16 selected combinations of Γ and Θ. As we have remarked earlier, the ratio L_E/L_T approaches unity in the classical limit ($\Theta >> 1$) and $75/13\pi^2 = 0.5845$ in the limit of complete Fermi degeneracy ($\Theta << 1$); the latter reflects the Wiedemann-Frantz relation (7.172).

For convenience in practical applications, simple analytic formulas parametrizing those computed results are provided as follows:

$$L_E = \frac{5}{9}\,\frac{\ln\left[1 + \frac{3}{4}(C\zeta)^{-9/10}\right]}{1 - 0.378X_b} \qquad (7.186)$$

Table 7.1 The generalized Coulomb logarithms, Eqs. (7.175) and (7.176), and the normalized interaction energy $u_{ex} = U_{int}/n(e^2/a)$ for hydrogen plasmas, at selected combinations of Γ and Θ. The values in parentheses are the fitted values in accord with Eqs. (7.186) and (7.187).

Γ	Θ	L_E	L_T	$-u_{ex}$
0.05	10	2.732 (2.630)	2.363 (2.298)	0.54483
0.1	10	2.710 (2.602)	2.256 (2.223)	0.75612
0.2	10	3.295 (3.175)	2.557 (2.637)	1.0918
0.3	10	5.250 (5.277)	3.784 (4.294)	1.5293
0.1	5	2.097 (2.602)	1.725 (2.223)	0.75009
0.3	5	2.506 (2.242)	1.853 (1.751)	1.3024
0.5	5	5.002 (5.158)	3.384 (3.884)	2.1899
0.1	1	0.9560 (0.9781)	0.7507 (0.7365)	0.75699
0.4	1	0.7184 (0.5747)	0.5024 (0.3880)	1.2285
0.7	1	0.7214 (0.5108)	0.4767 (0.3315)	1.5442
1.0	1	0.8273 (0.5528)	0.5301 (0.3511)	1.8822
1.0	0.1	0.03679 (0.03903)	0.02259 (0.02312)	1.1662
3.0	0.1	0.02696 (0.01742)	0.01614 (0.01024)	1.4295
5.0	0.1	0.02612 (0.01341)	0.01551 (0.00786)	1.6119
10.0	0.01	0.0005776 (0.0006390)	0.0003377 (0.0003736)	1.3290
30.0	0.01	0.0004450 (0.0002431)	0.0002580 (0.0001635)	1.5222

$$L_T = \frac{5}{9} \frac{\ln\left[1 + 0.5845 \times \frac{3}{4}(C\zeta)^{-9/10}\right]}{1 - 0.378X_b} \tag{7.187}$$

Here $C = 2\exp(\gamma) = 3.562\ldots$, and

$$X_b = r_s \tanh\left[\frac{2}{(9\pi)^{1/6}\sqrt{\Theta}}\right]. \tag{7.188}$$

The values of the parametrized expressions, Eqs. (7.186) and (7.187), are likewise listed in Table 7.1 for comparison. In the classical regime $\Theta > 1$, the fitting formulas reproduce the calculated values of L_E and L_T with good accuracy. Even in the quantum regime, a reasonable comparison is obtained. The simple formulas (7.186) and (7.187) can therefore account for the salient features in the Coulomb logarithms fairly accurately over wide ranges of plasma parameters.

The analytic formulas for the generalized Coulomb logarithms retain the following features.
(1) In the classical ($\Theta >> 1$) and weak-coupling ($\Gamma << 1$) limit, Eqs. (7.186) and (7.187) reproduce the first two terms on the right-hand side of Eq. (7.183).
(2) In the limit of complete Fermi degeneracy ($\Theta << 1$), L_E and L_T behave nearly proportionally to $\Theta^{3/2}$, as Eq. (7.175) suggests.
(3) Near the metal-insulator boundaries, L_E and L_T tend to diverge as the Coulomb coupling further increases. The condition for the divergence,

$$X_b = 2.65 \tag{7.189}$$

in Eqs. (7.186) and (7.187) qualitatively agrees with Eqs. (7.184) and (7.185). It remains to be seen whether Eq. (7.189) can truely account for the real metal-insulator transitions in dense plasmas.

C. Viscosity

Viscosity in a TCP arises mainly from momentum transfers between ions, because of the large difference between the masses of an electron and of an ion. As a first approximation, one may then adopt a classical OCP model in which the electrons provide an unpolarizable, uniform background of negative charges. The shear viscosity of a weakly coupled OCP has been calculated by Braginskii [1957] and by Braun [1967] through a solution to the Landau or the BGL collision terms. For a strongly coupled OCP, Vieillefosse and Hansen [1975] and Wallenborn and Baus [1978] have advanced theoretical estimates for the shear viscosity on the basis of the hydrodynamic and kinetic-theoretical arguments.

The simple OCP model described above, however, amounts to an oversimplification for those dense plasmas under present consideration. The polarization and screening effects of the electrons significantly influence the interionic correlations in such a TCP. In the treatment of shear viscosity, it is essential to incorporate the screening effects in the calculations of the interionic correlations [Tanaka & Ichimaru 1986b].

For the calculation of the coefficient of shear viscosity we may assume $\partial/\partial t = 0$ and $\langle \mathbf{E} \rangle = \langle \mathbf{B} \rangle = \mathbf{0}$ on the left-hand side of Eq. (7.14). Consider a plasma with a flow velocity $\mathbf{u}(z)$, a function of z alone, in the x direction; the density and temperature are assumed uniform and constant. If $d\mathbf{u}/dz$ is sufficiently small, we may set

$$F(X) = F^{(0)}(\mathbf{p})[1 + \varphi(\mathbf{w})] \tag{7.190}$$

for the nonequilibrium distribution function of the ions. Here $\mathbf{w} = \mathbf{v} - \mathbf{u}(z)$ is the peculiar velocity and

$$F^{(0)}(\mathbf{p}) = \frac{n_i}{(2\pi MT)^{3/2}} \exp\left(-\frac{Mw^2}{2T}\right) \tag{7.191}$$

is the Maxwellian distribution. The small quantity $\varphi(\mathbf{w})$ is to be determined in such a way as to satisfy the following constraints:

$$\int d\mathbf{p} F(X) = n_i \tag{7.192}$$

$$\int d\mathbf{p} M w_x w_z F(X) = P_{xz} \tag{7.193}$$

where P_{xy} denotes the x-z component of the pressure tensor. We thus set a solution

$$\varphi(\mathbf{w}) = P_{xz} \frac{M w_x w_z}{n_i T^2} \sum_{l=0}^{\infty} a_l S_{5/2}^{(l)}\left(\frac{Mw^2}{2T}\right), \tag{7.194}$$

where $a_0 = 1$ and $S_{5/2}^{(l)}(x)$ refer to the Sonine polynomials* of order 5/2. We here keep only the term $l = 0$, which corresponds to the single Sonine-polynomial approximation.

*The Sonine polynomials are defined through a series expansion of the generating function:

$$(1-s)^{-m-1} \exp\left(-\frac{xs}{1-s}\right) = \sum_{n=0}^{\infty} S_m^{(n)}(x) s^n .$$

In particular one has

$$S_m^{(0)} = 1 , \; S_m^{(1)}(x) = m + 1 - x .$$

Equation (7.190) is substituted directly in the left-hand side of Eq. (7.14) and through Eqs. (7.116) and (7.117) in the right-hand side of Eq. (7.14). Taking the moment of the resulting kinetic equation over the momentum space with a weighting function $w_x w_z$ (cf., Section 7.4A), we find

$$\frac{n_i T}{M}\frac{du}{dz} = -\frac{\pi P_{xz}}{2Mn_i T^2}\int\frac{d\mathbf{k}}{(2\pi)^3}\int d\mathbf{p}\int d\mathbf{p}'\frac{V(k)^2[1-G(k)]}{|\tilde{\varepsilon}(\mathbf{k},\mathbf{k}\cdot(\mathbf{w}+\mathbf{w}')/2|^2}$$

$$\times [k_x(w_z - w_z') + k_z(w_x - w_x')]^2\delta(\mathbf{k}\cdot\mathbf{w} - \mathbf{k}\cdot\mathbf{w}')F^{(0)}(\mathbf{p})F^{(0)}(\mathbf{p}') \qquad (7.195)$$

to the first order in P_{xz} and du/dz. Introducing new variables $\mathbf{W} = (\mathbf{w}+\mathbf{w}')/2$ and $\mathbf{s} = \mathbf{w} - \mathbf{w}'$, defining the coefficient of shear viscosity η via

$$P_{xz} = -\eta\frac{du}{dz} \qquad (7.196)$$

and performing the $\mathbf{s}$ integration, we obtain [Tanaka & Ichimaru 1986b]

$$\frac{1}{\eta} = -\frac{2M}{5\pi^2 T^4}\int_0^1 d(\cos\psi)\int_0^\infty dk\int_0^\infty dW\frac{W^2k^3V(k)^2[1-G(k)]}{|\tilde{\varepsilon}(k,kW\cos\psi)|^2}\exp\left(-\frac{Mw^2}{2T}\right), \qquad (7.197)$$

where ψ refers to the angle between $\mathbf{k}$ and $\mathbf{W}$.

For comparison of values of the viscosity in different parameter regimes, it is convenient to introduce a dimensionless shear viscosity defined by

$$\eta^* \equiv \frac{\eta}{Mn_i\omega_p a^2}, \qquad (7.198)$$

where $\omega_p = \sqrt{4\pi n_i(Ze)^2/M}$ and $a = (4\pi n_i/3)^{-1/3}$.

For the OCP in the weak-coupling limit, we may set $G(k) = 0$ and $\tilde{\varepsilon}(k,\omega) = 1$ in Eq. (7.197); $V(k)$ is that defined in Eq. (7.118). By introducing the upper and lower cutoffs at $T/(Ze)^2$ and $\sqrt{4\pi n_i(Ze)^2/T}$ in the k integration (which would diverge logarithmically without the cutoffs), we find

$$\eta_{\mathrm{L}}^* \equiv \frac{5\sqrt{\pi}}{6\sqrt{3}\Gamma^{5/2}\ln\Lambda}, \qquad (7.199)$$

where $\ln\Lambda = -\ln(\sqrt{3}\Gamma^{3/2})$ is a Coulomb logarithm in the weak-coupling limit. Since the derivation of Eq. (7.199) depends on the Landau collision term (cf., Section 2.4B), we shall call it the Landau value for the shear viscosity.

In the weak-coupling regime we may take account of the RPA dielectric screening in Eq. (7.198) by setting $G(k) = 0$ in Eq. (7.119). This corresponds to adopting the BGL collision term; the result thus obtained by Braun [1967] reads

$$\eta^*_{\mathrm{BGL}} \equiv \frac{\eta^*_{\mathrm{L}}}{1 + 0.346/\ln \Lambda}, \tag{7.200}$$

which we shall call the BGL value.

For an explicit calculation of the shear viscosity Eq. (7.197) in a strongly coupled plasma, either in the OCP model or in the electron-screened OCP model, the functional values of the static structure factors and the local field corrections need to be evaluated.* In the screened OCP, one approximates [Tanaka & Ichimaru 1986b]

$$V(k) = \frac{Z^2 v(k)}{\varepsilon_e(k,0)}, \tag{7.201}$$

$$G(k) = \frac{1}{\varepsilon_e(k,0)} + \frac{T}{n_i Z^2 v(k)}\left[1 - \frac{1}{S_{ii}(k)}\right] \tag{7.202}$$

in Eq. (7.198), where

$$\varepsilon_e(k,0) = 1 - \frac{v(k)\chi_e^{(0)}(k,0)}{1 + v(k)G_{ee}(k)\chi_e^{(0)}(k,0)} \tag{7.203}$$

is the static dielectric function of the electrons.

In Table 7.2, the values of the reduced shear viscosity computed in the OCP model with various theoretical schemes are listed. For small Γ, the results with Eq. (7.197) tend smoothly to the weak-coupling values, Eq. (7.199) obtained with the Landau collision term and Eq. (7.200) obtained with the BGL collision term. We remark in this connection that the values of η^* calculated in the single Sonine-polynomial approximation are smaller by a factor of 0.87 than those obtained by Braginskii [1957], who performed the calculations with the two Sonine-polynomial approximation.

In the strong-coupling regime ($\Gamma > 1$), the calculated values of η^* take on a minimum approximately at $\Gamma = 10$. Beyond that, η^* tends to increase owing to the diminishing contribution arising from the factor $1 - G(k)$ in the inte-

*Theoretical schemes for calculating these various structure factors and local field corrections in dense plasmas will be treated in Vol. II.

Table 7.2 Values of η^* calculated in various schemes for an OCP. L refers to Eq. (7.200); BGL, Eq. (7.201); VH [Vieillefosse & Hansen 1975]; WB [Wallenborn & Baus 1978]; TI [Tanaka & Ichimaru 1986b].

Γ	L	BGL	VH	WB	TI
0.01	13400	12700			14400
0.03	1160	1080			1270
0.1	92.8	83.0		86	101
0.2	25.6	21.6			25.9
0.5	9.84	5.77		3.7	4.92
1			0.349	1.0	1.57
2			0.194	0.29	0.565
5					0.181
10			0.0827	0.072	0.101
20			0.0781	0.097	0.153

grand of Eq. (7.197). The shear viscosity in a strongly coupled plasma has been evaluated also through a formulation including the dynamic local-field correction as in Eqs. (7.122)–(7.125) [Ichimaru & Tanaka 1986; Tanaka & Ichimaru 1987].

We proceed to calculate the shear viscosity in the screened OCP model for dense plasmas with $Z = 1$; the computed values are listed in Table 7.3. The values of η^* in this model are invariably larger than those computed in the simple OCP model at a same Γ. These increases may be interpreted in terms of the effective reduction in the ion-ion interaction owing to the electronic screening effects; by the same token the values of η^* increase with Θ. The contribution of electronic screening enhances the ionic shear viscosity in a hydrogen plasma by 9 to 66% at $\Gamma = 1$ for $0.1 \leq \Theta \leq 10$.

For convenience in practical applications, an analytic expression has been provided, parametrizing those computed results for $\Gamma \leq 3$. In terms of a generalized Coulomb logarithm $L(\Gamma,\Theta)$ defined via

$$\eta^* \equiv \frac{5\sqrt{\pi}}{6\sqrt{3}\,\Gamma^{5/2} L(\Gamma,\Theta)}\,, \tag{7.204}$$

the parametrization takes a form

$$L(\Gamma,\Theta) = \frac{-1.5\ln\Gamma + a(\Theta) + b(\Theta)\Gamma}{1 + c(\Theta)\Gamma^{5/2}} \tag{7.205}$$

where

Table 7.3 Values of η^* calculated on the basis of Eq. (7.197) for the electron-screened OCP with $Z = 1$ [Tanaka & Ichimaru 1986b].

Γ	$\Theta = 0.1$	$\Theta = 1$	$\Theta = 10$
0.1	104	116	119
0.2	27.0	30.9	32.3
0.5	5.23	6.40	6.94
1	1.71	2.29	2.61
2	0.632		

$$a(\Theta) = -\frac{1.0128 + 4.0142\Theta + 1.5071\Theta^2}{1 + 2.5902\Theta + \Theta^2}, \tag{7.206a}$$

$$b(\Theta) = \frac{2.5310 + 10.875\Theta + 3.7326\Theta^2}{1 + 3.0199\Theta + \Theta^2}, \tag{7.206b}$$

$$c(\Theta) = \frac{1.8143 + 9.6504\Theta + 5.8574\Theta^2}{1 + 2.0009\Theta + \Theta^2}. \tag{7.206c}$$

The numerical data in the simple OCP model calculations (Table 7.2) offer the boundary conditions in this parametrization at $\Theta = 0$; the fitting formula (7.205) reproduces those computed values in Tables 7.2 and 7.3 for $\Gamma \leq 3$ with digressions of less than 1%.

Two remarks are in order with regard to Eqs. (7.204)–(7.206). (i) In the limit of $\Gamma \to 0$, these expressions naturally approach the weak-coupling formulas, Eqs. (7.199) and (7.200). (ii) If a strong-coupling limit ($\Gamma >> 1$) is approached, Eq. (7.206) would take a form proportional to $\Gamma^{-3/2}$. This Γ dependence has been adopted on the basis of the OCP values of $L(\Gamma,\Theta)$ in the regime of $1 \leq \Gamma \leq 5$; if still larger values of Γ were considered, $L(\Gamma,\Theta)$ would decrease more rapidly. In such a strong-coupling regime, the present scheme based on static evaluation of the local field correction would become inaccurate; the range of applicability for those formulas should thus be limited to $\Gamma \leq 3$.

In Section 7.4B, we derived parametrized expressions for the generalized Coulomb logarithms in electric and thermal resistivities. The generalized Coulomb logarithm in shear viscosity Eq. (7.205) exhibits a markedly different large-Γ behavior from those in the resistivities. The difference stems from the fact that in the calculation of ionic viscosity one deals with the dynamic correlations between the ions as Eq. (7.197) illustrates, while only the static correlations between the ions take part in the treatment of the electronic transport problems in the resistivities.

7.5 RADIATIVE PROCESSES

The electromagnetic fluctuations in plasmas lead to radiative processes. Starting with a general consideration of the rates of excitations, we consider spontaneous and nonlinear-induced processes such as synchrotron radiation, bremsstrahlung, incoherent scattering, and inverse Compton processes.

A. Rates of Radiative Processes

We have noted that Eq. (7.15c) represents the law of energy conservation in particles and fields. In terms of the spectral representation Eq. (7.30) for the correlations between field fluctuations, the rate at which the energy contained in the electromagnetic-field fluctuations with wave vector **k** increases per unit volume in a plasma is then given by

$$\frac{\partial \mathcal{E}_{\mathbf{k}}}{\partial t} = -\int_{-\infty}^{\infty} d\omega \langle \mathbf{J} \cdot \mathbf{E}^*(\mathbf{k},\omega)\rangle . \tag{7.207}$$

Rates of radiative processes may be obtained through a separation of the transverse electromagnetic part in Eq. (7.207).

In the RPA, we may use Eqs. (7.39) and (7.41) to calculate

$$\left.\frac{\partial \mathcal{E}_{\mathbf{k}}}{\partial t}\right]^{\mathrm{RPA}} = -\int_{-\infty}^{\infty} d\omega \frac{4\pi}{i\omega} \mathsf{Z}^*(\mathbf{k},\omega) : \langle \mathbf{J}\mathbf{J}^*(\mathbf{k},\omega)\rangle^{(0)} \tag{7.208a}$$

$$+\sum_{\sigma}\int d\mathbf{p}\int_0^{\infty} d\tau \int_{-\infty}^{\infty} d\omega Z_{\sigma} e \left[\frac{\partial F_{\sigma}(\mathbf{p}')}{\partial \mathbf{p}'}\right] \cdot \mathbf{T}(\mathbf{k},\omega;p') \cdot \langle \mathbf{E}\mathbf{E}^*(\mathbf{k},\omega)\rangle$$

$$\times \exp\left[-i\varphi(\tau) - \eta\tau\right], \tag{7.208b}$$

where $\mathsf{Z}(\mathbf{k},\omega)$ is the inverse dispersion tensor defined by Eq. (7.42) and : implies a double contraction between tensors. The term (7.208a) describes the rate of *spontaneous emission* by the current fluctuations $\langle \mathbf{J}\mathbf{J}^*(\mathbf{k},\omega)\rangle^{(0)}$ produced by random motion of discrete particles. The term (7.208b), on the other hand, represents the rate of (linearly) *induced emission* by the presence of the electromagnetic-field fluctuations $\langle \mathbf{E}\mathbf{E}^*(\mathbf{k},\omega)\rangle$ in the plasma.

The dispersion tensor $\mathbf{Z}(\mathbf{k},\omega)$ contains

$$\det\left|\boldsymbol{\varepsilon}(\mathbf{k},\omega)-\left(\frac{ck}{\omega}\right)^2\mathbf{I}_T\right|$$

in its denominator, and therefore has poles at the solutions to Eq. (3.92). If the dielectric tensor is approximated by the form Eq. (3.94) applicable in an isotropic medium, the dispersion tensor can be split into longitudinal and transverse parts:

$$\mathbf{Z}(\mathbf{k},\omega)\approx\frac{\mathbf{I}_L}{\varepsilon(\mathbf{k},\omega)} \tag{7.209a}$$

$$+\frac{\mathbf{I}_T}{\varepsilon_T(\mathbf{k},\omega)-(ck/\omega)^2}\,. \tag{7.209b}$$

Here $\varepsilon(\mathbf{k},\omega)$ is the dielectric response function defined by Eq. (3.135), and $\varepsilon_T(\mathbf{k},\omega)$ refers to the transverse dielectric function defined by Eq. (3.96).

Solutions to Eqs. (3.54) and (3.99), respectively, determine the characteristic frequencies of the longitudinal and transverse excitations, ω_L and ω_T. In the vicinity of those characteristic frequencies, imaginary parts of Eq. (7.209) are expressed approximately as

$$\mathrm{Im}\mathbf{Z}(\mathbf{k},\omega)\approx-\frac{\pi\,\omega_L}{2\alpha_L}[\delta(\omega-\omega_L)-\delta(\omega+\omega_L)]\mathbf{I}_L \tag{7.210a}$$

$$-\frac{\pi\,\omega_T}{2\alpha_T}[\delta(\omega-\omega_T)-\delta(\omega+\omega_T)]\mathbf{I}_T \tag{7.210b}$$

where

$$\alpha_L=\frac{\omega_L}{2}\frac{\partial\varepsilon(k,\omega)}{\partial\omega}\Bigg]_{\omega=\omega_L} \tag{7.211a}$$

$$\alpha_T=\varepsilon_T(k,\omega_T)\,. \tag{7.211b}$$

Analogously those parts of the electromagnetic fluctuations associated with the characteristic solutions to Eqs. (3.54) and (3.99) may be singled out as

$$\langle \mathbf{EE}^*(\mathbf{k},\omega)\rangle \approx \langle |E^2|(\mathbf{k})\rangle_L \mathbf{l}_L \mathbf{l}_L \delta(\omega - \omega_L) \tag{7.212a}$$

$$+ \langle |E^2|(\mathbf{k})\rangle_T \mathbf{l}_T \mathbf{l}_T \delta(\omega - \omega_T) . \tag{7.212b}$$

Separation between the longitudinal and transverse parts such as Eq. (7.209) and (7.212) is an idealization of the actual excitation processes in real plasmas, and therefore should be used with caution. These representations, when applicable, serve a useful purpose in distinguishing between different excitation processes contained in the rate equation (7.207).

B. Synchrotron Radiation

We consider a plasma with relativistic electrons in a uniform magnetic field $\langle \mathbf{B}\rangle = B\hat{\mathbf{z}}$. The radiative process associated with spontaneous emission (7.208a) then gives rise to *synchrotron radiation.*

Substituting Eqs. (7.35) and (7.210b) in Eq. (7.208a) for the electrons, we find the rate of spontaneous emission for the transverse mode as

$$\left.\frac{\partial \mathcal{E}_{\mathbf{k}}}{\partial t}\right]_{\mathrm{T}}^{\mathrm{sp}} = (2\pi e)^2 n \sum_{\nu} \int d\mathbf{p} \left\{ [v_\perp J_\nu(z)]^2 + \left[\left(\frac{k_{\parallel}\omega}{kk_\perp} - \frac{kv_{\parallel}}{k_\perp} \right) J_\nu(z) \right]^2 \right\}$$

$$\times f_e(p_\perp, p_\parallel) \delta(\nu\Omega_e + k_\parallel v_\parallel - \omega_{\mathrm{T}}) . \tag{7.213}$$

Here z and Ω_e were defined in Eqs. (3.126) and (7.36); various angles are defined via $\cos\theta = v_\parallel/v$, $\cos\varphi = k_\parallel/k$, and $\psi = \theta - \varphi$, as depicted in Fig. 7.2.

Assuming ultrarelativistic ($\gamma >> 1$) electrons with $\theta > 1/\gamma$, for which the radiation emitted is confined in directions,

$$\psi < \frac{1}{\gamma}, \tag{7.214}$$

we may set

$$\omega_{\mathrm{T}} = ck = \frac{\nu\Omega_e}{\sin^2\theta},$$

$$\frac{\nu - z}{z} = \frac{1 + \gamma^2\psi^2}{2\gamma^2 \sin^2\theta} .$$

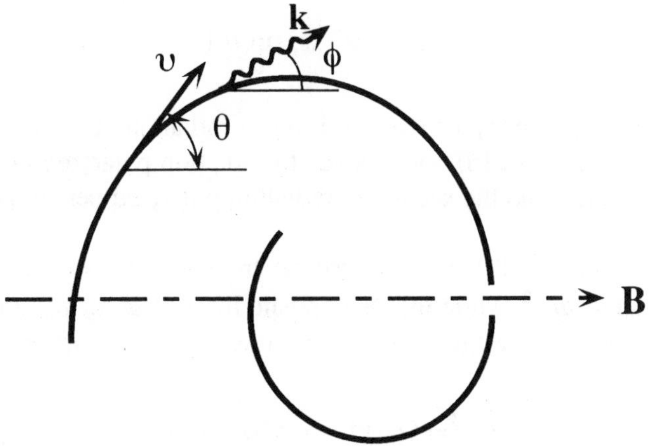

Figure 7.2 Angles of synchrotron orbital.

For the higher harmonics such that $\nu > z >> 1$, the Bessel functions may be asymptotically expressed by Airy integrals or by modified Bessel functions:

$$J_\nu(z) = \frac{1}{\pi}\left[\frac{2(\nu - z)}{3z}\right]^2 K_{1/3}\left[\frac{2^{3/2}(\nu - z)^{3/2}}{3\sqrt{z}}\right],$$

$$J_\nu'(z) = \frac{2(\nu - z)}{\sqrt{3}\,\pi z} K_{2/3}\left[\frac{2^{3/2}(\nu - z)^{3/2}}{3\sqrt{z}}\right].$$

Consequently the energy radiated by a single electron per infinitesimal frequency interval $d\omega$ per infinitesimal solid angle do is obtained from Eq. (7.213) as

$$\frac{d^2P}{do\,d\omega} = \frac{3e^2|\Omega_e|\gamma^2}{4\pi^3 c}\left(\frac{\omega}{\omega_c}\right)^2(1+\gamma^2\psi^2)^2$$

$$\times\left\{K_{2/3}^2\left[\frac{\omega}{2\omega_c}(1+\gamma^2\psi^2)^{3/2}\right] + \frac{\gamma^2\psi^2}{1+\gamma^2\psi^2}K_{1/3}^2\left[\frac{\omega}{2\omega_c}(1+\gamma^2\psi^2)^{3/2}\right]\right\} \quad (7.215)$$

[e.g., Jackson 1962], where

$$\omega_c = \frac{3}{2}\gamma^3 | \Omega_e \sin\theta | \tag{7.216}$$

marks a critical frequency of the synchrotron radiation. The first term in the curly bracket in Eq. (7.215) corresponds to radiation polarized in the plane of the electron orbital, and the second to radiation polarized perpendicular to that plane.

For frequencies well below the critical frequency ($\omega << \omega_c$), Eq. (7.215) is proportional to $\omega^{2/3}$, while for the opposite limit of $\omega >> \omega_c$, it decreases as $\sim\omega \exp(-\omega/\omega_c)$. Integration of Eq. (7.215) over the frequencies thus yields

$$\frac{dP}{do} \approx \frac{e^2}{c} | \Omega_e | \gamma^2 \omega_c \approx \gamma^3 \sigma_T c B^2 \sin\theta \tag{7.217}$$

where

$$\sigma_T = \frac{8\pi}{3}\left(\frac{e^2}{mc^2}\right)^2 = 6.553 \times 10^{-25} \ (\text{cm}^2) \tag{7.218}$$

is the cross section of *Thomson scattering*. Further integration with respect to the solid angle gives an estimate of the total power radiated:

$$P \approx \frac{dP}{do}\frac{\sin\theta}{\gamma} \approx \gamma^2 \sigma_T c B^2 \sin^2\theta \tag{7.219}$$

[e.g., Scheuer 1967].

C. Bremsstrahlung

Let us now consider radiative processes arising from scattering of electrons by the microscopic electric-field fluctuations produced by ions. Since these microscopic fields originate from the electrostatic fields of the individual ions, the processes under consideration correspond to the *bremsstrahlung*.

The current-density fluctuations relevant for substitution in Eq. (7.207) are those of the nonlinear fluctuations (7.26c) for the electrons in which the electric-field fluctuations $\delta\mathbf{E}(\mathbf{q},\omega')$ are those originating from the ions. Assuming a nonrelativistic plasma ($\omega >> \mathbf{k}\cdot\mathbf{v}$) without an external magnetic field, we thus obtain

$$\delta\mathbf{J}(\mathbf{k},\omega)^{\text{NL}} = -\frac{e^2}{m\omega}\sum_{\mathbf{q}}\int_{-\infty}^{\infty}\frac{d\omega'}{2\pi}\,\delta n_e(\mathbf{k}-\mathbf{q},\omega-\omega')\delta\mathbf{E}(\mathbf{q},\omega') \tag{7.220}$$

where $\delta n_e(\mathbf{k}-\mathbf{q},\omega-\omega')$ is the density fluctuation of the electrons.

In the notation adopted in Section 7.4B, the electric field fluctuations in Eq. (7.220) are expressed as

$$\delta\mathbf{E}(\mathbf{q},\omega') = -i\mathbf{q}\frac{Z}{e}\frac{v(q)}{\tilde{\varepsilon}(q,\omega')}\delta n_i(\mathbf{q},\omega') . \qquad (7.221)$$

The nonlinear transverse electric field associated with the current fluctuations Eq. (7.220) is also calculated as

$$\delta\mathbf{E}_{\mathrm{T}}(\mathbf{k},\omega)^{\mathrm{NL}} = -\frac{4\pi}{i\omega}\frac{\mathsf{I}_{\mathrm{T}}}{\varepsilon_{\mathrm{T}}(\mathbf{k},\omega)-(ck/\omega)^2}\cdot\delta\mathbf{J}(\mathbf{k},\omega)^{\mathrm{NL}} . \qquad (7.222)$$

Substitution of Eqs. (7.220) and (7.221) in Eq. (7.207) yields the rate of *free-free emission* or bremsstrahlung:

$$\frac{d^2P}{do d\omega} = \frac{4\pi(Ze)^2\sigma_{\mathrm{T}}c}{2\alpha_{\mathrm{T}}}\sum_{\mathbf{q}}\int_{-\infty}^{\infty}d\omega'\frac{|\hat{\mathbf{k}}\times\hat{\mathbf{q}}|^2}{q^2|\tilde{\varepsilon}(\mathbf{q},\omega)|^2}$$

$$\times[S_{ee}(\mathbf{k}-\mathbf{q},\omega-\omega')S_{ii}(\mathbf{q},\omega')+S_{ei}(\mathbf{k}-\mathbf{q},\omega-\omega')S_{ie}(\mathbf{q},\omega')] . \qquad (7.223)$$

In the derivation of this formula, the statistical average is approximately evaluated through pairwise factorizations on the product of four fluctuating quantities. Explicit calculations of the dynamic structure factors thus account for the bremsstrahlung processes.

D. Incoherent Scattering

The differential cross section of a plasma for the scattering of electromagnetic radiation (see Fig. 7.3) is likewise calculated through consideration of the nonlinear coupling processes (7.26c) for electrons.

We take the incident electromagnetic wave in the form of

$$\mathbf{E}(\mathbf{r},t) = \mathbf{E}_1\cos(\mathbf{k}_1\cdot\mathbf{r}-\omega_1 t) \qquad (7.224)$$

where $\omega_1 = k_1 c$ and $\mathbf{E}_1 = \hat{\eta}_1 E_1$ with $\hat{\eta}_1$ representing a unit polarization vector. Its Fourier components are

$$\mathbf{E}_1(\mathbf{k},\omega) = \hat{\eta}_1 E_1\pi\ \delta(\mathbf{k},\mathbf{k}_1)\delta(\omega-\omega_1)+\delta(\mathbf{k},-\mathbf{k}_1)\delta(\omega+\omega_1) \qquad (7.225)$$

where $\delta(\mathbf{k},\mathbf{k}_1)$ is the three-dimensional Kronecker's delta.

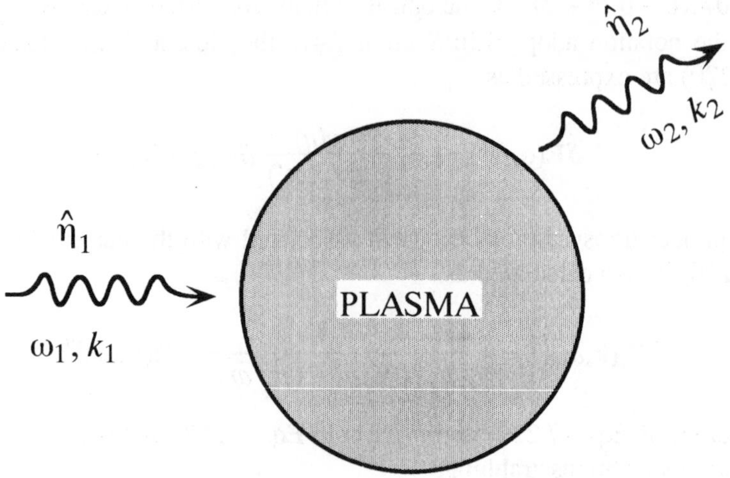

Figure 7.3 Incoherent scattering.

We assume that the frequencies, ω and ω', of the scattered and incident electromagnetic waves in (7.26c) are much greater than the plasma frequency ω_p and the cyclotron frequency $|\Omega_e|$ of the electrons. Since the latter frequencies represent the characteristic frequencies of the electron-density fluctuations $\delta N_e(\mathbf{k}-\mathbf{q},\omega-\omega';\mathbf{p}')$, the foregoing assumption implies

$$|\omega|, |\omega'| >> |\omega - \omega'| .$$

This also means that over the characteristic times, $|\omega|^{-1}$ and $|\omega'|^{-1}$, for the change in the transverse fields, the electrons cannot evolve significantly in the phase space, so that we may set $\mathbf{p}'=\mathbf{p}$. Finally we assume that the electrons are nonrelativistic so that $v << |\omega'/q| = c$; this observation permits a replacement

$$\mathsf{T}(\mathbf{q},\omega';\mathbf{p}) \to \mathsf{I} \tag{7.226}$$

in (7.26c).

We thus find after this simplification that the nonlinear electric-current density is calculated as

$$\delta \mathbf{J}(\mathbf{r},t)^{\mathrm{NL}} = \frac{e^2 E_1}{2mV} \hat{\eta}_1 \sum_{\mathbf{k}} \int_{-\infty}^{\infty} \frac{d\omega}{2\pi} \frac{i}{\omega} [\delta n(\mathbf{k}-\mathbf{k}_1, \omega - \omega_1)$$

$$+ \delta n(\mathbf{k}+\mathbf{k}_1, \omega + \omega_1)] \exp[i(\mathbf{k}\cdot\mathbf{r} - \omega t)] , \qquad (7.227)$$

where V refers to the scattering volume of the plasma. The radiation field produced by this current-density fluctuation at a distance R from the plasma may be described by a retarded vector potential. Choosing R sufficiently large as compared to the size of the plasma, we write the retarded potential for such scattered waves as

$$\mathbf{A}(t) = \frac{1}{cR} \int_V d\mathbf{r} \delta \mathbf{J} \left(\mathbf{r}, t - \frac{R}{c} + \frac{\hat{k}_2 \cdot \mathbf{r}}{c} \right)^{\mathrm{NL}} . \qquad (7.228)$$

Let us now suppose that the observation of the scattered wave is made in a fixed polarization direction $\hat{\eta}_2$ around the wave vector $\mathbf{k}_2$. The intensities of the scattered electromagnetic waves are then given by

$$E_2(t) = \hat{\eta}_2 \cdot \mathbf{E}_2(t) = \frac{1}{c} \hat{\eta}_2 \cdot [(\dot{\mathbf{A}} \times \hat{\mathbf{k}}_2) \times \hat{\mathbf{k}}_2] , \qquad (7.229a)$$

$$B_2(t) = | \hat{\eta}_2 \times \mathbf{B}_2(t) | = \frac{1}{c} | \hat{\eta}_2 \times (\dot{\mathbf{A}} \times \hat{\mathbf{k}}_2) | . \qquad (7.229b)$$

The energy flux associated with the scattered wave is calculated from the statistical average $\langle S \rangle$ of Poynting's vector; hence

$$\langle S \rangle = \frac{c}{4\pi} \langle E_2(t) B_2(t) \rangle$$

$$= \frac{\hat{\mathbf{k}}_2}{c} \frac{cE_1^2}{8\pi} \left(\frac{e^2}{mc^2} \right)^2 \frac{V}{R^2} (\hat{\eta}_1 \cdot \hat{\eta}_2)^2$$

$$\times \sum_{\mathbf{k}} \int_{-\infty}^{\infty} S(\mathbf{k},\omega) \delta \left(\mathbf{k} + \mathbf{k}_1, \frac{\omega + \omega_1}{2} \hat{\mathbf{k}}_2 \right) . \qquad (7.230)$$

Here $S(\mathbf{k},\omega)$ is the dynamic structure factor of the electrons.

In Eq. (7.230), the quantity $(cE_1^2/8\pi)$ is the time-averaged energy flux of the incident wave. Hence we obtain from Eq. (7.230) the differential cross section for the transfer of momentum $\hbar\mathbf{k}$ (corresponding to scattering into a solid angle do) and energy $\hbar\omega$ from the plasma to the electromagnetic wave as [e.g., Salpeter 1960, Rosenbluth & Rostoker 1962]

$$\frac{d^2Q}{do\,d\omega} = \frac{3V}{8\pi}\sigma_T(\hat{\eta}_1 \cdot \hat{\eta}_2)^2 S(\mathbf{k},\omega) \tag{7.231}$$

where $\mathbf{k} = \mathbf{k}_2 - \mathbf{k}_1$ and $\omega = \omega_2 - \omega_1$.

When no attention is paid to the polarization directions in the experiment, Eq. (7.231) is averaged over the two states of $\hat{\eta}_1$ and is summed over $\hat{\eta}_2$, with the result

$$\frac{d^2Q}{do\,d\omega} = \frac{3V}{8\pi}\sigma_T\left(1 - \frac{1}{2}\sin^2\theta\right)S(\mathbf{k},\omega), \tag{7.232}$$

where θ is the scattering angle between $\mathbf{k}_1$ and $\mathbf{k}_2$.

E. Inverse Compton Processes

Consider a situation in which an electromagnetic wave with frequency ω_1 and wave vector $\mathbf{k}_1$ is scattered by an ultrarelativistic electron ($\gamma >> 1$) with velocity $\mathbf{v}$, producing an elecgromagnetic wave with frequency ω_2 and wave vector $\mathbf{k}_2$ (see Fig. 7.4). Quantum effects in scattering may be negligible as we consider only those cases where the electron energy γmc^2 is much greater than the photon energy $\hbar\omega_1$. The incident and scattered waves satisfy the relation

$$\omega_2 - \omega_1 = \mathbf{v} \cdot (\mathbf{k}_2 - \mathbf{k}_1) \tag{7.233}$$

with $\omega_1 = ck_1$ and $\omega_2 = ck_2$. The frequency of the scattered wave is thus calculated as

$$\omega_2 = \omega_1 \frac{1 - (v/c)\cos\theta_1}{1 - (v/c)\cos\theta_2}. \tag{7.234}$$

Solving Eq. (7.6) for v/c and expanding the result with respect to γ^{-2}, one finds

$$\frac{v}{c} \approx 1 - \frac{1}{2\gamma^2}. \tag{7.235}$$

With the assumption $\theta_2^2 << 1$, one likewise has

$$\cos\theta_2 \simeq 1 - \frac{\theta_2^2}{2}. \tag{7.236}$$

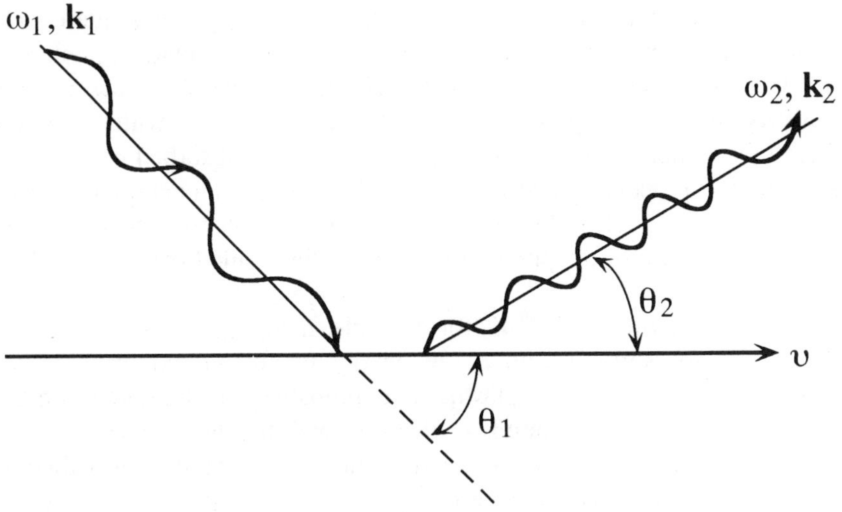

Figure 7.4 Inverse Compton scattering.

Substitution of these expansions in Eq. (7.234) yields

$$\omega_2 = 2\gamma^2\omega_1 \frac{1-(v/c)\cos\theta_1}{1+(\gamma\theta_2)^2}. \tag{7.237}$$

For those electromagnetic waves scattered into an opening angle θ_2 with respect to the direction of electron motion, satisfying

$$\theta_2^2 << \gamma^{-2}, \tag{7.238}$$

Eq. (7.237) becomes

$$\omega_2 = 2\gamma^2\omega_1\left(1-\frac{v}{c}\cos\theta_1\right). \tag{7.239}$$

The frequency ω_2 is distributed between 0 and $4\gamma^2\omega_1$, depending on the value of θ_1; it takes on the maximum value $4\gamma^2\omega_1$ at $\theta_1 = \pi$.

As the foregoing calculation has demonstrated, the electromagnetic wave scattered in the forward direction of the relativistic electron receives an enhancement of its frequency by a factor on the order of γ^2. Consequently, the bulk of energy in the scattered electromagnetic wave is confined in a narrow solid angle satisfying the condition (7.238); the energy of scattered photon becomes approximately γ^2 times as large as that of the incident photon. Ordinary Compton scattering, on the other hand, concerns the cases in which a relatively high-energy photon such as an X-ray collides with a low-energy electron; the photon energy decreases by an amount absorbed by recoil of the electron. The process treated in Eq. (7.236) produces an energy enhancement of the photon, in contrast to the ordinary Compton process, which is accompanied by a decrease of the photon energy; it is thus called the *inverse Compton process*.

In a plasma with $T \approx 10^9$ K, for example, many electrons have relativistic energies. If for some reason a low-energy photon—radio wave to visible light—is supplied in such a plasma, it is probable that the photon undergoes repeated scattering with relativistic electrons and thereby receives an enhancement of its energy into the X-ray domain. Such a process is called *Comptonization* of a low-energy photon; it plays an essential part in the interpretation of physical mechanisms involved in X-ray stars.

Finally let us remark that such an inverse Compton process is used profitably in *free-electron laser* experiments, such as those performed in the W.W. Hansen High Energy Physics Laboratory at Stanford [Elias et al. 1976; Deacon et al. 1977]. Here one deals with a static configuration of the magnetic field

$$\mathbf{B}(z) = \delta B\, \hat{\mathbf{x}} \cos(qz) + \hat{\mathbf{y}} \sin(qz) \tag{7.240}$$

called a *wiggler*. A beam of relativistic electrons is injected along the z axis of the wiggler field. The inverse Compton process relevant to this case is still described by Eq. (7.233), in which one now sets $\omega_1 = 0$ and $\mathbf{k}_1 = \mathbf{q}$. The photon produced in the z direction thus has a frequency

$$\omega_2 = 2\gamma^2 cq\,. \tag{7.241}$$

When a 24 MeV electron beam is injected into a wiggler of pitch length 3.2 cm (i.e., $q = 2\pi/3.2$ cm^{-1}), the wavelength of the output photon computed from Eq. (7.241) gives 7.2 μm.

PROBLEMS

7.1 Derive Eqs. (7.15) and explain the conservation properties of those equations.

7.2 Prove the relation Eq. (7.31).

7.3 Derive Eq. (7.52). In the derivation, it is useful to expand

$$\omega\varepsilon(\mathbf{k},\omega) = \omega'\varepsilon(\mathbf{k},\omega') + (\omega - \omega')\frac{\partial[\omega'\varepsilon(\mathbf{k},\omega')]}{\partial\omega'}$$

in conjunction with the definition Eq. (7.53).

7.4 Derive Eq. (7.58) with the use of the electrostatic approximation (7.47) to Eq. (7.43), and show that, when the momentum distribution is an undrifted Maxwellian, then

$$\mathcal{E}_{\mathbf{k}} = T\,.$$

7.5 Derive Eqs. (7.76).

7.6 Carry out a calculation analogous to Eq. (7.92) for the ion-acoustic mode in a Maxwellian TCP, and show

$$S_{ac}(\mathbf{k}) \approx 1\,.$$

7.7 For the TCP with $T_e >> T_i$, show that the energy density Eq. (7.57) in the ion-acoustic mode is given by

$$\mathcal{E}_k \approx T_e\,.$$

7.8 Show that a combination of Eq. (7.102) with Eq. (7.72) yields the Born-Green equation between the two- and three-particle radial distribution functions

7.9 Substituting the expression (7.72) in Eq. (7.63), show that for convergence of $g(0)$, $[1 - G(k)]$ should vanish in the limit, $k \to \infty$, faster than k^{-4}.

7.10 Assume that the local field correction in Eq. (7.72) is given by the STLS approximation Eq. (7.108), and solve the resultant set of self-consistent equations to calculate U_{int} in Eq. (7.65) at $\Gamma = 2$.

7.11 Assume that the local field correction in Eq. (7.72) is given by the MCA Eqs. (7.107), (7.109)–(7.111), and solve the resultant set of self-consistent equations to calculate U_{int} in Eq. (7.65) at $\Gamma = 2$.

7.12 Derive Eq. (7.138).

7.13 Derive Eq. (7.152).

7.14 Derive Eq. (7.183).

7.15 On a log$n$ vs. log$T$ plane for a hydrogen plasma, draw the relation specified by Eq. (7.189), and compare the result with the lines representing Eqs. (7.184) and (7.185).

7.16 Derive Eq. (7.223).

7.17 Derive Eq. (7.232) from Eq. (7.231).

7.18 Compute the wavelength of an output photon from Eq. (7.241) when a 43.5 MeV electron beam is injected into a wiggler of pitch length 3.2 cm.

CHAPTER **8**

Relaxations

In Chapter 2, we derived a collision term in the plasma kinetic theory. The fluctuation formalism described in the preceding chapter has provided another approach by which the transport processes in a plasma may be analyzed. In this chapter we extend such a fluctuation formalism to a consideration of relaxation processes based on the theory of Brownian motion.

Microscopically, each particle in a plasma is influenced by the fluctuating electric field produced by all other particles; its motion in the phase space may therefore resemble that of a Brownian particle. The behavior of such a particle is then described by the Fokker-Planck equation, the coefficients of which contain spectral functions of the microscopic electric fields; the rates of various relaxation processes are thereby determined. This approach thus makes it possible to express various transport coefficients in terms of the spectral functions of fluctuations.

This method of analyzing relaxation processes from a fluctuation-theoretic point of view has a definite advantage over a kinetic-theoretic approach, in that the formulation is applicable both to a plasma near thermodynamic equilibrium and to one in a turbulent state, as long as the relevant fluctuation spectra are known either experimentally or theoretically. In addition, one can investigate the extent to which each different frequency or wave vector regime of a fluctuation spectrum contributes to a given relaxation process through such a formalism.

In this chapter, we therefore calculate the Fokker-Planck coefficients in terms of the spectral functions of the electric-field fluctuations and the dielectric response function in a magnetic field. As concrete examples of the relaxation processes, we consider a temperature relaxation and a plasma diffusion across the magnetic field.

8.1 FOKKER-PLANCK EQUATIONS

In the theory of *Brownian motion*, it is essential to recognize the stochastic nature of the phenomenon and thereby to seek a description in terms of the transition probability in the phase space starting with a given initial distribution [Chandrasekhar 1943]. One assumes the existence of a time interval τ during which the phase-space coordinates for a Brownian particle change by infinitesimal amounts while fluctuations characteristic of interaction between such a Brownian motion and the surrounding medium take place very frequently. It is thus necessary once again to assume

$$\tau_1 >> \tau >> \tau_2 , \tag{8.1}$$

in conjunction with the hierarchical time scales considered in Sections 2.3B and 7.1A.

The time evolution of the distribution function $f(X;t)$ for the Brownian particles may then be described by an integral equation

$$f(X;t+\tau) = \int f(X-\Delta X;t) W_\tau(X-\Delta X;\Delta X) d(\Delta X) . \tag{8.2}$$

Here $X \equiv (\mathbf{r},\mathbf{p})$ are the phase-space coordinates for a particle, and $W_\tau(X;\Delta X)$ refers to a *transition probability* that X changes by an increment ΔX during the time interval τ. The probability function is normalized so that

$$\int W_\tau(X;\Delta X) d(\Delta X) = 1 . \tag{8.3}$$

In expecting that the integral equation (8.2) is applicable, we are actually supposing that the evolution of a Brownian motion depends only on the instantaneous values of its physical parameters and is entirely independent of its whole previous history. In probability theory, a stochastic process which has this characteristic, namely, that what happens at a given instant of time t depends only on the state of the system at time t, is said to be a *Markoff process.* That we should be able to idealize the motion of charged particles in a plasma as a Markoff process appears fairly reasonable, but far from obvious. In fact, discussions in Section 4.3A imply that plasma oscillations can have an infinite lifetime in the limit of long wavelengths; the behavior of those particles coupled with such long-lived excitations is expected to show a non-Markoffian character. Let us note, however, that the intensities of plasma oscillations become infinitesimally small as $k \to 0$ (see Eq. 7.93); their presence may thus have only a minor and perhaps negligible effect on the overall properties of relaxation processes in plasmas near thermodynamic equilibrium [Ichimaru & Rosenbluth 1970]. To an extent that effects of such long-lived collective oscillations are negligible in a given relaxation process, we can adopt a Markoffian hypothesis of Eq. (8.2).

Equation (8.2) is now expanded in a Taylor series with respect to τ and ΔX:

$$f(X;t) + \frac{\partial f}{\partial t}\tau + O\left(\frac{\tau^2}{\tau_1^2}\right)$$

$$= \int d(\Delta X)\left[f(X;t) - \Delta X \cdot \frac{\partial f}{\partial X} + \frac{1}{2}(\Delta X \Delta X) : \frac{\partial^2 f}{\partial X \partial X} - \cdots\right]$$

$$\times \left[W_\tau(X;\Delta X) - \Delta X \cdot \frac{\partial W_\tau}{\partial X} + \frac{1}{2}(\Delta X \Delta X) : \frac{\partial^2 W_\tau}{\partial X \partial X} - \cdots\right]. \tag{8.4}$$

Recalling the normalization (8.3), we may define and calculate the average values of the increments according to

$$\langle \Delta X \rangle = \int \Delta X W_\tau(X;\Delta X) d(\Delta X) \tag{8.5a}$$

$$\langle \Delta X \Delta X \rangle = \int \Delta X \Delta X W_\tau(X;\Delta X) d(\Delta X)\,. \tag{8.5b}$$

In terms of these averages, Eq. (8.4) can be rewritten to the lowest order in τ/τ_1 as

$$\frac{\partial f}{\partial t} = -\frac{\partial}{\partial X} \cdot \left[\frac{\langle \Delta X \rangle}{\tau} f\right] + \frac{1}{2}\frac{\partial^2}{\partial X \partial X} : \left[\frac{\langle \Delta X \Delta X \rangle}{\tau} f\right] + O\left(\frac{\langle \Delta X \Delta X \Delta X \rangle}{\tau}\right). \tag{8.6}$$

This is a *Fokker-Planck equation* in its general form; the factors $\langle \Delta X \rangle/\tau$ and $\langle \Delta X \Delta X \rangle/\tau$ are the *Fokker-Planck coefficients*.

In a treatment of relaxation processes through such a formalism, the Fokker-Planck coefficients are calculated by considering the effects of microscopic fluctuating forces on the motion of particles. In a homogeneous plasma, those coefficients as well as the distribution functions are independent of the spatial coordinates; Eq. (8.6) is then simplified as

$$\frac{\partial f}{\partial t} = -\frac{\partial}{\partial \mathbf{p}} \cdot \left[\frac{\langle \Delta \mathbf{p} \rangle}{\tau} f \right] + \frac{1}{2} \frac{\partial^2}{\partial \mathbf{p} \partial \mathbf{p}} : \left[\frac{\langle \Delta \mathbf{p} \Delta \mathbf{p} \rangle}{\tau} f \right]. \tag{8.7}$$

In obtaining this formula, we have truncated the series expansion by ignoring the terms with the coefficients $\langle \Delta \mathbf{p} \Delta \mathbf{p} \Delta \mathbf{p} \rangle / \tau$ and those of higher order in $\Delta \mathbf{p}$. For a weakly coupled plasma near thermodynamic equilibrium, such a truncation may be justified [Rosenbluth, MacDonald & Judd 1957]. The essential point is that in a Coulombic system the averages $\langle \Delta \mathbf{p} \rangle$ and $\langle \Delta \mathbf{p} \Delta \mathbf{p} \rangle$ contain terms that would diverge logarithmically and thus lead to Coulomb logarithms with an adoption of an appropriate cutoff for the integration. The averages $\langle \Delta \mathbf{p} \Delta \mathbf{p} \Delta \mathbf{p} \rangle$ and higher order contributions, on the other hand, do not exhibit such a divergent behavior. (See Problem 8.1.)

Logarithmically divergent terms leading to the Coulomb logarithms have been called the "dominant" terms by Chandrasekhar [1942]; these represent the major effects in relaxations for weakly coupled plasma. We shall consider examples of such a dominant-term approximation in Section 8.2.

The right-hand side of Eq. (8.7) should correspond to the collision term in a kinetic equation. The Fokker-Planck coefficients in the momentum space:

$$\mathbf{F}(\mathbf{p}) = \frac{1}{\tau} \langle \Delta \mathbf{p} \rangle , \tag{8.8}$$

$$\mathsf{D}(\mathbf{p}) = \frac{1}{\tau} \langle \Delta \mathbf{p} \Delta \mathbf{p} \rangle , \tag{8.9}$$

are then the basic quantities which determine the collision terms for plasmas in the Fokker-Planck formalism. $\mathbf{F}(\mathbf{p})$ and $\mathsf{D}(\mathbf{p})$ are, respectively, the friction and diffusion coefficients for the Brownian particles in the momentum space. Equations (7.121), (7.123), and (7.124) illustrate that the collision terms for plasmas are generally expressible in the Fokker-Planck form of Eq. (8.7).

8.2 FOKKER-PLANCK COEFFICIENTS

Effects of fluctuating electric fields on the motion of a charged particle in the absence of an external magnetic field have been extensively investigated by Hubbard and Thompson [Thompson & Hubbard 1960; Hubbard 1961b]. They were concerned with the first-order effects in the fluctuations on friction and diffusion in the velocity (i.e., momentum) space. Within such a framework their calculations are rigorous and provide an alternative method of deriving the BGL collision term when a theoretical expression for the spectral functions is substituted into the resulting Fokker-Planck coefficients. Let us therefore follow their approache in this section and calculate the Fokker-Planck coefficients.

A. Particle Motion in Fluctuating Fields

Consider the motion of a particle with electric charge Ze and mass m in a fluctuating electric field, $\mathbf{E}(\mathbf{r},t)$. Spectral distribution of the field fluctuations is assumed to be uniform in space and stationary in time. The equation of motion is

$$\frac{d\mathbf{p}(t)}{dt} = Ze\mathbf{E}[\mathbf{r}(t),t] \tag{8.10}$$

where $\mathbf{r}(t)$ and $\mathbf{p}(t)$ $[=md\mathbf{r}(t)/dt]$ are trajectories of a test particle in the phase space. The trajectories may be determined by integrating Eq. (8.10) as

$$\mathbf{r}(t) = \mathbf{r}(0) + \frac{\mathbf{p}(0)}{m}t + \frac{Ze}{m}\int_0^t dt'(t-t')\mathbf{E}[\mathbf{r}(t'),t'] , \tag{8.11}$$

$$\mathbf{p}(t) = \mathbf{p}(0) + \frac{Ze}{m}\int_0^t dt'\mathbf{E}[\mathbf{r}(t'),t'] . \tag{8.12}$$

The increments $\Delta\mathbf{r}(\tau)$ and $\Delta\mathbf{p}(\tau)$ of the particle coordinates during a time interval τ are readily calculated from Eqs. (8.11) and (8.12) as $\Delta\mathbf{r}(\tau) = \mathbf{r}(\tau) - \mathbf{r}(0)$ and $\Delta\mathbf{p}(\tau) = \mathbf{p}(\tau) - \mathbf{p}(0)$. We expand these quantities with respect to the fluctuating electric fields, to find

$$\Delta\mathbf{r}(\tau) = \frac{\mathbf{p}}{m}\tau + \frac{Ze}{m}\int_0^\tau ds s\mathbf{E}[\mathbf{r}_0(\tau - s),\tau - s] + \left(\frac{Ze}{m}\right)^2 \int_0^\tau ds s \int_0^{\tau - s} ds' s'$$
$$\times \mathbf{E}[\mathbf{r}_0(\tau - s - s'),\tau - s - s'] \cdot \frac{\partial}{\partial \mathbf{r}_0} \mathbf{E}[\mathbf{r}_0(\tau - s),\tau - s] + \cdots , \tag{8.13}$$

$$\Delta\mathbf{p}(\tau) = Ze\int_0^\tau ds\mathbf{E}[\mathbf{r}_0(s),s]$$
$$+ \frac{(Ze)^2}{m}\int_0^\tau ds \int_0^s ds' s' \mathbf{E}[\mathbf{r}_0(s - s'),s - s'] \cdot \frac{\partial}{\partial \mathbf{r}_0} \mathbf{E}[\mathbf{r}_0(s),s] + \cdots \tag{8.14}$$

where $\mathbf{p} \equiv \mathbf{p}(0)$ is the initial momentum, and

$$\mathbf{r}_0(t) = \mathbf{r}(0) + \frac{\mathbf{p}}{m}t \tag{8.15}$$

is the unperturbed orbit of the test particle.

When a uniform magnetic field is applied in the z direction, we take additional account of the spiral motion of the particle as described by Eqs. (3.115) and (3.116). Equations (8.13)–(8.15) are now replaced by

$$\begin{aligned}\Delta\mathbf{r}(\tau) = {} & \frac{1}{\Omega}\mathsf{H}(\tau)\cdot\frac{\mathbf{p}}{m} + \frac{c}{B}\int_0^\tau ds\mathsf{H}(s)\cdot\mathbf{E}[\mathbf{r}_0(\tau-s),\tau-s] \\ & + \frac{c^2}{B^2}\int_0^\tau ds\mathsf{H}(s)\cdot\left[\left\{\int_0^{\tau-s} ds'\mathsf{H}(s')\cdot\mathbf{E}[\mathbf{r}_0(\tau-s-s'),\tau-s-s']\right\}\right. \\ & \left.\cdot\frac{\partial}{\partial\mathbf{r}_0}\mathbf{E}[\mathbf{r}_0(\tau-s),\tau-s]\right] + \cdots,\end{aligned} \tag{8.16}$$

$$\begin{aligned}\Delta\mathbf{p}(\tau) = {} & [\mathsf{B}(\tau)-\mathsf{I}]\cdot\mathbf{p} + Ze\int_0^\tau ds\mathsf{B}(s)\cdot\mathbf{E}[\mathbf{r}_0(\tau-s),\tau-s] \\ & + \frac{Zec}{B}\int_0^\tau ds\mathsf{B}(s)\cdot\left[\left\{\int_0^{\tau-s} ds'\mathsf{H}(s')\cdot\mathbf{E}[\mathbf{r}_0(\tau-s-s'),\tau-s-s']\right\}\right. \\ & \left.\cdot\frac{\partial}{\partial\mathbf{r}_0}\mathbf{E}[\mathbf{r}_0(\tau-s),\tau-s]\right] + \cdots,\end{aligned} \tag{8.17}$$

$$\mathbf{r}_0(t) = \mathbf{r}(0) + \frac{1}{\Omega}\mathsf{H}(t)\cdot\frac{\mathbf{p}}{m}. \tag{8.18}$$

These equations form the basis for evaluating various relaxation rates in the following sections.

B. Diffusion Coefficients in Momentum Space

According to Eq. (8.10), the position and momentum of a test particle fluctuate as the microscopic electric fields; their instantaneous displacements are given by Eqs. (8.13) and (8.14). The diffusion tensor in the momentum space is then calculated through definition (8.9) with τ satisfying (8.1). To the lowest order in the fluctuations, we thus obtain

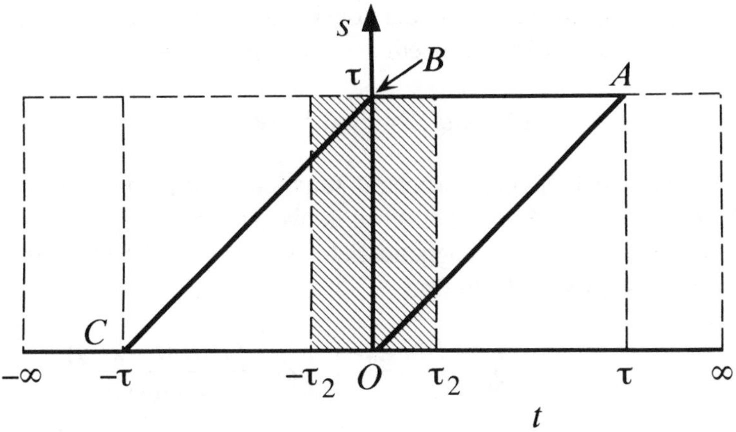

Figure 8.1 Domain of integration in Eq.(8.19).

$$\mathsf{D}(\mathbf{p}) = \frac{(Ze)^2}{\tau} \int_0^{\tau} ds \int_{s-\tau}^{s} dt \langle \mathbf{E}[\mathbf{r}_0(s), s] \mathbf{E}[\mathbf{r}_0(s-t), s-t] \rangle \tag{8.19}$$

where one of the integration variables has been changed to t via $t = s - s'$. The domain of the double integration in Eq. (8.19) is illustrated in Fig. 8.1 as the parallelogram ABCO. By the definition of the correlation time τ_2, the integrand of Eq. (8.19) takes on considerable magnitude only in the shaded domain of Fig. 8.1. Neglecting the errors stemming from the contributions in the small triangular domains outside the parallelogram in the vicinity of O and B, we may extend the upper and lower limits of the t integration in Eq. (8.19) to $+\infty$ and $-\infty$; the errors incurred are on the order of τ_2/τ and thus negligible in light of (8.1).

We may then express the microscopic electric fields in the spectral representation as developed in Eqs. (7.28)–(7.31). In terms of the spectral tensor $\langle \mathbf{EE}^*(\mathbf{k},\omega) \rangle$ of the field fluctuations,* Eq. (8.19) is thus rewritten as

$$\mathsf{D}(\mathbf{p}) = 2\pi (Ze)^2 \sum_{\mathbf{k}} \langle \mathbf{EE}^*(\mathbf{k}, \mathbf{k}\cdot\mathbf{v}) \rangle \, . \tag{8.20}$$

This equation gives a general relationship between the diffusion tensor and the spectral tensor of the electric-field fluctuations.

*In this chapter we use Fourier decompositions of field variables with the periodic boundary conditions appropriate to a cube of a unit volume.

When the field fluctuations under consideration are primarily of longitudinal nature, a condition which we here assume, the spectral tensor takes a form

$$\langle \mathbf{EE}^*(\mathbf{k},\omega)\rangle = \frac{\mathbf{kk}}{k^2}\langle | E^2 |(\mathbf{k},\omega)\rangle\,, \tag{8.21}$$

where $\langle | E^2 |(\mathbf{k},\omega)\rangle$ is a power spectral function of the longitudinal electric-field fluctuations (cf., Eq. 7.212a). In a weakly coupled, multicomponent plasma without an external magnetic field, we may evoke the superposition calculations of Section 7.2C, to find

$$\langle | E^2 |(\mathbf{k},\omega)\rangle = \sum_\sigma n_\sigma \left(\frac{4\pi Z_\sigma e}{k}\right)^2 \int d\mathbf{v}\frac{f_\sigma(\mathbf{v})\delta(\omega - \mathbf{k}\cdot\mathbf{v})}{|\,\varepsilon(\mathbf{k},\omega)\,|^2}\,. \tag{8.22}$$

Here $\varepsilon(\mathbf{k},\omega)$ is the (longitudinal) dielectric response function calculated as in Eq. (3.136), in terms of the velocity distribution functions $f_\sigma(\mathbf{v})$ normalized to unity (cf., Eq. 2.54). The diffusion tensor Eq. (8.20) now takes a form

$$\mathsf{D}(\mathbf{p}) = 2\pi(Ze)^2\sum_\sigma\sum_\mathbf{k}\int d\mathbf{v}'\left(\frac{4\pi Z_\sigma e}{k^2}\right)^2 \mathbf{kk}\,\frac{n_\sigma f_\sigma(\mathbf{v}')}{|\,\varepsilon(\mathbf{k},\mathbf{k}\cdot\mathbf{v})\,|^2}\,\delta[\mathbf{k}\cdot(\mathbf{v}-\mathbf{v}')]\,. \tag{8.23}$$

In many cases, the relaxation processes are considered for plasmas near thermodynamic equilibrium. It is then useful to evaluate the Fokker-Planck coefficients for a plasma with an isotropic Maxwellian distribution, Eq. (1.1). We assume that no magnetic field is applied to the plasma. Based on an argument similar to the one leading to Eq. (3.94), we then note that the diffusion tensor must be expressible as

$$\mathsf{D}(\mathbf{p}) = D_{||}(p)\frac{\mathbf{pp}}{p^2} + D_\perp(p)\left(\mathsf{I} - \frac{\mathbf{pp}}{p^2}\right). \tag{8.24}$$

We refer $D_{||}(p)$ and $D_\perp(p)$, respectively, to the parallel and perpendicular diffusion coefficients; these depend on the magnitude of $\mathbf{p}$ only.

The *parallel diffusion coefficient* is thus calculated as

$$D_{||}(p) = \frac{\sqrt{2}}{\pi}(Ze)^2\sum_\sigma \frac{4\pi n_\sigma (Z_\sigma e)^2}{\sqrt{T_\sigma/m_\sigma}}\int_{-1}^{1} d\mu\,\mu^2 \exp\left(-\frac{v^2}{2\sqrt{T_\sigma/m_\sigma}}\mu^2\right)$$
$$\times\left\{\ln\frac{k_\mathrm{m}}{k_\mathrm{D}} + \frac{1}{4}\ln\frac{[1+(k_\mathrm{D}/k_\mathrm{m})^2X]^2 + (k_\mathrm{D}/k_\mathrm{m})^4Y^2}{X^2+Y^2} - \frac{X}{2|\,Y\,|}\left[\frac{\pi}{2} - \tan^{-1}\frac{X}{|\,Y\,|}\right]\right\}. \tag{8.25}$$

Here the dielectric response function is written in terms of the W function Eq. (4.3) as

$$\varepsilon(\mathbf{k},\mathbf{k}\cdot\mathbf{v}) = 1 + \sum_{\sigma} \frac{k_{\sigma}^2}{k^2} W\left(\frac{\mathbf{k}\cdot\mathbf{v}}{k\sqrt{T_{\sigma}/m_{\sigma}}}\right)$$

$$= 1 + \frac{k_{\mathrm{D}}^2}{k^2}(X + iY) \tag{8.26}$$

with

$$k_{\mathrm{D}}^2 = \sum_{\sigma} k_{\sigma}^2 = \sum_{\sigma} \frac{4\pi n_{\sigma}(Z_{\sigma}e)^2}{T_{\sigma}},$$

$$X = \frac{1}{k_{\mathrm{D}}^2} \sum_{\sigma} k_{\sigma}^2 \mathrm{Re}\, W\left(\frac{\mathbf{k}\cdot\mathbf{v}}{k\sqrt{T_{\sigma}/m_{\sigma}}}\right), \tag{8.27a}$$

$$Y = \frac{1}{k_{\mathrm{D}}^2} \sum_{\sigma} k_{\sigma}^2 \mathrm{Im}\, W\left(\frac{\mathbf{k}\cdot\mathbf{v}}{k\sqrt{T_{\sigma}/m_{\sigma}}}\right). \tag{8.27b}$$

Note that X and Y are independent of the magnitude k of the wave vector and are functions of only its direction cosine, $\mu = \cos\theta$, where θ is the angle between $\mathbf{k}$ and $\mathbf{p}$.

In the derivation of Eq. (8.25), we have set the upper bound of the k integration at k_{m}. As this upper bound approaches infinity, the first term, $\ln(k_{\mathrm{m}}/k_{\mathrm{D}})$, in the curly bracket of Eq. (8.25) diverges logarithmically, while the other two terms remain finite. The former corresponds to the "dominant" term of Chandrasekhar [1942]. Since the value of k_{m} may be estimated as the inverse of an average distance of the closest approach between two colliding particles (see Eq. 2.71), the logarithmic term represents the major effect for a weakly coupled plasma, where $\ln(k_{\mathrm{m}}/k_{\mathrm{D}}) >> 1$. Hence we may identify

$$\ln\left(\frac{k_{\mathrm{m}}}{k_{\mathrm{D}}}\right) = \ln\Lambda, \tag{8.28}$$

a Coulomb logarithm. It must be remarked, however, that such a dominant-term approximation will break down if both X and Y take on infinitesimally small values at certain μ and p.

Within the validity of such a dominant-term approximation, we thus carry out the remaining μ integration explicitly to obtain

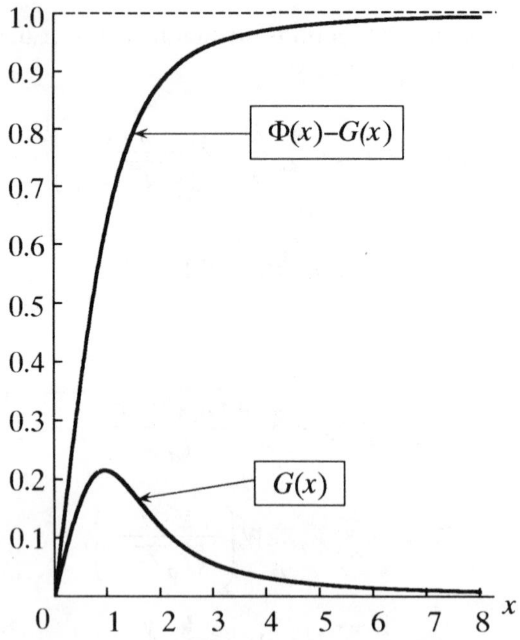

Figure 8.2 Graphs of $G(x)$ and $\Phi(x)$-$G(x)$.

$$D_{||}(p) = \frac{8\pi(Ze)^2}{v} \sum_{\sigma} n_\sigma (Z_\sigma e)^2 \ln \Lambda \, G\left(\frac{v}{\sqrt{2T_\sigma/m_\sigma}}\right). \tag{8.29}$$

Here the function $G(x)$ is defined in terms of the error function

$$\Phi(x) = \frac{2}{\sqrt{\pi}} \int_0^x dy \exp(-y^2) \tag{8.30}$$

as

$$G(x) = \frac{\Phi(x) - x\Phi'(x)}{2x^2}. \tag{8.31}$$

These functions are plotted in Fig. 8.2 and are tabulated in Spitzer [1962].

The *perpendicular diffusion coefficient* $D(p)$ is determined from Eq. (8.24) as

$$D_\perp(p) = \frac{1}{2}\left[\mathrm{Tr}\,\mathbf{D}(\mathbf{p}) - D_{||}(p)\right]. \tag{8.32}$$

In the dominant-term approximation, we thus find from Eq. (8.23)

$$D_{\perp}(p) = \frac{4\pi(Ze)^2}{v}\sum_{\sigma} n_{\sigma}(Z_{\sigma}e)^2 \ln\Lambda\left[\Phi\left(\frac{v}{\sqrt{2T_{\sigma}/m_{\sigma}}}\right) - G\left(\frac{v}{\sqrt{2T_{\sigma}/m_{\sigma}}}\right)\right]. \quad (8.33)$$

In the presence of a uniform magnetic field in the z direction, one calculates the diffusion tensor in an analogous way, starting with Eq. (8.17) [Ichimaru & Rosenbluth 1970]. In the electrostatic approximation (8.21), the Cartesian components of the diffusion tensor are written as

$$\mathsf{D} = \begin{bmatrix} D_{x}(p_{\perp},p_{\shortparallel}) & 0 & 0 \\ 0 & D_{x}(p_{\perp},p_{\shortparallel}) & 0 \\ 0 & 0 & D_{z}(p_{\perp},p_{\shortparallel}) \end{bmatrix} \quad (8.34)$$

where

$$D_{x}(p_{\perp},p_{\shortparallel}) = 2\pi(Ze)^2\sum_{\mathbf{k}}\sum_{n=-\infty}^{\infty}\frac{k_{\perp}^2}{4k^2}\left[J_{n-1}^2(z) + J_{n+1}^2(z)\right]\langle|E^2|(\mathbf{k},n\Omega + k_{\shortparallel}v_{\shortparallel})\rangle \quad (8.35)$$

$$D_{z}(p_{\perp},p_{\shortparallel}) = 2\pi(Ze)^2\sum_{\mathbf{k}}\sum_{n=-\infty}^{\infty}\frac{k_{\shortparallel}^2}{k^2}J_{n}^2(z)\langle|E^2|(\mathbf{k},n\Omega + k_{\shortparallel}v_{\shortparallel})\rangle\left(z = \frac{k_{\perp}v_{\perp}}{\Omega}\right). \quad (8.36)$$

With knowledge of the spectral functions of fluctuations, these formulas enable one to calculate the diffusion coefficients in a magnetic field.

C. Friction Coefficients in Momentum Space

When a particle is singled out in a plasma as a test charge, an induced electric field arises due to polarization of the medium. Such an effect has been considered in Section 3.2C; according to Eq. (3.46), the induced field acting on the test particle with momentum $\mathbf{p}$ and charge Ze is given by

$$\mathbf{E}_{\text{ind}}(\mathbf{p}) = 4\pi Ze\sum_{\mathbf{k}}\frac{\mathbf{k}}{k^2}\text{Im}[\varepsilon(\mathbf{k},\mathbf{k}\cdot\mathbf{v})]^{-1}. \quad (8.37)$$

Motion of the test charge is therefore retarded both by the induced field (8.37) and by the spontaneously fluctuating fields $\mathbf{E}(\mathbf{r},t)$ considered in Eq. (8.10).

At this stage it might be argued that the diffusion coefficients treated in the preceding subsection should have been calculated not from $\mathbf{E}(\mathbf{r},t)$ alone, but from the total field

$$\mathbf{E}_{\text{tot}} = \mathbf{E}_{\text{ind}}(\mathbf{p}) + \mathbf{E}(\mathbf{r},t). \quad (8.38)$$

We would then have

$$\mathbf{D}(\mathbf{p}) = (Ze)^2\mathbf{E}_{\text{ind}}(\mathbf{p})\mathbf{E}_{\text{ind}}(\mathbf{p})\tau + \text{Eq. (8.20)}\,, \tag{8.39}$$

assuming the translational invariance so that $\langle \mathbf{E}[\mathbf{r}_0(t),t]\rangle = 0$ along the unperturbed orbit (8.15) of the particle.

It is possible to show, however, that the contribution of the first term on the right-hand side of Eq. (8.39) is in fact negligibly small, so that the results of the previous subsection remain intact. The argument, due to Hubbard [1961b], proceeds as follows: Since the momentum of the test charge undergoes a change of order p in a time of order τ_1, one must have

$$Ze|\,\mathbf{E}_{\text{tot}}\,| \sim \frac{p}{\tau_1}\,,$$

so that

$$(Ze)^2|\,\mathbf{E}_{\text{ind}}(\mathbf{p})\mathbf{E}_{\text{ind}}(\mathbf{p})\,|\tau \lesssim (Ze)^2|\,\mathbf{E}_{\text{tot}}\,|^2\tau \sim \frac{p^2\tau}{\tau_1^2}\,. \tag{8.40}$$

On the other hand, $D(\mathbf{p})$ must be of the order p^2/τ_1; the collision terms should describe the effects pertaining to the relaxation time τ_1. In light of Eqs. (8.1) and (8.40), we find that the first term on the right-hand side of Eq. (8.39) is indeed negligible.

The coefficients of friction defined by Eq. (8.8) may be calculated as the statistical average of Eq. (8.38). Since the statistical average of the first term on the right-hand side of Eq. (8.14) vanishes, the second term represents the leading contribution arising from the fluctuating fields. The induced polarization field, the first term in Eq. (8.38), then constitutes a contribution on the same order as that of the second fluctuation term. Since $\tau >> \tau_2$, we may thus employ a technique of integration used in the calculation of Eq. (8.19) for an evaluation of the latter fluctuation term, to obtain

$$\mathbf{F}(\mathbf{p}) = Ze\mathbf{E}_{\text{ind}}(\mathbf{p}) + (Ze)^2\frac{\partial}{\partial \mathbf{p}}\cdot\int_0^\infty ds'\langle \mathbf{E}[\mathbf{r}_0(s),s]\mathbf{E}[\mathbf{r}_0(s)+\mathbf{v}s',s+s']\rangle\,. \tag{8.41}$$

Substitution of Eq. (8.37) in this first term and use of the spectral expression Eq. (8.20) in the second term yield

$$\mathbf{F}(\mathbf{p}) = 4\pi(Ze)^2\sum_{\mathbf{k}}\frac{\mathbf{k}}{k^2}\,\text{Im}[\varepsilon(\mathbf{k},\mathbf{k}\cdot\mathbf{v})]^{-1} + \frac{1}{2}\frac{\partial}{\partial \mathbf{p}}\cdot\mathbf{D}(\mathbf{p})\,. \tag{8.42}$$

For plasmas with an isotropic Maxwellian distribution Eq. (1.1) in the absence of an external magnetic field, we may follow the calculations resulting in Eqs. (8.29) and (8.33), to find that Eq. (8.43) is evaluated in the dominant-term approximation as

$$\mathbf{F}(\mathbf{p}) = -\frac{4\pi (Ze)^2 \mathbf{v}}{v} \sum_\sigma \frac{n_\sigma (Z_\sigma e)^2}{(T_\sigma/m_\sigma)} \left(1 + \frac{m}{m_\sigma}\right) \ln \Lambda \, G\left(\frac{v}{\sqrt{2T_\sigma/m_\sigma}}\right). \quad (8.43)$$

For a plasma in a magnetic field, the calculation of the friction coefficients proceeds quite similarly to that described for Eqs. (8.35) and (8.36). Expressing $\mathbf{F}(\mathbf{p})$ as a summation of two terms, the polarization term $\mathbf{F}_p(\mathbf{p})$ and the fluctuation term $\mathbf{F}_f(\mathbf{p})$, we find in the electrostatic approximation [Ichimaru & Rosenbluth 1970]

$$\mathbf{F}_p(\mathbf{p}) = 2(Ze)^2 \sum_\mathbf{k} \int_{-\infty}^{\infty} d\omega \int_{-\infty}^{\infty} dt \frac{\mathbf{k}}{k^2} \operatorname{Im}[\varepsilon(\mathbf{k},\omega)]^{-1} \exp -i[\mathbf{k}\cdot\Delta\mathbf{r}_0(t) - \omega t] \quad (8.44)$$

$$\mathbf{F}_f(\mathbf{p}) = i\frac{Zec}{2B} \sum_\mathbf{k} \int_{-\infty}^{\infty} d\omega \int_{-\infty}^{\infty} dt [\mathbf{k}\cdot\mathsf{H}(t)\cdot\mathbf{k}] \frac{\mathbf{k}}{k^2} \langle |E^2|(\mathbf{k},\omega)\rangle$$

$$\times \exp i[\mathbf{k}\cdot\Delta\mathbf{r}_0(t) - \omega t] \quad (8.45)$$

where $\Delta\mathbf{r}_0(t)$ has been defined in terms of the orbital (8.18).

Now that the Fokker-Planck coefficients have been evaluated in Eqs. (8.23) and (8.42), it is not difficult to obtain an explicit form of the collision term through Eq. (8.7), which is now written as

$$\frac{\partial f}{\partial t} = \frac{\partial}{\partial \mathbf{p}} \cdot \left\{ \frac{1}{2} \mathsf{D}(\mathbf{p}) \cdot \frac{\partial}{\partial \mathbf{p}} f(\mathbf{p}) - 4\pi (Ze)^2 \sum_\mathbf{k} \operatorname{Im}[\varepsilon(\mathbf{k}, \mathbf{k}\cdot\mathbf{v})]^{-1} f(\mathbf{p}) \right\}. \quad (8.46)$$

Substitution of Eqs. (3.136) and (8.23) into Eq. (8.46) then yields the BGL collision term Eq. (2.68).

8.3 TEMPERATURE RELAXATIONS

As an example of application of the Fokker-Planck coefficients calculated in the previous section, we consider in this section the equipartition rates in energies between electrons and ions and between the parallel and perpendicular degrees of freedom for plasmas in a magnetic field [Spitzer 1962; Ichimaru & Rosenbluth 1970].

A. Relaxation of Particle Energy

Consider the kinetic energy of a test particle, $w = p^2/2m$. As in the cases of Eqs. (8.8) and (8.9), its time rate of change may be defined and calculated as

$$\frac{dw}{dt} = \frac{\langle \Delta w(\tau) \rangle}{\tau} = \frac{\langle |\mathbf{p} + \Delta\mathbf{p}(\tau)|^2 - |\mathbf{p}|^2 \rangle}{2m\tau}$$

$$= \mathbf{v} \cdot \mathbf{F}(\mathbf{p}) + \mathrm{Tr}\mathbf{D}(\mathbf{p})/2m \,. \tag{8.47}$$

Each term in Eq. (8.47) can readily be calculated with the aid of Eqs. (8.35), (8.36), (8.44), and (8.45):

$$\frac{\mathrm{Tr}\mathbf{D}(\mathbf{p})}{2m} = \frac{\pi (Ze)^2}{m} \sum_{\mathbf{k}} \sum_{\nu} \int_{-\infty}^{\infty} d\omega \left\{ \frac{k_\perp^2}{2k^2} [J_{\nu-1}^2(\xi) + J_{\nu+1}^2(\xi)] + \frac{k_{\parallel}^2}{k^2} J_\nu^2(\xi) \right\}$$

$$\times \langle |E^2|(\mathbf{k},\omega) \rangle \delta(\nu\Omega + k_{\parallel}v_{\parallel} - \omega) \tag{8.48}$$

$$\mathbf{v} \cdot \mathbf{F}_{\mathrm{f}}(\mathbf{p}) = -\frac{\pi (Ze)^2}{m} \sum_{\mathbf{k}} \sum_{\nu} \int_{-\infty}^{\infty} d\omega \left[\frac{k_\perp^2}{2k^2} \left\{ [J_{\nu-1}^2(\xi) + J_{\nu+1}^2(\xi)] \right. \right.$$

$$\left. - \frac{\omega}{\Omega} [J_{\nu-1}^2(\xi) + J_{\nu+1}^2(\xi)] \right\} \langle |E^2|(\mathbf{k},\omega) \rangle$$

$$\left. - \frac{k_{\parallel}^2}{k^2} J_\nu^2(\xi) \omega \frac{\partial}{\partial \omega} \langle |E^2|(\mathbf{k},\omega) \rangle \right] \delta(\nu\Omega + k_{\parallel}v_{\parallel} - \omega) \,, \tag{8.49}$$

$$\mathbf{v} \cdot \mathbf{F}_\sigma(\mathbf{p}) = 4\pi (Ze)^2 \sum_{\mathbf{k}} \sum_{\nu} \int_{-\infty}^{\infty} d\omega \frac{\omega}{k^2} J_\nu^2(\xi) \mathrm{Im}[\varepsilon(\mathbf{k},\omega)]^{-1} \delta(\nu\Omega + k_{\parallel}v_{\parallel} - \omega) \,. \tag{8.50}$$

In obtaining Eqs. (8.49) and (8.50), we have carried out the averages with respect to differences in the azimuthal angles between **k** and **p**. Summing Eqs. (8.48)–(8.50), we obtain

$$\frac{d\omega}{dt} = \frac{\pi (Ze)^2}{m} \sum_{\mathbf{k}} \sum_{\nu} \int_{-\infty}^{\infty} d\omega \left\{ \frac{k_\perp^2}{2k^2} [J_{\nu-1}^2(\xi) - J_{\nu+1}^2(\xi)] \frac{\omega}{\Omega} \langle |E^2|(\mathbf{k},\omega) \rangle \right.$$

$$\left. + \frac{k_{\parallel}^2}{k^2} J_\nu^2(\xi) \frac{\partial}{\partial \omega} [\omega \langle |E^2(\mathbf{k},\omega)| \rangle] \frac{4m\omega}{k^2} J_\nu^2(\xi) \mathrm{Im}[\varepsilon(\mathbf{k},\omega)]^{-1} \right\} \delta(\nu\Omega + k_{\parallel}v_{\parallel} - \omega) \,. \tag{8.51}$$

The rate of temperature relaxation may be obtained by integrating the product between Eq. (8.51) and the Maxwellian (1.1) over the velocities. After lengthy but straightforward algebra with the aid of some of the well-known identities on the Bessel functions, we obtain

$$\frac{dT}{dt}=\frac{\sqrt{2\pi}\,(Ze)^2}{3m\sqrt{T/m}}\sum_{\mathbf{k}}\sum_{\nu}\int_{-\infty}^{\infty}d\omega\,\frac{\omega^2}{|k_{\parallel}|k^2}$$

$$\times\left\{\langle|E^2|(\mathbf{k},\omega)\rangle+\frac{4T}{\omega}\operatorname{Im}[\varepsilon(\mathbf{k},\omega)]^{-1}\right\}\Lambda_\nu(\beta)\exp\left[-\frac{(n\Omega-\omega)^2}{2k_{\parallel}^2(T/m)}\right], \tag{8.52}$$

where β and $\Lambda_\nu(\beta)$ have been defined by Eqs. (4.12) and (4.13).

Equation (8.52) contains important physical implications: Generally, the spectral functions of fluctuations for systems in thermodynamic equilibria can be calculated with the aid of the fluctuation-dissipation theorem (Section 3.1B); for a classical plasma, this theorem gives

$$\langle|E^2|(\mathbf{k},\omega)\rangle=-\frac{4T}{\omega}\operatorname{Im}[\varepsilon(\mathbf{k},\omega)]^{-1}. \tag{8.53}$$

A conclusion that follows naturally from Eqs. (8.52) and (8.53) is that $dT/dt=0$ for a plasma in thermodynamic equilibrium.

Alternatively, we may choose to turn the arguments around and regard the calculations in this section as another way of deriving the fluctuation-dissipation theorem. The calculations up to Eq. (8.52) have been with regard to the average rate of change in energy for a test particle when the plasma dielectric function and the spectrum of the (longitudinal) electric-field fluctuations are given by $\varepsilon(\mathbf{k},\omega)$ and $\langle|E^2|(\mathbf{k},\omega)\rangle$. For thermodynamic equilibrium, we then require that this rate of change vanish; the relation (8.53) thus follows from this requirement.

B. Relaxation between Electrons and Ions

Let us proceed to consider a nonequilibrium TCP in which the temperatures of the electrons and the ions are different. For simplicity, we assume $Z=1$ and $n_e=n_i=n$. The relaxation time τ_{ei} between the temperatures of the electrons and of the ions is defined through the equation

$$\frac{dT_e}{dt}=-\frac{T_e-T_i}{\tau_{ei}}. \tag{8.54}$$

This rate may thus be obtained through a detailed analysis of Eq. (8.52). For the two-component Maxwellian plasma under consideration, the dielectric function $\varepsilon(\mathbf{k},\omega)$ takes the form

$$\varepsilon(\mathbf{k},\omega)=1+\sum_{\sigma=e,i}\frac{k_\sigma^2}{k^2}\left\{1+\sum_{\nu}\frac{\omega}{\omega-\nu\Omega_\sigma}\left[W\left(\frac{\omega-\nu\Omega_\sigma}{|k_{\|}|\sqrt{T_\sigma/m_\sigma}}\right)-1\right]\Lambda_\nu(\beta_0)\right\}. \tag{8.55}$$

The spectral function $\langle|E^2|(\mathbf{k},\omega)\rangle$ of the electric-field fluctuations can be calculated with the aid of the superposition technique of the dressed test particles (Section 7.2C). We thus find

$$\langle|E^2|(\mathbf{k},\omega)\rangle=\frac{16\pi^2e^2}{k^2|\varepsilon(\mathbf{k},\omega)|^2}\sum_{\sigma=e,i}S_\sigma^{(0)}(\mathbf{k},\omega), \tag{8.56}$$

$$S_\sigma^{(0)}(\mathbf{k},\omega)=\frac{N}{\sqrt{2\pi T_\sigma/m_\sigma}\,|k_{\|}|}\sum_{\nu}\Lambda_\nu(\beta_\sigma)\exp\left[-\frac{(\nu\Omega_\sigma-\omega)^2}{2k_{\|}^2(T_\sigma/m_\sigma)}\right]. \tag{8.57}$$

Equations (8.56) and (8.57) are now substituted in Eq. (8.52) to yield

$$\frac{1}{\tau_{ei}}=\frac{16\pi^2ne^4\sqrt{m_em_i}}{3(T_eT_i)^{3/2}}\sum_{\mathbf{k}}\sum_{\nu,\nu'}\int_{-\infty}^{\infty}d\omega\,\frac{\omega^2}{k_{\|}^2k^4|\varepsilon(\mathbf{k},\omega)|^2}$$

$$\times\Lambda_\nu(\beta_e)\Lambda_{\nu'}(\beta_i)\exp\left[-\frac{(\omega-\nu\Omega_e)^2}{2k_{\|}^2(T_e/m_e)}-\frac{(\omega-\nu'\Omega_i)^2}{2k_{\|}^2(T_i/m_i)}\right]. \tag{8.58}$$

In the limit of $B\to 0$, we may go over to the continuum (4.16) for the ν and ν' summation in Eq. (8.58). In the dominant-term approximation of Section 7.2B, the result reproduces the Spitzer [1962] value

$$\frac{1}{\tau_{ei}^{(0)}}=\frac{8\sqrt{2\pi}\,ne^4}{3m_em_i(T_e/m_e+T_i/m_i)^{3/2}}\ln\Lambda, \tag{8.59}$$

where the Coulomb logarithm has been given by Eq. (8.28).

When the magnetic field is finite, it is essential in the calculation of Eq. (8.58) to take explicit account of discreteness in the cyclotron motions of particles represented by the harmonic numbers, ν and ν'. Assuming $m_i \gg m_e$, we note that the domain of the $\mathbf{k}$ space such that

$$|k_{\|}|\sqrt{\frac{T_e}{m_e}}>k_\perp\sqrt{\frac{T_i}{m_i}}, \tag{8.60a}$$

$$|\Omega_e| > k_\perp \sqrt{\frac{T_i}{m_i}} \tag{8.60b}$$

covers most of the important range in Eq. (8.58). We may then replace the electronic polarizability by its static value in the the dielectric function contained in Eq. (8.58). The summation over the ion-harmonic numbers (ν') and the ω integration in Eq. (8.58) can then be performed exactly with the aid of a Kramers-Kronig relation analogous to Eq. (3.40a). Hence Eq. (8.58) becomes

$$\frac{1}{\tau_{ei}} = \frac{8\sqrt{2\pi}\,ne^4\sqrt{m_e}}{3m_iT_e^{3/2}} \int_0^\infty dk_\perp \int_{k_\perp\sqrt{m_eT_i/m_iT_e}}^\infty dk_{\|} \frac{k_\perp\Lambda_0(\beta_e)}{k_{\|}k^2[\varepsilon_e(\mathbf{k})]^2}$$

$$+\frac{8\sqrt{2\pi}\,ne^4\sqrt{m_e}}{3m_iT_e^{3/2}} \sum_\nu{}' \int_0^\infty dk_\perp \int_0^\infty dk_{\|} \frac{k_\perp\Lambda_\nu(\beta_e)}{k_{\|}k^2[\varepsilon_e(\mathbf{k})]^2} \exp\left[-\frac{\nu^2\Omega_e^2}{2k_{\|}^2(T_{\|}/m_{\|})}\right] \tag{8.61}$$

[Ichimaru & Rosenbluth 1970], where $\varepsilon_e(k)$ is the static dielectric constant defined by Eq. (4.52), and the prime means omission of the term with $\nu = 0$ in the summation. In this derivation we have transformed the **k** summation into integration via

$$\sum_{\mathbf{k}} \to \frac{1}{2\pi^2}\int_0^\infty k_\perp dk_\perp \int_{k_\perp\sqrt{m_eT_i/m_iT_e}}^\infty dk_{\|}\,. \tag{8.62}$$

The lower limit of the $k_{\|}$ integration in Eq. (8.62) derives from (8.60a). In the second term on the right-hand side of Eq. (8.61), this limit has been set equal to zero by going over to $m_e/m_i \to 0$, as no anomaly arises there. This term reduces to the Spitzer value, Eq. (8.59), when the continuum limit of Eq. (4.16) is applied to the summation; such a procedure may be justified for the domain $\beta_e >> 1$, which contributes the bulk of the effect.

The first term in Eq. (8.61), on the other hand, exhibits a singular behavior in the limit of $m_e/m_i \to 0$; the $k_{\|}$ integration diverges logarithmically for small $k_{\|}$. Keeping the mass ratio finite, we thus calculate this term in terms of the modified Bessel functions of the second kind as

$$\frac{1}{\tau_{ei}^*} = \frac{2\sqrt{2\pi}\,ne^4}{3m_em_i(T_e/m_e)^{3/2}} \exp(\alpha)[(1+\alpha)K_0(\alpha) - \alpha K_1(\alpha)] \ln\left(\frac{m_i}{m_e}\right) \tag{8.63}$$

where

$$\alpha \equiv k_e^2\left(\frac{T_e}{m_e\Omega_e^2}\right) = \frac{\omega_e^2}{\Omega_e^2}\,. \tag{8.64}$$

When the magnetic field is strong so that $\Omega_e^2 >> \omega_e^2$ (i.e., $\alpha << 1$), Eq. (8.63) reduces to

$$\frac{1}{\tau_{ei}^*} = \frac{2\sqrt{2\pi}\, ne^4}{3m_e m_i (T_i/m_e)^{3/2}} \ln\left(\frac{\Omega_e^2}{\omega_e^2}\right) \ln\left(\frac{m_i}{m_e}\right). \tag{8.65}$$

Equation (8.63) or (8.65) is the relaxation rate which appears in a magnetized plasma in addition to the usual term of Eq. (8.59). It is then instructive to compare between the magnitudes of those two contributions:

$$\frac{1/\tau_{ei}^*}{1/\tau_{ei}^{(0)}} = \frac{\ln|\Omega_e/\omega_e| \ln(m_i/m_e)}{2\ln\Lambda}. \tag{8.66}$$

Assuming $\ln\Lambda \approx 15$ and $\Omega_i \approx \omega_i$, we find this ratio to be approximately 1.1 for a deuterium plasma. The additional temperature-relaxation process of Eq. (8.63) or (8.65) is here seen to be as effective as the ordinary process of Eq. (8.59) in a strongly magnetized plasma.

The physical origin of such an anomalous relaxation process may be traced to the spiral motion of the electrons along the magnetic lines of force. These electrons naturally tend to couple strongly with long-wavelength fluctuations (i.e., small $k_{||}$) along the magnetic field. In addition, when such fluctuations are characterized by slow variation in time (i.e., small ω), the contact time or the rate of energy exchange between the electrons and the fluctuations will be further enhanced. In a TCP with $m_i >> m_e$, such low-frequency fluctuations are provided mostly by the thermal motion of the ions. The foregoing coupling can, therefore, be an effective mechanism of energy exchange between the electrons and the ions. In the limit of $m_e/m_i \to 0$, the frequency spectrum of the ionic thermal fluctuations consists of a δ function at $\omega = 0$; the contact time and hence the relaxation rate would then diverge.

The reasoning above, emphasizing the special role that the fluctuations with small $k_{||}$ and small ω play, may lead one to suspect that collective modes with similar characteristics will likewise give rise to significant contributions in the energy exchange processes. For a plasma in a strong magnetic field, the electron plasma waves and the ion acoustic waves have the frequencies $(k_{||}/k)\omega_e$ and $k_{||}\sqrt{T_e/m_i}$, respectively; potentially, they might provide a significant energy-exchange mechanism. A detailed analysis through evaluation of the spectral strengths in these collective modes, however, indicates [Ichimaru & Rosenbluth 1970] that the fluctuations are not strong enough to be able to affect the result, Eq. (8.63).

C. Relaxation between Parallel and Perpendicular Motions to the Magnetic Field

For a plasma in a magnetic field, it is sometimes possible to define two temperatures for the particles of a given species, associated with the parallel and perpendicular degrees of freedom with respect to the magnetic field. Interaction between identical particles accounts for relaxation between these two temperatures.

The relaxation processes in these circumstances may be analyzed by dealing separately with $w_\perp = p^2/2m$ and $w_{||} = p_{||}^2/2m$. Time rates of change for these quantities are multiplied by the anisotropic Maxwellian (6.22) and are integrated over the velocities. We thus obtain [Ichimaru & Rosenbluth 1970]

$$\frac{dT_\perp}{dt} = \sqrt{\frac{\pi}{2}} \frac{e^2}{m(T_{||}/m)^{3/2}} \sum_{\mathbf{k}} \sum_{\nu} \int_{-\infty}^{\infty} d\omega \frac{\nu\Omega}{|k_{||}|k^2} \left\{ \nu\Omega \left(\frac{T_{||}}{T_\perp} - 1 \right) \langle |E^2|(\mathbf{k},\omega) \rangle \right.$$

$$\left. + \omega \left[\langle |E^2|(\mathbf{k},\omega)\rangle + \frac{4T_{||}}{\omega} \mathrm{Im}[\varepsilon(\mathbf{k},\omega)]^{-1} \right] \right\} \Lambda_\nu(\beta) \exp\left[-\frac{(\nu\Omega - \omega)^2}{2k_{||}^2 (T_{||}/m)} \right], \tag{8.67}$$

$$\frac{dT_{||}}{dt} = \frac{\sqrt{2\pi}\, e^2}{m(T_{||}/m)^{3/2}} \sum_{\mathbf{k}} \sum_{\nu} \int_{-\infty}^{\infty} d\omega \frac{\omega - \nu\Omega}{|k_{||}|k^2} \left\{ \nu\Omega \left(\frac{T_{||}}{T_\perp} - 1 \right) \langle |E^2|(\mathbf{k},\omega) \rangle \right.$$

$$\left. + \omega \left[\langle |E^2|(\mathbf{k},\omega)\rangle + \frac{4T_{||}}{\omega} \mathrm{Im}[\varepsilon(\mathbf{k},\omega)]^{-1} \right] \right\} \Lambda_\nu(\beta) \exp\left[-\frac{(\nu\Omega - \omega)^2}{2k_{||}^2 (T_{||}/m)} \right]. \tag{8.68}$$

The relaxation rates are again written as functionals of the spectral densities of fluctuations and the dielectric response functions. In thermodynamic equilibrium, $T_\perp = T_{||}$; both Eqs. (8.67) and (8.68) vanish by virtue of the fluctuation-dissipation theorem.

As an example of the nonequilibrium situation, we consider a case in which the ions are characterized by the anisotropic distribution, Eq. (6.22). For simplicity, however, the electrons are treated only as providing a static dielectric background to the ions; the relaxation is thus between $T_\perp$ and $T_{||}$ for the ions. The ionic relaxation time τ_i is defined through the equation

$$\frac{dT_\perp}{dt} = -\frac{1}{2}\frac{dT_{||}}{dt} = -\frac{T - T_{||}}{\tau_i}. \tag{8.69}$$

It can be calculated in a way similar to τ_{ei} in the previous subsection. Within the dominant-term approximation, we find

$$\frac{1}{\tau_i} = \frac{8\sqrt{\pi}\, ne^4}{15\sqrt{m_i}(T_{\rm eff})^{3/2}} \ln \Lambda \,, \tag{8.70}$$

where the effective ion temperature $T_{\rm eff}$ has been defined through an integral relation

$$\frac{1}{(T_{\rm eff})^{3/2}} \equiv \frac{15}{4}\int_{-1}^{1} d\mu \, \frac{\mu^2(1-\mu^2)}{[(1-\mu^2)T_\perp + \mu^2 T_{||}]^{3/2}} \,. \tag{8.71}$$

when $T_{||} = T_\perp$, $T_{\rm eff}$ becomes equal to either of these temperatures. In this relaxation process, no anomalous term corresponding to Eq. (8.63) appears.

8.4 SPATIAL DIFFUSION ACROSS A MAGNETIC FIELD

In the previous section, we studied the rates of temperature relaxation in a plasma with a magnetic field, based on the study of a charged particle in fluctuating electric fields. Thus it has been shown that the relaxation rate between the electron and ion temperatures contains, in addition to the usual Spitzer rate, an extra term which exhibits a different functional dependence on the magnetic field and which diverges logarithmically as the mass ratio of an electron to an ion approaches zero. The physical origin of this anomalous term has been identified as a strong coupling between the spiral motion of the electrons and the long-wavelength, low-frequency fluctuations produced by the ions.

In this section, the foregoing approach is extended and applied to the problem of spatial diffusion of a plasma across a magnetic field [Ichimaru & Tange 1974]. We thus calculate the first and second moments of the guiding-center displacements of a test particle under the action of microscopic electric fields. Originally, such calculations were carried out by Longmire and Rosenbluth [1956] based on a collision-theoretic consideration, followed by Spitzer [1960] and by Taylor [1961].

From a fluctuation-theoretic point of view, a new feature emerges in the treatment of spatial diffusion: In the formulation of the dielectric response function such as Eq. (8.55) and the spectral functions of fluctuations such as Eqs. (8.56) and (8.57), effects of inhomogeneity or the spatial gradient,

$$q \equiv \frac{1}{n_e}\frac{dn_e}{dX} = \frac{1}{n_i}\frac{dn_i}{dX} \,, \tag{8.72}$$

should be accurately taken into account. These problems have been considered by Tange [1974].

A. Motion of Guiding Centers and Spatial Diffusion

As Eq. (8.72) implies, the plasma under consideration has a density gradient in the x direction with a uniform magnetic field B in the z direction. We calculate the flux $\Gamma_1(2)$ of the guiding centers of the particles "1" scattered in those fluctuating electric fields $\mathbf{E}(\mathbf{r},t)$ which originate from the particles "2"; the two types of particles, designated by subscripts 1 and 2, need not be different. According to the stochastic formula (8.6), the flux may be calculated as

$$\Gamma_1(2) = n_1(X)\frac{\langle \Delta X_1 \rangle}{\tau} - \frac{1}{2}\frac{\partial}{\partial X}\left[n_1(X)\frac{\langle (\Delta X_1)^2 \rangle}{\tau} \right] \tag{8.73}$$

where $n_1(X)$ is the density of the guiding centers for particles "1." In this section, we let X denote the x coordinate of a guiding center. The first and second moments, $\langle \Delta X_1 \rangle$ and $\langle (\Delta X_1)^2 \rangle$, of the guiding-center displacements in the x direction for a particle "1" may be evaluated with knowledge of the stochastic motion of the guiding center in the microscopic electric fields. With the aid of Eqs. (8.16) and (8.17) for the instantaneous position $\mathbf{r}(t) = \mathbf{r}(0) + \Delta\mathbf{r}(t)$ and velocity $\mathbf{v}(t) = \mathbf{v} + \Delta\mathbf{v}(t)$ of a particle with charge Ze and mass m, the position $\mathbf{R}(t)$ of the guiding center is obtained as

$$\begin{aligned} \mathbf{R}(t) &= \mathbf{r}(t) - \frac{1}{\Omega}\hat{\mathbf{z}} \times \mathbf{v}(t) \\ &= \mathbf{R}(0) - \frac{c}{B}\hat{\mathbf{z}} \times \int_0^t dt' \mathbf{E}[\mathbf{r}(t'),t'] \\ &\quad + \hat{\mathbf{z}}\left\{ v_{\parallel}(0)t + \frac{Ze}{m}\int_0^t dt' \int_0^{t'} dt'' E_{\parallel}[\mathbf{r}(t''),t''] \right\}. \end{aligned} \tag{8.74}$$

We thus have

$$\frac{\langle \Delta X_1 \rangle}{\tau} = \frac{c}{\tau B}\int_0^\tau dt \langle E_y[\mathbf{r}_1(t),t] \rangle, \tag{8.75}$$

$$\frac{\langle (\Delta X_1)^2 \rangle}{\tau} = \frac{c^2}{\tau B^2}\int_0^\tau dt \int_0^\tau dt' \langle E_y[\mathbf{r}_1(t),t] E_y[\mathbf{r}_1(t'),t'] \rangle. \tag{8.76}$$

It is clear from the definition of Eq. (8.73) that the electric-field fluctuations in Eqs. (8.75) and (8.76) must be those produced by the particles "2."

For an application of these formulas, we shall analyze the diffusion phenomena in an electron-ion plasma with $Z = 1$. The electron and ion fluxes calculated according to Eq. (8.73) may generally be different; an average electric field will then be induced in the x direction to make the diffusion ambipolar [e.g., Hoh 1962]. The most important case to be considered in these circumstances is that with "1" = "e" and "2" = "i."

B. Calculation of the Second Moment

We begin with a calculation of the second moment, Eq. (8.76), in an inhomogeneous plasma. Let us first note that this moment takes on a finite value in a homogeneous limit. From symmetry considerations, we then find that the first-order contributions of the spatial inhomogeneity, Eq. (8.72), vanish; the next leading terms therefore arise from the second-order contributions of the inhomogeneity. We shall not delve into such higher-order calculations. From a collision-theoretic point of view, such higher-order terms were calculated by Longmire and Rosenbluth [1956].

In the longitudinal approximation to the microscopic fluctuations, the procedure in Section 8.3 leads to

$$\frac{\langle(\Delta X_1)^2\rangle}{\tau} = \frac{\pi c^2}{B^2}\sum_{\mathbf{k}}\sum_{\nu}\int_{-\infty}^{\infty} d\omega \langle | E^2 |(\mathbf{k},\omega)\rangle_2 \frac{k_{\perp}^2}{k^2} J_\nu^2(z_1)\delta(\nu\Omega_1 + k_{\parallel}v_{\parallel} - \omega) \tag{8.77}$$

where $\langle | E^2 |(\mathbf{k},\omega)\rangle_2$ denotes that part of the field fluctuations produced by the dynamically screened particles "2."

For a Maxwellian plasma, we may average Eq. (8.77) with respect to a Maxwellian distribution for particles "1"; hence

$$\frac{\langle(\Delta X_1)^2\rangle}{\tau} = \frac{\pi c^2}{B^2}\sqrt{\frac{m_1}{2\pi T_1}}\sum_{\mathbf{k}}\sum_{\nu}\int_{-\infty}^{\infty} d\omega \langle | E^2 |(\mathbf{k},\omega)\rangle_2 \frac{k_{\perp}^2}{| k_{\parallel} | k^2}$$

$$\times \Lambda_\nu(\beta_1)\exp\left[-\frac{(n\Omega_1 - \omega)^2}{2k_{\parallel}^2(T_1/m_1)}\right] \tag{8.78}$$

where β and $\Lambda_\nu(\beta)$ have been defined by Eqs. (4.12) and (4.13). The spectral function of the electric-field fluctuations may be written as

$$\langle | E^2 |(\mathbf{k},\omega)\rangle_2 = \frac{16\pi^2(Z_2 e)^2}{k^2 | \varepsilon(\mathbf{k},\omega) |^2} S_2^{(0)}(\mathbf{k},\omega)\,. \tag{8.79}$$

Here

$$S_2^{(0)}(\mathbf{k},\omega) = \frac{n_2}{\sqrt{2\pi T_2/m_2}\,|k_{\parallel}|} \sum_{\nu} \Lambda_{\nu}(\beta_2) \exp\left[-\frac{(\nu\Omega_2 - \omega)^2}{2k_{\parallel}^2(T_2/m_2)} \right] \quad (8.80)$$

and $\varepsilon(\mathbf{k},\omega)$ is the dielectric response function given by Eq. (8.55); n_2 in Eq. (8.90) may contain a spatial inhomogeneity. Equation (8.79) is clearly an adequate approximation for the case when particles "2" are ions with the Larmor radii larger than the Debye length.

C. Calculation of the First Moment

As Eq. (8.75) implies, the first moment $\langle \Delta X_1 \rangle$ is proportional to the average force on a particle "1" in the y direction due to scattering in the fields produced by the particles "2." Such a force arises from two separate sources: one due to the field fluctuations and another due to the polarization. These may be calculated, respectively, as

$$\frac{\langle \Delta X_1 \rangle_{\phi}}{\tau} = i\frac{e^2}{2B^2} \sum_{\mathbf{k}} \int_{-\infty}^{\infty} d\omega \int_{-\infty}^{\infty} dt \frac{k_y}{k^2} \mathbf{k} \cdot \mathbf{H}(t) \cdot \mathbf{k} \langle |E^2|(\mathbf{k},\omega) \rangle_2 \exp\{ i[\mathbf{k} \cdot \Delta\mathbf{r}(t) - \omega t] \}$$

of the particles "2." Physically, such a density gradient acts in a number of ways to produce nonvanishing contributions in $\langle \Delta X_1 \rangle$: For instance, it gives rise to a net drift motion of the particles "2" in the y direction with velocity (see Eq. 6.50) (8.81)

$$\frac{\langle \Delta X_1 \rangle_{\pi}}{\tau} = -\frac{2Z_1 ec}{B} \sum_{\mathbf{k}} \int_{-\infty}^{\infty} d\omega \int_{-\infty}^{\infty} dt \frac{k_y}{k^2} \frac{\mathrm{Im}\,\varepsilon(\mathbf{k},\omega)_2}{|\varepsilon(\mathbf{k},\omega)|^2} \exp\{ i[\mathbf{k} \cdot \Delta\mathbf{r}(t) - \omega t] \} \quad (8.82)$$

where $\mathrm{Im}\,\varepsilon(\mathbf{k},\omega)_2$ denotes the imaginary part of that portion of $\varepsilon(\mathbf{k},\omega)$ which arises from the polarizability of the particles "2." For a Maxwellian plasma,

$$\mathrm{Im}\,\varepsilon(\mathbf{k},\omega)_2 = \frac{4\pi^2(Z_2 e)^2\omega}{T_2 k^2} S_2^{(0)}(\mathbf{k},\omega) \quad (8.83)$$

where $S_2^{(0)}(\mathbf{k},\omega)$ is given by Eq. (8.80).

of the particles "2." Physically, such a density gradient acts in a number of ways to produce nonvanishing contributions in $\langle \Delta X_1 \rangle$: For instance, it gives rise to a net drift motion of the particles "2" in the y direction with velocity (see Eq. 6.50)

$$v_2 = \frac{T_2}{m_2\Omega_2}\frac{1}{n_2}\frac{\partial n_2}{\partial X}.$$

In the frame of reference comoving with the average motion of the particles "2," a test particle "1" acquires a net drift velocity $-v_2$ in the y direction superposed on its own cyclotron motion. A frictional force is thus exerted onto "1" and thereby produces $\langle \Delta X_1 \rangle$. A second instance of the friction effects stems from finiteness of Larmor radii with the particles "1." In the presence of $\partial n_2/\partial X$, these gyrating particles then perceive different densities of the particles "2" between one and the other sides of their cyclotron orbitals. The frictional forces at the two opposite sides of the orbitals do not exactly cancel each other. The resulting net friction points in the y direction and thus contributes to $\langle \Delta X_1 \rangle$.

All of these effects arising from density gradients in "2" have been formulated on the dielectric responses and the electric-field fluctuations [Tange 1974] and thereby in the calculations of $\langle \Delta X_1 \rangle$ [Ichimaru & Tange 1974]. Details of the theories are somewhat beyond the scope of this volume; the interested reader is referred to these articles. In the subsequent section, the results with $\langle \Delta X_1 \rangle$ will be used for the calculations of the diffusion coefficients.

D. Diffusion Coefficients

A complete expression for the diffusion equation may be obtained by substituting Eqs. (8.75) and (8.76) into Eq. (8.73). The result is

$$\Gamma_1(2) = n_1 \frac{\pi m_1 c^2 \Omega_1}{2B^2}\sqrt{\frac{m_1}{2\pi T_1}} \sum_{\nu}\sum_{\mathbf{k}} \int_{-\infty}^{\infty} d\omega \, \frac{1}{|k_{\parallel}| k^2} \Lambda_\nu(\beta_1) \exp\left[-\frac{(\nu\Omega_1 - \omega)^2}{2k_{\parallel}^2 (T_1/m_1)}\right]$$

$$\times \left\{ -q k_{\perp}^2 \left(\frac{1}{m_1\Omega_1} - \frac{1}{m_2\Omega_2}\right) \langle |E^2|(\mathbf{k},\omega)\rangle_2 - \nu\omega\left(\frac{1}{T_1} - \frac{1}{T_2}\right) \frac{\partial \langle |E^2|(\mathbf{k},\omega)\rangle_2}{\partial X}\right.$$

$$\left. + q\omega\left(\frac{1}{T_1} - \frac{1}{T_2}\right)\left[\sum_{\nu'} \nu' \langle |E^2|(\mathbf{k},\omega)\rangle_{2,\nu'} + \left(\frac{\varepsilon}{\varepsilon_y} + \frac{\varepsilon^*}{\varepsilon_y^*}\right)\langle |E^2|(\mathbf{k},\omega)\rangle_2\right]\right\}. \tag{8.84}$$

Here $\varepsilon \equiv \varepsilon(\mathbf{k},\omega)$,

$$\langle |E^2|(\mathbf{k},\omega)\rangle_{\sigma,\nu} \equiv \left|\frac{4\pi Z_\sigma e}{k\varepsilon}\right|^2 n_\sigma \sqrt{\frac{m_\sigma}{2\pi T_\sigma}} \frac{1}{|k_{||}|} \Lambda_\nu(\beta_\sigma)$$

$$\times \exp\left[-\frac{(\nu\Omega_\sigma - \omega)^2}{2k_{||}^2(T_\sigma/m_\sigma)}\right], \tag{8.85}$$

$$\langle |E^2|(\mathbf{k},\omega)\rangle_\sigma = \sum_\nu \langle |E^2|(\mathbf{k},\omega)\rangle_{\sigma,\nu}\,, \tag{8.86}$$

$$\frac{\partial}{\partial X}\langle |E^2|(\mathbf{k},\omega)\rangle_\sigma = q\left(1 + \frac{1-\varepsilon}{\varepsilon} + \frac{1-\varepsilon^*}{\varepsilon^*}\right)\langle |E^2|(\mathbf{k},\omega)\rangle_\sigma\,, \tag{8.87}$$

$$\frac{1}{\varepsilon_y} = \frac{1}{\varepsilon^2}\sum_\sigma\sum_\nu \frac{k_\sigma^2}{k^2}\left[W\left(\frac{\omega - \nu\Omega_\sigma}{|k_{||}|\sqrt{T_\sigma/m_\sigma}}\right) - 1\right]\frac{\beta_\sigma\Omega_\sigma - \nu\omega}{\omega - \nu\Omega_\sigma}\Lambda_\nu(\beta_\sigma)\,. \tag{8.88}$$

Note that $\Gamma_1(2) = 0$ when "1" = "2"; collisions between like particles do not contribute to the guiding-center diffusion.

The diffusion coefficient $D_\perp$ across the magnetic field may be defined through the equation

$$\Gamma_1 = -D_\perp \frac{\partial n}{\partial X}\,. \tag{8.89}$$

Equivalently, we find it useful to introduce an effective collision frequency ν_D pertaining to the diffusion process by

$$\nu_D \equiv \left(\frac{m_1}{T_1}\right)\Omega_1^2 D_\perp\,. \tag{8.90}$$

This expression finds its analogy in a problem of one-dimensional random walk [Chandrasekhar 1943], in which a particle suffers displacements along a straight line with an average step of $\sqrt{2T_1/m_1\Omega_1^2}$, an average Larmor radius. Substitution of Eq. (8.87) in Eq. (8.84) thus yields

$$\nu_D = \sqrt{\frac{\pi}{2}} \left(\frac{Z_1 e}{m_1}\right)^2 \left(\frac{m_1}{T_1}\right)^{3/2} \sum_{\nu} \sum_{\mathbf{k}} \int_{-\infty}^{\infty} d\omega \, \frac{m_1 \Omega_1}{2 |k_{\parallel}| k^2} \Lambda_\nu(\beta_1)$$

$$\times \exp\left[-\frac{(\nu\Omega_1 - \omega)^2}{2k_{\parallel}(T_1/m_1)}\right] \left\{ k_\perp^2 \left(\frac{1}{m_1\Omega_1} - \frac{1}{m_2\Omega_2}\right) \langle | E^2 |(\mathbf{k},\omega)\rangle_2 \right.$$

$$+ \omega \left(\frac{1}{T_1} - \frac{1}{T_2}\right) \left[\nu \left(-1 + \frac{1}{\varepsilon} + \frac{1}{\varepsilon^*}\right) - \left(\frac{\varepsilon}{\varepsilon_y} + \frac{\varepsilon^*}{\varepsilon_y^*}\right)\right] \langle | E^2 |(\mathbf{k},\omega)\rangle_2$$

$$\left. - \omega \left(\frac{1}{T_1} - \frac{1}{T_2}\right) \sum_{\nu'} \nu' \langle | E^2 |(\mathbf{k},\omega)\rangle_{2,\nu'} \right\}. \tag{8.91}$$

If the effects of the magnetic field are neglected in the calculation of the collision frequency Eq. (8.91), the treatment above should correspond to a standard theory of classical diffusion. We thus consider the limit of $B \to 0$ in Eq. (8.91) and go over to the continuum for the ν summation as prescribed by Eq. (4.16). The result of such a calculation is

$$\nu_D^{(0)} = \frac{8\sqrt{2\pi}\, n Z_1^2 Z_2^2 e^4}{3 m_1 m_2 (T_1/m_1 + T_2/m_2)^{3/2}} \ln \Lambda \left\{ \left(\frac{m_2}{m_1}\right)\left(1 + \frac{m_1 T_2}{m_2 T_1}\right) - \frac{m_1 + m_2}{2 m_1} \left(1 + \frac{Z_1 T_2}{Z_2 T_1}\right) \right\} \tag{8.92}$$

where $\ln \Lambda$ is the Coulomb logarithm. When $T_1 = T_2$, Eq. (8.92) reproduces the Longmire-Rosenbluth [1956] result; when $T_1 \neq T_2$, it agrees with the Taylor [1961] calculation. For the electron-ion TCP with $Z = 1$ and $m_e \ll m_i$, this frequency takes the form

$$\nu_D^{(0)} = \frac{4\sqrt{2\pi}\, n e^4}{3 m_e^2 (T_e/m_e + T_i/m_i)^{3/2}} \left(1 + \frac{T_i}{T_e}\right) \ln \Lambda \,. \tag{8.93}$$

As we remarked in Section 8.3C, an anomalous contribution of the magnetic field to the collisional processes exists even for a plasma in thermodynamic equilibrium. Such a term can likewise be singled out in Eq. (8.91). Let us study the physical consequences arising from it for the electron-ion TCP.

Following the arguments leading to Eq. (8.63), we obtain from the $\nu = 0$ terms in Eq. (8.91)

$$\nu_D^* = \frac{\sqrt{2\pi}\, ne^4}{\sqrt{m_e} T_e^{3/2}} \left(1 + \frac{T_i}{T_e}\right) \ln\left(\frac{m_i}{m_e}\right) \left\{ \frac{1}{2}\left(1 + \frac{T_i}{T_e}\right) \exp(\alpha_D) K_0(\alpha_D) - \frac{T_i}{2T_e} \exp(\alpha_e) K_0(\alpha_e) \right\}, \qquad (8.94)$$

where

$$\alpha_D \equiv \left(1 + \frac{T_e}{T_i}\right) \frac{\omega_e^2}{\Omega_e^2}, \qquad (8.95)$$

$$\alpha_e \equiv \frac{\omega_e^2}{\Omega_e^2}. \qquad (8.96)$$

Equation (8.94) is the expression for the collision frequency between the electrons and the ions pertaining to the anomalous diffusion of the electrons across the magnetic field.

The anomalous part $D_\perp^*$ of the diffusion coefficient may be calculated with the aid of Eq. (8.90). When $\alpha_D > \alpha_e >> 1$, we may expand $K_0(\alpha_D)$ and $K_0(\alpha_e)$ in Eq. (8.94) in asymptotic series and keep only the leading terms, to obtain

$$D_\perp^* = \frac{e^2 k_e}{8 m_e |\Omega_e|} \sqrt{\frac{T_i}{T_e}} \left(1 + \frac{T_i}{T_e}\right) \ln\left(\frac{m_i}{m_e}\right) \left(\sqrt{1 + \frac{T_i}{T_e}} - \sqrt{\frac{T_i}{T_e}}\right). \qquad (8.97)$$

Note that this diffusion coefficient is proportional to B^{-1}; later in Chapter 9 we shall explore a possible connection between Eq. (8.97) and the *Bohm diffusion coefficient* [Bohm 1949; Hoh 1962] in terms of enhanced fluctuations above a thermal level in a turbulent plasma.

The other limiting cases of $\alpha_e < \alpha_D << 1$ may be similarly obtained by series expansions of $K_0(\alpha_D)$ and $K_0(\alpha_e)$ for small α_D and α_e; we find

$$D_\perp^* = \frac{\sqrt{2\pi}\, ne^4}{m_e^{3/2} \sqrt{T_e} \Omega_e^2} \left(1 + \frac{T_i}{T_e}\right) \ln\left(\frac{m_i}{m_e}\right) \times \left\{ \frac{1}{2}\left(1 + \frac{T_i}{T_e}\right) \ln\left[\frac{\Omega_e^2}{\omega_e^2}\left(1 + \frac{T_e}{T_i}\right)^{-1}\right] - \frac{T_i}{2T_e} \ln\left(\frac{\Omega_e^2}{\omega_e^2}\right) \right\}. \qquad (8.98)$$

This diffusion coefficient, which is valid only with a very strong magnetic field, is therefore proportional to $B^{-2} \ln B$.

We emphasize that these anomalous terms arise as a consequence of electron fluctuations in the low-frequency domain (i.e., $\nu = 0$) such that $\omega < |\Omega_e|$. In many cases of turbulent plasmas, we expect that such low-frequency fluctuations are especially enhanced above the thermal levels. In these circumstances, these anomalous terms should grow accordingly, resulting in an enhanced diffusion. A normal term such as Eq. (8.92), however, is not related to such an enhancement.

PROBLEMS

8.1 Consider events of Coulomb scattering as illustrated in Fig. 1.2, with $\Delta\mathbf{p}$ denoting the increment in momentum before and after scattering. Calculate the quantities

$$\Delta\mathbf{p} \equiv \int \Delta\mathbf{p}\,dQ \text{ and } \Delta\mathbf{p}\Delta\mathbf{p} \equiv \int (\Delta\mathbf{p}\Delta\mathbf{p})\,dQ$$

with the differential cross section Eq. (1.11), and show that these will diverge logarithmically as the lower limit of the χ integration approaches zero. Also show that

$$\{\Delta\mathbf{p}\Delta\mathbf{p}\Delta\mathbf{p}\} \equiv \int (\Delta\mathbf{p}\Delta\mathbf{p}\Delta\mathbf{p})\,dQ$$

does not contain such a divergent term.

8.2 In the collision term of Eq. (7.121), write explicit expressions for the coefficients of friction and diffusion, $\mathbf{F}(\mathbf{p})$ and $\mathbf{D}(\mathbf{p})$.

8.3 Derive Eq. (8.43).

8.4 Show that Eq. (8.46) reduces to the BGL collision term Eq. (2.68) by substitution of Eqs. (3.136) and (8.23) in it.

8.5 Describe the steps in obtaining Eq. (8.53) through the fluctuation-dissipation theorem.

8.6 Derive Eqs. (8.56) and (8.57).

8.7 Carry out the summation and integration for Eq.(8.58) in the limit of $B \to 0$ and derive Eq. (8.59).

8.8 Evoke the conservation of momentum for collisions in the guiding-center motion of Eq. (8.74), and show that collisions between like particles do not contribute to the guiding-center diffusion.

CHAPTER **9**

Plasma Turbulence

In Chapter 7, we studied the theory of fluctuations in plasmas near thermodynamic equilibrium. In such a plasma, the density fluctuations associated with the collective modes are at a thermal level; the plasma may be said to be in a quiescent state. Motion of a single particle is not appreciably affected by the presence of such weak fluctuations. The discreteness parameters, such as the one defined in Eq. (1.54), play a special part in characterizing the relaxation processes in such a quiescent plasma; in Chapter 2, the BBGKY hierarchy was solved through a systematic expansion in powers of the plasma parameters. The transport processes are thus viewed in terms of the ordinary Coulomb collisions.

In most cases, plasmas found in nature are significantly away from thermodynamic equilibrium. Such is especially the case in weakly coupled plasmas ($\Gamma \ll 1$), where the usual relaxations are extremely slow; nonequilibrium states may be maintained over quite a long time. A collective mode may thus remain unstable; we studied about those instabilities in Chapter 6. In such an unstable plasma, fluctuations are enhanced greatly over a thermal level; the plasma may eventually go over to a state of turbulence. The enhanced fluctuations are expected to affect the rates of relaxation drastically; the so-called anomalous transports of the conserved quantities may take place in such a plasma. The resulting states may be either weakly or strongly turbulent, depending on the degree of deviation at which the plasmas are maintained away from equilibrium. Clearly, the Klimontovich formalism or the BBGKY hierarchy should be able to produce solutions applicable to any of the three plasma states—namely, quiescent, weakly turbulent, and strongly turbulent states—although the truncation scheme for a quiescent plasma is no longer applicable to those plasmas in a turbulent state. It is a purpose of the present chapter to elucidate some of the fundamental issues in the treatment of plasma turbulence.

For a theoretical treatment of plasma turbulence, it may be meaningful to distinguish between the following two classes of problems: One is what may be called an initial value problem. Here the plasma is isolated from external energy sources; initially, however, it is characterized by physical conditions such that a certain collective mode can grow exponentially. Experimentally, a situation corresponding to this case may, for example, be realized in the initial stage of a beam-plasma interaction experiment when a pulsed beam of charged particles is injected into a quiescent plasma. The theoretical problem then is to analyze the time development, or an approach to equilibrium, for the combined system of particles and oscillations, starting with such unstable initial conditions. Since the amount of an extra free energy available in the initial state to drive the plasma-wave instability is limited, the resulting plasma is most likely to stay in a weakly turbulent state. The advent of the quasilinear theory of plasma oscillations has marked an important step toward a solution to turbulence problems in this category. This approach takes explicit account of a feedback action of the growing oscillations upon the single-particle distribution functions and treats the wave-wave interaction in a perturbation-theoretic way.

Many significant contributions have been made toward the understanding of nonlinear interactions between various collective modes in plasmas. The treatments developed here are basically perturbation-theoretic; iterative solutions to evolution equations for nonlinear processes in the fluctuations, such as (7.26c), lead to a microscopic description on the nonlinear coupling mechanisms. An important consequence derived from such an analysis is a nonlinear Landau damping of the plasma waves; this damping stems from a resonant coupling between a particle and a beating of two plasma waves as described by the relation (7.233). Combining those linear and nonlinear processes in the evolution of the fluctuations, one obtains the wave-kinetic equations.

Consider a second class of turbulence problems, which arises when a system maintains connection with external sources and sinks of free energy; this is thus an open-system problem. In such a system, if one waits long enough, the plasma may reach a new kind of stationary state: a turbulent stationary state. Experimental examples pertaining to this case may be found in various plasma phenomena, both in laboratories and in astrophysical settings. The plasma may be either strongly or weakly turbulent, depending on the abilities of the external source to feed energy into the plasma and in turn of the plasma to process such extra energy internally. In either of these states, a steady flow of energies associated with mode transformation takes place in the turbulent plasma. The specific transport and relaxation processes relevant directly to the internal structures of those turbulent plasmas include particle and energy transports, resistivities, and dissipation mechanisms in magnetic-field turbulence.

In the description of plasma turbulence, the concept of discreteness or the plasma-coupling parameters, elucidated in Chapter 1, is essential. A quantity of primary interest in the description of plasma turbulence is the energy spectrum $\mathcal{E}_{\mathbf{k}}$ contained in the wave vector **k** mode of the fluctuations (cf., Section 7.2E). For a plasma in thermodynamic equilibrium at temperature T, $\mathcal{E}_{\mathbf{k}}$ in a given collective mode takes on an equipartition value around T. Generally for a quiescent plasma, $\mathcal{E}_{\mathbf{k}}$ is written in a form proportional to the discreteness parameters; these are the fluctuations arising from random thermal motion of the discrete particles; hence $\mathcal{E}_{\mathbf{k}}$ vanishes in the fluid limit of Section 1.4C. *A strongly turbulent state of a plasma may be viewed as that in which the energy spectra of fluctuations remain finite even in the fluid limit.* Correlations at distances on the order of the wavelengths therefore persist with "macroscopic" intensities in such a state. It therefore becomes essential for a theory of plasma turbulence to deal directly with a state with strong correlations, rather than to start with a Vlasov description of the plasma state which would ignore the interparticle correlations altogether.

Put another way, a theory of plasma turbulence calls for an approach beyond the RPA. We recall that, when plasmas in strong Coulomb-coupling (i.e., $\Gamma > 1$) were treated, e.g., in Chapters 3 and 7, the local field corrections were introduced to account for such effects of nonlinear coupling between fluctuations. The strongly coupled plasmas are usually in a state of thermodynamic equlibrium, due to their relatively frequent interparticle collisions; an account for the non-RPA effects has been necessitated because of the nonideal nature brought about by $\Gamma > 1$. In a turbulent plasma, scarcity of ordinary Coulomb collisions due to weakness in the Coulomb coupling (i.e., $\Gamma << 1$) acts to keep the plasma in a nonequilibrium turbulent state with an enhanced level of fluctuations, thereby requiring a treatment beyond the RPA. It is in this sense that we may look upon plasma turbulence as a manifestation of strong coupling effects in a nonequilibrium plasma. It is therefore useful to formulate turbulence problems also in terms of the dynamic local-field corrections in the theory of strongly coupled plasmas, which renormalize interactions between particles by the strong correlations.

9.1 WAVE-KINETIC EQUATIONS

The wave-kinetic equations are concerned with evolution of the energy spectrum $\mathcal{E}_{\mathbf{k}}$ of the fluctuations associated with a collective mode in the plasma.* For an OCP, the rate of change in the spectral density is calculated from the longitudinal part of Eq. (7.207) as

$$\frac{\partial \mathcal{E}_{\mathbf{k}}}{\partial t} = -iZe \int_{-\infty}^{\infty} d\omega \, \omega \, \langle n\Phi^*(\mathbf{k},\omega) \rangle . \tag{9.1}$$

Equation (7.26) is the fundamental equation describing microscopic evolution of fluctuations in phase space. In the absence of an external magnetic field, it reads

$$\delta N(\mathrm{k};\mathbf{p}) = \delta N^{(0)}(\mathrm{k};\mathbf{p}) \tag{9.2a}$$

$$+ iZeg_0(\mathrm{k};\mathbf{p})\mathbf{k} \cdot \frac{\partial F}{\partial \mathbf{p}} \Phi(\mathrm{k}) \tag{9.2b}$$

$$+ iZeg_0(\mathrm{k};\mathbf{p}) \int d^4\mathrm{q}\mathbf{q} \cdot \frac{\partial}{\partial \mathbf{p}} \delta N(\mathrm{k}-\mathrm{q};\mathbf{p})\Phi(\mathrm{q}) . \tag{9.2c}$$

Here $F(\mathbf{p})$ is a single-particle distribution normalized to the average number density n as in Eq. (2.22a),

$$g_0(\mathrm{k};\mathbf{p}) \equiv \frac{i}{\omega - \mathbf{k}\cdot\mathbf{v} + i\eta} \tag{9.3}$$

represents a free-particle propagator, and a four-vector notation

$$\mathrm{k} \equiv (\mathbf{k},\omega) \, , \, \mathrm{q} \equiv (\mathbf{q},\chi) \, , \, \mathrm{q}' \equiv (\mathbf{q}',\chi') \, , \, \text{etc.} , \tag{9.4}$$

is employed. In conjunction with this four-vector notation, we introduce a short-hand

$$\sum_{\mathbf{k}}{}' \int_{-\infty}^{\infty} \frac{d\omega}{2\pi} \to \int d^4\mathrm{k} \tag{9.5}$$

for summation and integration.

*In this chapter, we perform Fourier expansion of physical quantities with the periodic boundary conditions appropriate to a cube of a unit volume, unless specified otherwise.

The density fluctuations in Eq. (9.1) are given by

$$\delta n(\mathbf{k}) = \int d\mathbf{p}\,\delta N(\mathbf{k};\mathbf{p})$$

and the fluctuations in the potential field $\Phi(\mathbf{k})$ are connected with the density fluctuations via the Poisson equation.

A. The Quasilinear Approach

To the extent that the fluctuations are weak and the nonlinear term (9.2c) negligible, one may adopt an approximation whereby only the linear terms (9.2a) and (9.2b) are retained. Equation (9.1) in such a linear approximation takes the form

$$\left.\frac{\partial \mathcal{E}_{\mathbf{k}}}{\partial t}\right]^{\mathrm{L}} = -\frac{4\pi (Ze)^2}{k^2}\int d\mathbf{p}\int_{-\infty}^{\infty} d\omega\,\omega F(\mathbf{p})\delta(\omega - \mathbf{k}\cdot\mathbf{v})\mathrm{Im}\frac{1}{\varepsilon_0(\mathbf{k})}$$

$$-\frac{k^2}{4\pi}\int_{-\infty}^{\infty} d\omega\,\omega\,\langle|\Phi|^2(\mathbf{k})\rangle\,\mathrm{Im}\varepsilon_0(\mathbf{k}) \tag{9.6}$$

with the dielectric response function $\varepsilon_0(\mathbf{k})$ given by

$$\varepsilon_0(\mathbf{k}) = 1 - V(k)\chi_0(\mathbf{k}) \tag{9.7a}$$

$$\chi_0(\mathbf{k}) = \int d\mathbf{p} g_0(\mathbf{k};\mathbf{p}) i\mathbf{k}\cdot\frac{\partial F(\mathbf{p})}{\partial \mathbf{p}} \tag{9.7b}$$

and $V(k) = 4\pi(Ze)^2/k^2$ referring to the Coulomb interaction as usual.

The contributions arising from a collective mode $\omega \approx \omega_{\mathbf{k}}$ (cf. Section 7.2E) are separated in Eq. (9.6) to yield

$$\left.\frac{\partial \mathcal{E}_{\mathbf{k}}}{\partial t}\right]^{\mathrm{L}} = \left.\frac{\partial \mathcal{E}_{\mathbf{k}}}{\partial t}\right]^{\mathrm{SP}} - 2\gamma_{\mathbf{k}}\mathcal{E}_{\mathbf{k}} \tag{9.8}$$

where

$$\left.\frac{\partial \mathcal{E}_{\mathbf{k}}}{\partial t}\right]^{\mathrm{SP}} = \left(\frac{2\pi Ze\omega_{\mathbf{k}}}{k}\right)^2 \frac{1}{\alpha_{\mathbf{k}}}\int d\mathbf{p} F(\mathbf{p})\delta(\omega_{\mathbf{k}} - \mathbf{k}\cdot\mathbf{v}) \tag{9.9}$$

and the decay rate $\gamma_{\mathbf{k}}$ has been calculated as Eq. (7.89); $\alpha_{\mathbf{k}}$ is a rewriting of α_{L} in Eq. (7.211a). The first term on the right-hand side of Eq. (9.8), i.e., Eq.

(9.9), represents the rate of spontaneous emission for the plasma wave by the Cherenkov process. The second term describes the balance between the induced emission and absorption, giving rise to linear Landau damping ($\gamma_k > 0$) or growth ($\gamma_k < 0$).

It is instructive to substitute an isotropic Maxwellian for the momentum distribution in Eq. (9.8); we find

$$\left.\frac{\partial \mathcal{E}_{\mathbf{k}}}{\partial t}\right]^{\mathrm{L}} = \left(\frac{2\pi Ze\omega_{\mathbf{k}}}{k}\right)^2 \frac{1}{\alpha_{\mathbf{k}}}\left(1 - \frac{\mathcal{E}_{\mathbf{k}}}{T}\right)\int d\mathbf{p}F(\mathbf{p})\delta(\omega_{\mathbf{k}} - \mathbf{k}\cdot\mathbf{v}) . \qquad (9.10)$$

This implies that $\partial\mathcal{E}_{\mathbf{k}}/\partial t]^{\mathrm{L}} < 0$ if $\mathcal{E}_{\mathbf{k}} > T$ and vice versa. Thermalization of the energy spectrum in the collective mode toward an equipartition value T is thus indicated.

For characterization of a state of plasma turbulence, let us introduce a dimensionless parameter,

$$\mathcal{W} \equiv \frac{\sum_{\mathbf{k}}' \mathcal{E}_{\mathbf{k}}}{nT} . \qquad (9.11)$$

The range of the **k** summation here is confined to that where the collective mode is well defined. Equation (9.11) thus measures the ratio of the turbulence energy to the kinetic energy.

In a plasma near thermodynamic equilibrium, $\mathcal{E}_{\mathbf{k}} \approx T$ and $k < k_{\mathrm{D}}$, so that

$$\mathcal{W} \approx \frac{k_{\mathrm{D}}^3}{n} \approx \frac{1}{N_{\mathrm{D}}} , \qquad (9.12)$$

an inverse of the Debye number introduced in Eq. (1.23). In light of Eq. (1.54), we find that the turbulence level in Eq. (9.12) is of the first order in the discreteness, corresponding to the thermal fluctuation.

The evolution of the single-particle distribution function in the presence of turbulent fluctuations is described by Eq. (7.114). In the linear approximation to fluctuations which collects (9.2a) and (9.2b), the resultant collision term takes a Fokker-Planck form,

$$\left.\frac{\partial F(\mathbf{p})}{\partial t}\right]_{\mathrm{C}} = \frac{\partial}{\partial \mathbf{p}} \cdot [\mathbf{b}(\mathbf{p})F(\mathbf{p})] + \frac{\partial}{\partial \mathbf{p}} \cdot \left[\mathbf{D}(\mathbf{p}) \cdot \frac{\partial}{\partial \mathbf{p}} F(\mathbf{p})\right], \qquad (9.13)$$

with the coefficients of friction and diffusion given by

$$\mathbf{b}(\mathbf{p}) = (2\pi Ze)^2 \sum_{\mathbf{k}}' \frac{\omega_{\mathbf{k}}}{\alpha_{\mathbf{k}}} \frac{\mathbf{k}}{k^2} \delta(\omega_{\mathbf{k}} - \mathbf{k} \cdot \mathbf{v}) \tag{9.14a}$$

$$\mathsf{D}(\mathbf{p}) = (2\pi Ze)^2 \sum_{\mathbf{k}}' \frac{\mathbf{kk}}{\alpha_{\mathbf{k}} k^2} \mathcal{E}_{\mathbf{k}} \delta(\omega_{\mathbf{k}} - \mathbf{k} \cdot \mathbf{v}) \,. \tag{9.14b}$$

The friction stems physically from the reaction to the Cherenkov emission processes described in Eq. (9.9). The diffusion is a consequence of the recoils in particles at the induced emission and absorption processes.

The rate of increase in the plasma kinetic energy is obtained through an integral

$$\int d\mathbf{p}\varepsilon \left. \frac{\partial F(\mathbf{p})}{\partial t} \right]_{\mathrm{C}} = - \sum_{\mathbf{k}}' \int d\mathbf{p} \left(\frac{2\pi Ze\omega_{\mathbf{k}}}{k} \right)^2 \frac{1}{\alpha_{\mathbf{k}}} \delta(\omega_{\mathbf{k}} - \mathbf{k} \cdot \mathbf{v})$$

$$\times \left[\frac{\mathcal{E}_{\mathbf{k}}}{\omega_{\mathbf{k}}} \mathbf{k} \cdot \frac{\partial F(\mathbf{p})}{\partial \mathbf{p}} + F(\mathbf{p}) \right]. \tag{9.15}$$

This equation may be interpreted as describing the rate of *turbulent heating* due to enhanced fluctuations when $\mathcal{E}_{\mathbf{k}} > T$.

The theory presented here has a structure such that the fluctuations in the collective mode evolve linearly by Eq. (9.8) in accord with the solution to Eq. (3.54). The spectral distribution of the fluctuations then acts to change the single-particle distributions via Eq. (9.13), which in turn affect the evolution of the collective mode through Eq. (3.54). Hence it is called a *quasilinear approach*. In such a treatment of coupling between the fluctuations and the single-particle distributions, the hierarchy (7.10) is assumed. The quasilinear theory of plasma oscillations [Drummond & Pines 1962; Vedenov, Velikhov & Sagdeev 1962; Rutherford & Frieman 1963] has marked an important step toward solution of the problems associated with a weakly turbulent plasma.

B. Weak Turbulence

Let us consider a way to take into account the nonlinear term (9.2c) in the wave-kinetic equations. Introducing the potential-field fluctuations via the Poisson equation,

$$\Phi(\mathbf{k}) = \frac{4\pi Ze}{k^2} \delta n(\mathbf{k}) \,, \tag{9.16}$$

we transform Eq. (9.2) to the form,

$$\varepsilon_0(\mathrm{k})\Phi(\mathrm{k}) = \Phi^{(0)}(\mathrm{k}) + \Phi^{\mathrm{NL}}(\mathrm{k})\,. \tag{9.17}$$

Here the nonlinear contributions, $\Phi^{\mathrm{NL}}(\mathrm{k})$, are expanded perturbation-theoretically as [Kadomtsev 1965]

$$\Phi^{\mathrm{NL}}(\mathrm{k}) = Ze\int d^4\mathrm{q}\,\upsilon^{(2)}(\mathrm{k},\mathrm{q})\Phi(\mathrm{k}-\mathrm{q},\mathrm{q})$$

$$+(Ze)^2\int d^4\mathrm{q}\int d^4\mathrm{q}'\,\upsilon^{(3)}(\mathrm{k},\mathrm{q},\mathrm{q}')\Phi(\mathrm{k}-\mathrm{q}-\mathrm{q}')\Phi(\mathrm{q})\Phi(\mathrm{q}')\,. \tag{9.18}$$

Various nonlinear polarizabilities are introduced in conjunction with such expansions:

$$\upsilon^{(2)}(\mathrm{k},\mathrm{q}) = -V(k)\int d\mathbf{p}\,g_0(\mathrm{k};\mathbf{p})\mathbf{q}\cdot\frac{\partial}{\partial\mathbf{p}}\,g_0(\mathrm{k}-\mathrm{q};\mathbf{p})\cdot\frac{\partial F(\mathbf{p})}{\partial\mathbf{p}} \tag{9.19a}$$

$$\upsilon^{(3)}(\mathrm{k},\mathrm{q},\mathrm{q}') = -iV(k)\int d\mathbf{p}\,g_0(\mathrm{k};\mathbf{p})\mathbf{q}\cdot\frac{\partial}{\partial\mathbf{p}}\,g_0(\mathrm{k}-\mathrm{q};\mathbf{p})\,,$$

$$\times\,\mathbf{q}'\cdot\frac{\partial}{\partial\mathbf{p}}\,g_0(\mathrm{k}-\mathrm{q}-\mathrm{q}';\mathbf{p})(\mathbf{k}-\mathbf{q}-\mathbf{q}')\cdot\frac{\partial F(\mathbf{p})}{\partial\mathbf{p}} \tag{9.20}$$

and

$$\chi^{(2)}(\mathrm{k},\mathrm{q}) \equiv \upsilon^{(2)}(\mathrm{k},\mathrm{q}) + \upsilon^{(2)}(\mathrm{k},\mathrm{k}-\mathrm{q}) = \chi^{(2)}(\mathrm{k},\mathrm{k}-\mathrm{q})\,. \tag{9.19b}$$

These polarizabilities are Hermitian in the sense that

$$\chi^{(2)}(-\mathrm{k},-\mathrm{q}) = \chi^{(2)}(\mathrm{k},\mathrm{q})^*\,. \tag{9.21}$$

The nonlinear contributions to the growth rate of the energy spectrum are then calculated in a standard way (cf., Eqs. 7.207 and 9.1) from the rate of dissipation in the nonlinear (longitudinal) current; to the second order in the spectral density, we thus find

$$\left.\frac{\partial\mathcal{E}_{\mathbf{k}}}{\partial t}\right]^{\mathrm{NL}} = -\int d\omega\langle\mathbf{J}^{\mathrm{NL}}\cdot\mathbf{E}^*(\mathbf{k})\rangle$$

$$= \sum_{\mathbf{q}}\frac{V(q)\omega_k}{\alpha_k\alpha_q}\Big\{\mathrm{Im}[\upsilon^{(3)}(\mathrm{k},\mathrm{q},-\mathrm{q}) - \upsilon^{(3)}(\mathrm{k},\mathrm{q},\mathrm{k})]$$

$$+\,\mathrm{Im}[\chi^{(2)}(\mathrm{k},\mathrm{k}-\mathrm{q})\chi^{(2)}(\mathrm{k}-\mathrm{q},\mathrm{k})]\mathrm{Re}\frac{1}{\varepsilon_0(\mathrm{k}-\mathrm{q})}\Big\}\,\mathcal{E}_{\mathbf{q}}\mathcal{E}_{\mathbf{k}} \tag{9.22a}$$

$$-\frac{\pi}{2}\sum_{\mathbf{q}}\frac{V(q)\omega_k\omega_{k-q}\mathrm{Re}[\chi^{(2)}(\mathrm{k,k-q})\chi^{(2)}(\mathrm{k-q,k})]}{\alpha_k\alpha_q\alpha_{k-q}}\delta(\omega_k-\omega_q-\omega_{k-q})\mathcal{E}_{\mathbf{q}}\mathcal{E}_{\mathbf{k}}$$

(9.22b)

$$+\frac{\pi}{4}\sum_{\mathbf{q}}\frac{V(q)V(|\mathbf{k}-\mathbf{q}|)\omega_k^2|\chi^{(2)}(\mathrm{k,q})|^2}{V(k)\alpha_k\alpha_q\alpha_{k-q}}\delta(\omega_k-\omega_q-\omega_{k-q})\mathcal{E}_{\mathbf{q}}\mathcal{E}_{\mathbf{k-q}}\,.$$

(9.22c)

The term (9.22a), depending on the imaginary parts of the nonlinear polarizabilities, represents that part of the growth rate due to the *nonlinear wave-particle interactions*; the resonant denominators in the free-particle propagators, Eq. (9.3), contribute to the processes. The terms (9.22b) and (9.22c), on the other hand, describe the contributions of those *three-wave processes* to the growth rate which satisfy the condition

$$\omega_{\mathbf{k}}=\omega_{\mathbf{q}}+\omega_{\mathbf{k-q}}\,. \tag{9.23}$$

Summation between the linear and nonlinear contributions, Eqs. (9.8) and (9.22), yields the total growth rate in the perturbation approximation, which may be expressed in the form,

$$\frac{\partial\mathcal{E}_{\mathbf{k}}}{\partial t}=\left.\frac{\partial\mathcal{E}_{\mathbf{k}}}{\partial t}\right]^{\mathrm{SP}}-2\gamma_{\mathbf{k}}\mathcal{E}_{\mathbf{k}}$$

$$-\sum_{\mathbf{q}}A(\mathbf{k},\mathbf{q})\mathcal{E}_{\mathbf{q}}\mathcal{E}_{\mathbf{k}}+\sum_{\mathbf{q}}B(\mathbf{k},\mathbf{q})\mathcal{E}_{\mathbf{q}}\mathcal{E}_{\mathbf{k-q}}\,. \tag{9.24}$$

A stationary solution to this equation is then written as

$$\mathcal{E}_{\mathbf{k}}=\frac{\partial\mathcal{E}_{\mathbf{k}}/\partial t]^{\mathrm{SP}}+\sum_{\mathbf{q}}{}'B(\mathbf{k},\mathbf{q})\mathcal{E}_{\mathbf{q}}\mathcal{E}_{\mathbf{k-q}}}{2\gamma_{\mathbf{k}}+\sum_{\mathbf{q}}{}'A(\mathbf{k},\mathbf{q})\mathcal{E}_{\mathbf{q}}}\,. \tag{9.25}$$

Equation (9.25) makes it possible to order the level of plasma turbulence with respect to powers in the plasma discreteness parameters such as Eq. (1.54). To begin, we note that

$$\left.\frac{\partial\mathcal{E}_{\mathbf{k}}}{\partial t}\right]^{\mathrm{SP}}\sim\Delta^1\,, \tag{9.26}$$

$$A(\mathbf{k},\mathbf{q}) \sim B(\mathbf{k},\mathbf{q}) \sim \Delta^0 . \tag{9.27}$$

The latter coupling coefficients remain finite in the fluid limit of Section 1.4C.

Let us now introduce a formal definition of the three states for plasmas: quiescent, weakly turbulent, and strongly turbulent. A *quiescent state* is defined as that in which γ_k is positive (i.e., linearly stable) and $| \gamma_k/\omega_k | \Delta^0$. Equation (9.25) then yields

$$\mathcal{E}_{\mathbf{k}} = \frac{1}{2\gamma_{\mathbf{k}}} \frac{\partial \mathcal{E}_{\mathbf{k}}}{\partial t} \Big]^{\mathrm{SP}} . \tag{9.28}$$

The thermal fluctuations, Eq. (9.12), are special examples of this general relation. The spectral density of fluctuations associated with a collective mode in such a quiescent stationary plasma is maintained by a balance between the rate of spontaneous emission and that of the linear Landau damping.

A *weakly turbulent state* is defined as that in the vicinity of the critical condition ($\gamma_k = 0$), so that $| \gamma_k/\omega_k | \leq \sqrt{\Delta}$. Equation (9.25) then yields $\mathcal{W} \approx \sqrt{\Delta}$, some enhancement over the thermal level of Eq. (9.12). Each term on the right-hand side of the wave-kinetic equation (9.24) is on the first order in the discreteness parameters; nonlinear terms of orders higher than those terms retained need not be considered.

A *strongly turbulent state* is defined as that in which γ_k is negative (i.e., linearly unstable) and $| \gamma_k/\omega_k | \Delta^0$. Equation (9.25) then yields $\mathcal{W} \approx \Delta^0$; fluctuations thus remain finite even in the fluid limit. It then turns out that the truncation on the right-hand side of Eq. (9.24) may be premature; those higher-order terms ignored in the expansion may contribute as strongly as those retained. The coupling coefficients, $A(\mathbf{k},\mathbf{q})$ and $B(\mathbf{k},\mathbf{q})$, should thus take on renormalized values, accounting for the summation over those higher-order contributions. We shall consider some such renormalization schemes in Section 9.1E.

C. Nonlinear Landau Damping

For plasma oscillations with the dispersion relation as given by Eq. (4.30), the three-wave processes specified by Eq. (9.23) cannot take place; these are waves of the *nondecay* type. The *nonlinear decay rate* defined by

$$\gamma_{\mathbf{k}}^{\mathrm{NL}} \equiv \frac{1}{2} \sum_{\mathbf{q}}{}' A(\mathbf{k},\mathbf{q}) \mathcal{E}_{\mathbf{q}} \tag{9.29}$$

is then calculated as

$$\gamma_{\mathbf{k}}^{\mathrm{NL}} = -\frac{\omega_{\mathbf{k}}}{2} \sum_{\mathbf{q}}{}' V(q)\mathcal{E}_{\mathbf{q}} \mathrm{Im} \left\{ \upsilon^{(3)}(\mathbf{k},\mathbf{q},-\mathbf{q}) + \upsilon^{(3)}(\mathbf{k},\mathbf{q},\mathbf{k}) \right.$$

$$\left. + \frac{\chi^{(2)}(\mathbf{k},\mathbf{k}-\mathbf{q}) + \chi^{(2)}(\mathbf{k}-\mathbf{q},\mathbf{k})}{\varepsilon_0(\mathbf{k}-\mathbf{q},\omega_{\mathbf{k}}-\omega_{\mathbf{q}})} \right\} \tag{9.30}$$

since $\alpha_{\mathbf{k}} = 1$ for such a plasma oscillation.

The plasma oscillations are long-lived elementary excitations in the long-wavelength regime, $k < k_{\mathrm{D}}$. In light of their dispersion relation (4.30), one then assumes

$$\omega' \equiv \omega_{\mathbf{k}} - \omega_{\mathbf{q}} \approx 0\,, \tag{9.31a}$$

$$|\mathbf{k}'| \equiv |\mathbf{k}-\mathbf{q}| < k_{\mathrm{D}}\,, \tag{9.31b}$$

so that $|\varepsilon_0(\mathbf{k}',\omega')| >> 1$. Equation (9.30) under these circumstances becomes

$$\gamma_{\mathbf{k}}^{\mathrm{NL}} = -\frac{32\pi^3(Ze)^4}{m^2\omega_{\mathbf{k}}^5} \sum_{\mathbf{q}}{}' \frac{(\mathbf{k}\cdot\mathbf{q})^2}{k^2q^2}\mathcal{E}_{\mathbf{q}}$$

$$\times \int d\mathbf{p}(\mathbf{k}\cdot\mathbf{v})^2\delta(\omega' - \mathbf{k}'\cdot\mathbf{v})\mathbf{k}'\cdot\frac{\partial F(\mathbf{p})}{\partial\mathbf{p}}\,. \tag{9.32}$$

We may interpret this decay rate as arising from a resonant coupling between a particle with momentum **p** and a *beating* of two plasma oscillations at a frequency ω' and a wave vector $\mathbf{k}'$, that is,

$$\omega' = \mathbf{k}'\cdot\mathbf{v}\,. \tag{9.33}$$

Hence it is referred to as a *nonlinear Landau damping*; the decay rate Eq. (9.32) depends on the spectral distribution of the plasma oscillations. Because of (9.31a), the bulk of the particles can satisfy the resonance condition (9.33); the nonlinear Landau damping can be effective even when the linear Landau damping is not. A similarity between Eq. (9.33) and (7.233) should be remarked; the nonlinear Landau damping is related closely to the Compton scattering of the plasma waves by the particles.

Note that Eq. (9.32) vanishes in the long-wavelength limit as k^2, reflecting conservation of the total momentum due to (9.31b). One likewise derives a conservation law of the total energy in the plasma oscillations:

$$\sum_{\mathbf{k}} \gamma_{\mathbf{k}}^{\mathrm{NL}} \mathcal{E}_{\mathbf{k}} = 0 , \tag{9.34}$$

stemming from Eq. (9.31a). The nonlinear Landau damping of the plasma waves therefore describes a diffusive redistribution of the fluctuation spectrum in the wave vector space.

D. Ion-Acoustic Wave Turbulence

Ion-acoustic waves with the dispersion relation Eq. (4.53) likewise are of a nondecay type, and hence it is assumed that the wave-particle coupling provides the leading contribution to nonlinear decay processes. The nonlinear Landau damping has been incorporated in Kadomtsev's theory of ion-acoustic wave turbulence in a current-carrying plasma [Kadomtsev 1965].

We consider the ion-acoustic wave in a TCP with one-dimensional velocity distributions given by Eqs. (6.8) under a condition such that $T_i << T_e$. The second term in the square bracket of Eq. (6.12), representing the linear Landau damping due to ions, is negligible in these circumstances; the condition for instability is given by

$$\gamma^{\mathrm{L}} \equiv \sqrt{\frac{\pi m_e}{8 m_i}}\, k[s - v_{\mathrm{d}} \cos \theta] < 0 . \tag{9.35}$$

Here s refers to the sound velocity Eq. (4.50b) and θ is the angle between $\mathbf{k}$ and $\mathbf{v}_{\mathbf{d}}$. The inequality in (9.35) defines a Cherenkov cone in the $\mathbf{k}$ space as depicted in Fig. 9.1.

The rate of nonlinear Landau damping for the ion-acoustic wave arising from resonant coupling (9.33) with the ions is calculated as

$$\gamma_{\mathbf{k}}^{\mathrm{NL}} = -\frac{2\pi s^2 T_e}{n^2 m_i \omega_k^2} \sum_{\mathbf{q}} (\mathbf{k}\cdot\mathbf{q})^2 \mathcal{E}_{\mathbf{q}} \int d\mathbf{p} \left\{ \left(\frac{\mathbf{k}\cdot\mathbf{v}}{\omega_k}\right)^2 \left(1 - 3\frac{\mathbf{k}\cdot\mathbf{k}'}{|\mathbf{k}'|^2}\frac{\omega'}{\omega_k}\right) \right.$$
$$\left. + \left(\frac{\mathbf{k}\cdot\mathbf{v}}{\omega_k}\right)^3 \left(3 - 4\frac{\mathbf{k}\cdot\mathbf{k}'}{|\mathbf{k}'|^2}\frac{\omega'}{\omega_k}\right) - \left(\frac{\mathbf{k}\cdot\mathbf{k}'}{|\mathbf{k}'|^2}\frac{\omega'}{\omega_k}\right)^2 \right\} \delta(\omega' - \mathbf{k}'\cdot\mathbf{v})\mathbf{k}' \cdot \frac{\partial F_i(\mathbf{p})}{\partial \mathbf{p}} . \tag{9.36}$$

For simplicity we deal with quantities integrated with respect to solid angle do over the Cherenkov cone of Fig. 9.1:

$$\mathcal{E}_k \equiv \int do \mathcal{E}_{\mathbf{k}} , \tag{9.37a}$$

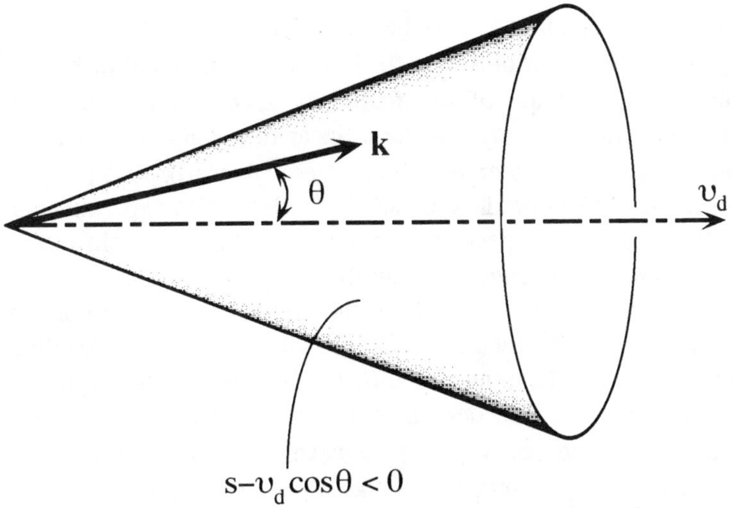

Figure 9.1 Cherenkov cone for the ion-acoustic wave instability.

$$\theta_0^2 \equiv \frac{1}{\mathcal{E}_k} \int do \sin^2\theta \cos^2\theta \, \mathcal{E}_{\mathbf{k}} . \tag{9.37b}$$

This θ_0^2 is assumed independent of k. In the limit of $T_i/T_e \rightarrow 0$, the momentum distribution $F_i(\mathbf{p})$ of the ions may be approximated by a three-dimensional delta function in the calculation of Eq. (9.36). Hence we find

$$\gamma^{\rm NL} = -\frac{\theta_0^2 T_i}{(2\pi)^2 n m_i s T_e}\left[k^5 \frac{\partial \mathcal{E}_k}{\partial k} + 3k^4 \mathcal{E}_k\right]. \tag{9.38}$$

The stationary state is determined by a condition that the total growth rate vanish, i.e.,

$$\gamma^{\rm L} + \gamma^{\rm NL} = 0 . \tag{9.39}$$

Solution of this differential equation yields the *Kadomtsev spectrum* of the ion-acoustic wave turbulence,

$$\mathcal{E}_k = \frac{\sqrt{\pi m_e}}{8 m_i} \frac{(2\pi)^2 n T_e^2}{\theta_0^2 T_i} \frac{v_{\rm d} - s}{s} \frac{1}{k^3} \ln\left(\frac{k_0}{k}\right). \tag{9.40}$$

The constant of integration k_0 may be chosen in the vicinity of k_e by physical reasoning. A spectral distribution in the **k** space proportional to $k^{-3} \ln (k_e/k)$ is thus indicated. The shape of the frequency spectrum is then obtained as $\omega^{-1} \ln (\omega_i/\omega)$, since we may assume a linear dispersion, $\omega = sk$. This shape has been experimentally observed [Hamberger & Jancarik 1972].

Concrete theoretical prediction and comparison with experiments notwithstanding, the Kadomtsev spectrum appears to contain a certain inconsistency in its derivation. As Kadomtsev himself recognizes, the magnitude of Eq. (9.40) is "very large." This stems from the inefficiency of the nonlinear Landau damping (9.38) to balance the linear growth rate (9.35) through Eq. (9.39); the nonlinear Landau damping is inefficient because of the assumptions that $T_i/T_e \to 0$ and $\theta_0^2 << 1$. The derivation of the spectrum is based on a weak turbulence theory. The result (3.40), however, is of the zeroth order in the discreteness, a scaling appropriate rather to a strong turbulence. In such a circumstance, it becomes essential to take into account the strong modification of particle orbitals due to those enhanced fluctuations, or the renormalization of single-paticle propagators, which we consider in the next subsection.

E. Toward Strong Turbulence

In the dynamic theory of strongly coupled plasmas ($\Gamma > 1$) treated in Sections 3.2 and 7.3, the dynamic local-field corrections have played an important part in a description of the modified interactions brought about by the strong interparticle correlations. Let us consider how such an effect may be accounted for in a theory of strong plasma turbulence.

We take the point of view that turbulence is a nonequilibrium realization of strong coupling effects in plasmas, where fluctuations or correlations cannot be treated as weak perturbations. A most salient effect in a turbulent plasma is a statistical modification of single-particle orbits by fluctuating fields. The idea is basically due to Dupree [1966], who has argued that the dominant nonlinear effect of low-frequency instabilities is an incoherent scattering of particles by waves, which causes particle diffusion and appears in the theory as an enhanced viscosity. Kono and Ichikawa [1973] have shown a way to carry out a partial summation of the most secular terms in each order of the perturbation-theoretic expansion with respect to fluctuations.

Another aspect of strong-correlation effects in a turbulent plasma is modification of effective interactions between particles. The dynamic aspects should correspond physically to creation and destruction of clumps [Dupree 1972], bunched particles, and coherent waves [Goldman 1984]. It has been shown [Ichimaru 1977a] that such an effect can be described through local field corrections in the theory of strongly coupled plasmas. Equations (7.123)–(7.125) can then be employed for evolution of the single-particle distribution in the presence of such a strong dynamic correlation.

Let us therefore revisit the fundamental equation (9.2) for the evolution of turbulent fluctuations and derive the renormalized single-particle propagator $g(\mathrm{k};\mathbf{p})$ through a partial summation of higher-order correlation diagrams. The dynamic local-field correction $G(\mathrm{k})$ is then obtained via

$$\begin{aligned}\tilde{\varepsilon}(\mathrm{k}) &= 1 - V(k)\int d\mathbf{p}\, g(\mathrm{k};\mathbf{p})i\mathbf{k}\cdot\frac{\partial F(\mathbf{p})}{\partial \mathbf{p}} \\ &= 1 - V(k)\chi(\mathrm{k}) \\ &= 1 - V(k)[1 - G(\mathrm{k})]\chi_0(\mathrm{k}) \end{aligned} \tag{9.41}$$

with $\chi_0(\mathrm{k})$ defined in Eq. (9.7b).

The function $\tilde{\varepsilon}(\mathrm{k})$ acts to screen the Coulomb field between an external test charge and a plasma particle. The dielectric response function for the plasma, which has been defined as the screening function for the Coulomb field between two external test charges, is then given by

$$\varepsilon(\mathrm{k}) = 1 - \frac{V(k)\chi_0(\mathrm{k})}{1 + V(k)G(\mathrm{k})\chi_0(\mathrm{k})} \tag{9.42}$$

[Hedin & Lundqvist 1971; Kukkonen & Overhauser 1979; Ichimaru & Iyetomi 1983].

The integral equation (9.25) for the spectral distribution of turbulence is now duly modified by the use of $\varepsilon(\mathrm{k})$ and $g(\mathrm{k};\mathbf{p})$ in place of $\varepsilon_0(\mathrm{k})$ and $g_0(\mathrm{k};\mathbf{p})$ in Eqs. (9.17)–(9.19). The use of $G(\mathrm{k})$ in Eqs. (7.123)–(7.125) then sets the kinetic equation for $F(\mathbf{p})$ in the turbulent plasma. Thus we have a complete formulation of particles and fluctuations in a turbulent plasma.

The renormalized propagator is calculated in the following way by the use of diagrams [Ichimaru 1977a]. Let thin straight, thick straight, and shaded double lines, respectively, correspond to the free-particle propagator $g_0(\mathrm{k};\mathbf{p})$, the renormalized propagator $g(\mathrm{k};\mathbf{p})$, and the screened propagator $g(\mathrm{k};\mathbf{p})/\tilde{\varepsilon}(\mathrm{k})$. A wavy line with both ends terminated by vertices represents the contribution of the internal fluctuations $\langle|\Phi^2|(\mathrm{k})\rangle$; an open wavy line describes the potential field $\Phi(\mathrm{k})$ stemming from a summed contribution of the external and induced potentials. When a wavy line with wave vector $\mathbf{q}$ is terminated at a filled-circle vertex, the vertex gives rise to a differential operator,

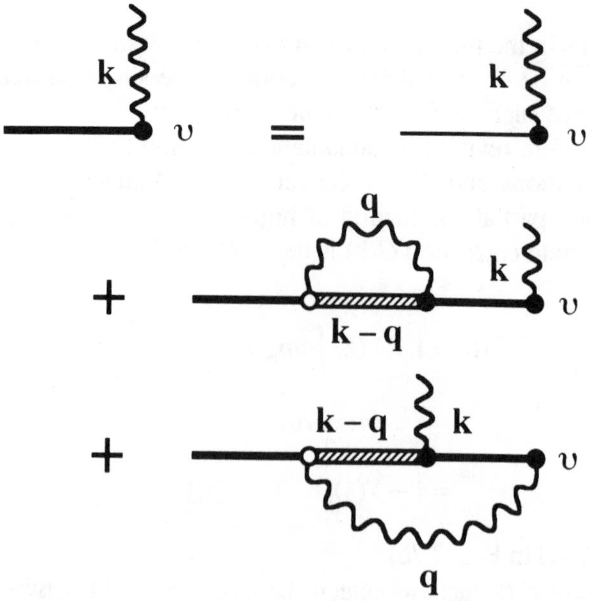

Figure 9.2 Diagrammatic representation of a renormalization scheme for a dielectric propagator.

$$iZe\mathbf{q} \cdot \frac{\partial}{\partial \mathbf{p}} . \tag{9.43}$$

When a wavy line with wave vector **q** and a straight line with wave vector **k** − **q** merge at an open-circle vertex, this contributes an operator of (9.42) in which **q** is replaced by

$$\mathbf{Q} = \mathbf{q} + \frac{q^2}{|\mathbf{k}-\mathbf{q}|^2}(\mathbf{k}-\mathbf{q}) , \tag{9.44}$$

reflecting the symmetrization in Eq. (7.100) or in Eq. (9.19b). Involvement of such an open-circle vertex is related to conservation of the total momentum in an OCP; **Q** defined in (9.44) vanishes in the long-wavelength limit, $k \to 0$.

In terms of these diagrams, the renormalized propagator is calculated in accord with the summation scheme as depicted in Fig. 9.2. It yields the equation:

$$g(\mathrm{k};\mathbf{p})\mathbf{k}\cdot\frac{\partial}{\partial\mathbf{p}} = g_0(\mathrm{k};\mathbf{p})\mathbf{k}\cdot\frac{\partial}{\partial\mathbf{p}}$$

$$+2\pi(Ze)^2\int d^4\mathrm{q}\langle|\Phi^2|(\mathrm{q})\rangle g(\mathrm{k};\mathbf{p})\mathbf{Q}\cdot\frac{\partial}{\partial\mathbf{p}}\frac{g(\mathrm{k}-\mathrm{q};\mathbf{p})}{\tilde{\varepsilon}(-\mathrm{q})}$$

$$\times\left[\mathbf{q}\cdot\frac{\partial}{\partial\mathbf{p}}g(\mathrm{k};\mathbf{p})\mathbf{k}\cdot\frac{\partial}{\partial\mathbf{p}}+\mathbf{k}\cdot\frac{\partial}{\partial\mathbf{p}}g(-\mathrm{q};\mathbf{p})\mathbf{q}\cdot\frac{\partial}{\partial\mathbf{p}}\right]. \tag{9.45}$$

The dynamic local-field correction $G(\mathrm{k})$ is then obtained by substitution of Eq. (9.45) in Eq. (9.41). The first two terms on the right-hand side of (9.45) contain all the processes related to the most secular terms [Kono & Ichikawa 1973]; in addition they include screening of the intermediate propagators and enable one to sum the next most secular terms without vertex corrections. The last term of (9.45) takes account of partial effects for the vertex corrections.

In formulation of the strong-turbulence problems, the imaginary parts of the dynamic local-field corrections should play a leading role: These provide new channels of dissipation other than those arising from the resonant denominators of the free-particle propagators Eq. (9.3). Such a distinction between the dissipation channels has been elucidated in Eqs. (7.124a) and (7.124b). The contribution (7.124b) stemming from the imaginary part of $G(\mathrm{k})$ has been essential in a generalized viscoelastic theory of glass transition for strongly coupled, classical OCPs [Ichimaru & Tanaka 1986]. It will be equally essential in a nonequilibrium problem such as strong plasma turbulence.

Finally we find it instructive to derive a simplified expression for the propagator out of Eq. (9.45). To do so, we note that the polarizabilities, $\chi_0(\mathrm{k})$ and $\chi(\mathrm{k})$, defined in Eqs. (9.7b) and (9.41), may be interpreted as a momentum average of operators,

$$\chi_0(\mathrm{k}) = n\left\langle g_0(\mathrm{k};\mathbf{p})i\mathbf{k}\cdot\frac{\partial}{\partial\mathbf{p}}\right\rangle \tag{9.46a}$$

$$\chi(\mathrm{k}) = n\left\langle g(\mathrm{k};\mathbf{p})i\mathbf{k}\cdot\frac{\partial}{\partial\mathbf{p}}\right\rangle. \tag{9.46b}$$

Then, within the range of validity for such a momentum-average approximation, we may write

$$ng_0(\mathrm{k};\mathbf{p})i\mathbf{k}\cdot\frac{\partial F(\mathbf{p})}{\partial\mathbf{p}} \approx \chi_0(\mathrm{k})F(\mathbf{p}), \tag{9.47a}$$

$$ng(\text{k};\mathbf{p})i\mathbf{k}\cdot\frac{\partial F(\mathbf{p})}{\partial \mathbf{p}} \approx \chi(\text{k})F(\mathbf{p})\,. \tag{9.47b}$$

Applying these approximations successively to the terms in Eq. (9.45), we obtain

$$G(\text{k}) = \frac{H(\text{k})}{1+H(\text{k})} \tag{9.48}$$

with

$$H(\text{k}) = 2\pi(Ze)^2\int d^4\text{q}\,\langle|\Phi^2|(\text{q})\rangle\frac{\chi(\text{k}-\text{q})}{\tilde{\varepsilon}\,(\text{k}-\text{q})}K(\mathbf{k},\mathbf{q})$$

$$\times\frac{q^2(\mathbf{k}-\mathbf{q})}{k^2|\,\mathbf{k}-\mathbf{q}\,|^2}\cdot[\mathbf{q}\chi(\text{k})-\mathbf{k}\chi(-\text{q})] \tag{9.49}$$

and $K(\mathbf{k},\mathbf{q})$ in Eq. (7.100). These expressions should be of use in explicit treatments for specific cases of turbulence problems.

9.2 TRANSPORTS

Transport processes in a turbulent plasma differ in an essential way from those in a quiescent plasma. Let us note that the frequencies of Coulomb collisions, such as Eqs. (1.28), (8.92), and (8.94), are of the first order in the discreteness parameters of Section 1.4C, apart from a weak dependence in the Coulomb logarithm; these collision frequencies therefore vanish in the fluid limit. Analogous dependence will be observed in the Coulomb resistivity of a quiescent plasma, the Spitzer resistivity [Spitzer 1962], which we shall consider in Section 9.2B. In fact such a linear dependence on the discreteness parameters is a basic feature inherent in the collision term of a quiescent plasma such as Eq. (2.68).

Empirically there exist various transport coefficients which remain *finite* in the fluid limit. A well known example is the *coefficient of Bohm diffusion* [Bohm 1949],

$$D_B \sim \frac{cT}{eB}\,. \tag{9.50}$$

The *pseudo-classical diffusion coefficient* such as $\sim |\gamma_k|/k_\perp^2$ is another example, where $1/k_\perp$ and $|\gamma_k|$ are the characteristic length and growth rate of turbulent fluctuations [Kadomtsev 1965]. In the ensuing subsections we shall encounter the Buneman resistivity [Hamberger & Jancarik 1972] and Parker's reconnection velocity of magnetic lines of force [Parker 1973], both of which belong to the class of transport processes that would not vanish in the fluid limit.

As Eq. (1.23) illustrates, N_D takes on a huge number in a high-temperature, tenuous plasma. Hence those transports that remain finite in the fluid limit can become the dominant processes in such a plasma. These processes are distinctly different from the classical collision processes, and are called *anomalous transports*.

A. Description by Fluctuation Spectra

A formulation of the electric resistivity may be obtained by following the process of momentum transfer in the kinetic equation (7.14). An equation relevant to the resistivity is derived from Eq. (7.146) as

$$\mathbf{E} + \frac{1}{c}\mathbf{u}_\sigma \times \mathbf{B} = -\frac{2\pi}{Z_\sigma e n_\sigma}\int d^4\mathrm{k}\,\frac{\mathbf{k}}{\omega}\langle \mathbf{J}_\sigma \cdot \mathbf{E}(\mathrm{k})\rangle . \tag{9.51}$$

Here $\mathbf{u}_\sigma$ and $\mathbf{J}_\sigma$ are the flow velocity and the current density carried by the particles of the σ species, and the four-vector notation of Eqs. (9.4) and (9.5) is in use.

The current-density fluctuations induced by the electric-field fluctuations are expressed as

$$\delta\mathbf{J}_\sigma(\mathrm{k}) = \sigma(\mathrm{k}) \cdot \delta\mathbf{E}(\mathrm{k}) . \tag{9.52}$$

This defines the conductivity tensor $\sigma(\mathrm{k})_\sigma$ of the σ species (cf., Eq. 3.83). Substituting Eq. (9.52) in (9.51), we find a general expression for the dc resistivity as

$$\rho_{E\sigma} = -\frac{2\pi}{(Z_\sigma e n_\sigma)^2}\int d^4\mathrm{k}\,\frac{k_{\parallel}}{\omega}\,\sigma(\mathrm{k})_\sigma : \langle \mathbf{E}\mathbf{E}^*(\mathrm{k})\rangle . \tag{9.53}$$

Here $k_{\parallel}$ is the component of the wave vector in the direction of the flow velocity. In evaluating Eq. (9.53), one selects the particle species that has the most effective contribution in the dissipative processes under consideration.

When the longitudinal approximation

$$\delta \mathbf{E}(\mathbf{k}) = -i \frac{4\pi \mathbf{k}}{k^2} \sum_{\sigma} Z_\sigma e \delta n_\sigma(\mathbf{k}) \tag{9.54}$$

is applicable, we may rewrite the right-hand side of Eq. (9.51) in terms of density-density correlation functions with the aid of the continuity relation. We thus find

$$\rho_{E\sigma} = \sum_{\sigma'} \frac{8\pi^2 Z_{\sigma'}}{Z_\sigma (n_\sigma u_\sigma)^2} \int d^4\mathrm{k} \frac{\mathbf{k} \cdot \mathbf{u}_\sigma}{k^2} \mathrm{Im} S_{\sigma\sigma'}(\mathrm{k}) \,, \tag{9.55}$$

where

$$S_{\sigma\sigma'}(\mathrm{k}) = \frac{1}{2\pi} \int d\mathbf{r} \int_{-\infty}^{\infty} dt \langle \delta n_\sigma(\mathbf{r},t) \delta n_{\sigma'}(\mathbf{0},0) \rangle \exp(-i\mathbf{k} \cdot \mathbf{r} + i\omega t) \,. \tag{9.56}$$

Imaginary parts of these correlation functions vanish for $\sigma = \sigma'$; scattering between like particles does not directly contribute to Eq. (9.55), due to conservation of the total momenta.

B. The Spitzer Resistivity

In the cases of weak fluctuations, we may use the superposition principle of dressed test-particles (Section 7.2C). For a hydrogenic plasma with $Z = 1$, assuming that the electrons have a drifted Maxwellian distribution

$$F_e(\mathbf{p}) = n(2\pi m_e T_e)^{-3/2} \exp\left[-\frac{|\mathbf{p} - m_e \mathbf{u}|^2}{2m_e T_e}\right] \tag{9.57a}$$

while the ions as a whole are at rest with

$$F_i(\mathbf{p}) = n(2\pi m_i T_i)^{-3/2} \exp\left[-\frac{p^2}{2m_i T_i}\right], \tag{9.57b}$$

we find from Eq. (9.55) [Tange & Ichimaru 1974]

$$\rho_E = \frac{4\sqrt{2\pi}\, e^2 \sqrt{m_e}}{3T_e^{3/2}} \left[2 \ln \Lambda - \left(1 + \frac{T_i}{T_e}\right) \ln\left(1 + \frac{T_e}{T_i}\right) + \cdots\right]. \tag{9.58}$$

Here

$$\ln \Lambda = \left(1 + \frac{T_i}{T_e}\right) \ln \left(\frac{k_{\max}}{k_e}\right) \tag{9.59}$$

is another version of the Coulomb logarithm, which now arises by setting the upper bound of the **k** summation in Eq. (9.51) $k_{\max}$. The leading term of Eq. (9.58) is exactly 1.97 times the Spitzer-Härm value [Spitzer 1962]. The origin of this discrepancy is the adoption of Eq. (9.57a), which amounts to a single Sonine-polynomial approximation in the solution to the kinetic equation.

We note that Eq. (9.58) is of the first order in the discreteness parameters, characteristic of the ordinary processes of Coulomb collisions.

C. Anomalous Electric Resistivities

In a turbulent situation such as those created by excitations of collective modes Eq. (3.54), the density fluctuations induced by the electric-field fluctuations play a greater part in controlling the transport processes than the spontaneous fluctuations do; Eq. (9.58) depends rather on the latter. In those circumstances involving the induced fluctuations, one writes

$$Z_\sigma e \delta n_\sigma(\mathrm{k}) = \frac{i}{4\pi} v(k) \chi_\sigma(\mathrm{k}) \mathbf{k} \cdot \delta \mathbf{E}(\mathrm{k}) \,, \tag{9.60}$$

which defines polarizabilities $\chi_\sigma(\mathrm{k})$ with $v(k) = 4\pi e^2/k^2$. Since the spectral function $\langle | E^2 | (\mathrm{k}) \rangle_{\mathrm{L}}$ of the longitudinal electric-field fluctuations should exhibit peak structures associated with a collective mode

$$\varepsilon(\mathrm{k}) = 1 - v(k) \sum_\sigma \chi_\sigma(\mathrm{k}) = 0 \,, \tag{9.61}$$

we find that Eq. (9.55) now becomes

$$\rho_{E\sigma} = \frac{2\pi}{(n_\sigma u_\sigma)^2} \int d^4\mathrm{k} \frac{\mathbf{k} \cdot \mathbf{u}_\sigma}{k^2} \operatorname{Im} \chi_\sigma(\mathrm{k}) \langle | E^2 | (\mathrm{k}) \rangle_{\mathrm{L}} \,. \tag{9.62}$$

The electric resistivity is here expressed as a functional of the spectral density of longitudinal fluctuations associated with the collective mode Eq. (9.61).

For the ion-acoustic wave turbulence treated in Section 9.1D, one considers the electric resistivity arising from scattering of electrons by the ion-acoustic fluctuations, for which

$$\mathrm{Im}\chi_e(\mathbf{k}) \simeq -\sqrt{\frac{\pi}{2}}\frac{n}{T_e}\frac{\omega - \mathbf{k}\cdot\mathbf{u}}{k\sqrt{T_e/m_e}}\,. \tag{9.63}$$

Substitution of (9.63) in Eq. (9.62) yields a resistivity

$$\rho_E = \frac{(2\pi)^{3/2}\sqrt{m_e}}{nT_e^{3/2}}\sum_{\mathbf{k}}\frac{\mathbf{k}\cdot\mathbf{u}(\mathbf{k}\cdot\mathbf{u} - \omega_\mathbf{k})}{ku^2(k^2 + k_e^2)}\mathcal{E}_\mathbf{k} \tag{9.64}$$

due to Sagdeev [1967]. In Eq. (9.64), $\mathbf{u}$ is the drift velocity of the electrons and $\mathcal{E}_\mathbf{k}$ refers to the energy-spectral density of the ion-acoustic fluctuations.

Hamberger and Jancarik [1972] performed turbulent heating experiments in a toroidal discharge of a Stellarator type (major radius 32.5 cm, minor radius 5 cm, main magnetic field 2 to 3 kG, working gas: H_2 or A, plasma density 10^{11} to 10^{13} cm^{-3}, initial electron temperature 2 to 5 eV), with an application of electric fields with strength ~100 to 500 V/cm. Three different turbulent regimes A, B, and C were distinguished whose occurrence depended on the electron drift velocities

$$\mathrm{A}:\ u_{\max} < 2\times 10^8\ \mathrm{cm/s},$$
$$\mathrm{B}:\ u_{\max} > (2\ \text{to}\ 3)\times 10^8\ \mathrm{cm/s},$$
$$\mathrm{C}:\ u_{\max} \sim 5\times 10^9\ \mathrm{cm/s}\,,$$

and corresponded, respectively, to the excitation of ion-acoustic, two-stream, and beam-plasma instabilities.

It was particularly observed that the correlation times of the fluctuations and electric resistivities assume plateau values in the regime B independent of the drift velocity. The effective collision frequency ν_C defined via the Drude relation $\rho_E = 4\pi\,\nu_c/\omega_e^2$ takes on a magnitude

$$\tilde{\nu}_C = \frac{1}{2\pi}\left(\frac{m_e}{m_i}\right)^{1/3}\omega_e \approx 10^{-2}\omega_e\,. \tag{9.65}$$

This offers another example of the anomalous transports, since Eq. (9.65) remains finite in the fluid limit of Section 1.4C.

The saturated value of electric resistivity observed at Eq. (9.65) may be accounted for in terms of the Buneman [1959] instability (Section 6.3C). Here one assumes a two-beam situation,

$$f_e(\mathbf{v}) = \delta(\mathbf{v} - \mathbf{u})\,, \tag{9.66a}$$

$$f_i(\mathbf{v}) = \delta(\mathbf{v})\,, \tag{9.66b}$$

so that the RPA dielectric function is given by

$$\varepsilon(\mathrm{k}) = 1 - \frac{\omega_e^2}{(\omega - \mathbf{k}\cdot\mathbf{u})^2} - \frac{\omega_i^2}{\omega^2}. \tag{9.67}$$

The dispersion relatin obtained from Eq. (9.61) has been depicted in Fig. 6.4 with $u = v_d$.

The maximum wave number for the Buneman instability takes place at Eq. (6.18); the maximum growth rate Eq. (6.19) is found at Eq. (6.20) with the assumption $m_e << m_i$. The frequency (6.20) has been identified in the regime B of the Hamberger-Jancarik [1972] experiments.

In the vicinity of the maximum growth rate, we note

$$\frac{1}{u}\,\mathrm{Im}\chi_e(\mathrm{k}) \approx \frac{k_{||}k^2}{2\pi e^2\omega_e^2}\,\mathrm{Im}\,\omega_{\max}\,, \tag{9.68}$$

so that Eq. (9.62) becomes

$$\rho_E \simeq \frac{\sqrt{3}}{2\omega_e}\left(\frac{m_e}{2m_i}\right)^{1/3}\int d^4\mathrm{k}\,\frac{k_{||}^2}{(ne)^2}\langle|\,E^2\,|(\mathrm{k})\rangle_{\mathrm{L}}\,. \tag{9.69}$$

The energy density in turbulence, normalized by nT, is given by Eq. (9.11); we thus find

$$\rho_E \simeq \frac{4\pi\sqrt{3}}{\omega_e}\left(\frac{m_e}{2m_i}\right)^{1/3}\left(\frac{T_e}{m_e u^2}\right)\mathcal{W}. \tag{9.70a}$$

Since $u \approx \sqrt{T_e/m_e}$ in Buneman-instability experiments, we may likewise estimate

$$\rho_E \simeq \frac{4\pi\sqrt{3}}{\omega_e}\left(\frac{m_e}{2m_i}\right)^{1/3}\mathcal{W}. \tag{9.70b}$$

The Poisson equation coupled with Parseval's theorem (cf., Eq. 7.68) yields

$$\langle\delta n^2\rangle \int d^4\mathrm{k}\,\frac{k_{||}^2}{8\pi e^2}\langle|\,E^2\,|(\mathrm{k})\rangle_{\mathrm{L}} \tag{9.71}$$

for the Buneman instability. Thus Eq. (9.69) is integrated as

$$\rho_E \frac{4\pi\sqrt{3}}{\omega_{pe}}\left(\frac{m_e}{2m_i}\right)^{1/3}\frac{\langle\delta n^2\rangle}{n^2}\,. \tag{9.70c}$$

The saturated resistivity in the regime B of the Hamberger-Jancarik expriments may thus be reconciled with the Buneman-instability concept if an equipartition relation such as

$$\mathcal{W} \approx \frac{\langle \delta n^2 \rangle}{n^2} \approx 0.1 \tag{9.72}$$

is assumed in Eqs. (9.70) [Ichimaru 1975a].

D. Magnetic-Field Reconnections

Rapid reconnection of magnetic lines of force plays an essential part in the mechanisms of solar flares [Dungey 1953; Sweet 1969] and geomagnetic activity [Dungey 1961; Dessler 1968a, b]. On the basis of observations of solar flares and in particular the absence of intense small-scale magnetic fields throughout the turbulent solar chromosphere, Parker [1973] pointed out that the reconnection rates, or merging speed, of two oppositely directed magnetic fields (see Fig. 9.3) are on the general order of the Alfvén speed c_A of Eq. (4.107). It was postulated that the plasma between the opposite fields escapes via instabilities, permitting rapid close approach of the fields, and that the merging rates are "universally" on the general order of

$$u \approx 0.1 c_A \tag{9.73}$$

or faster, in most cases.

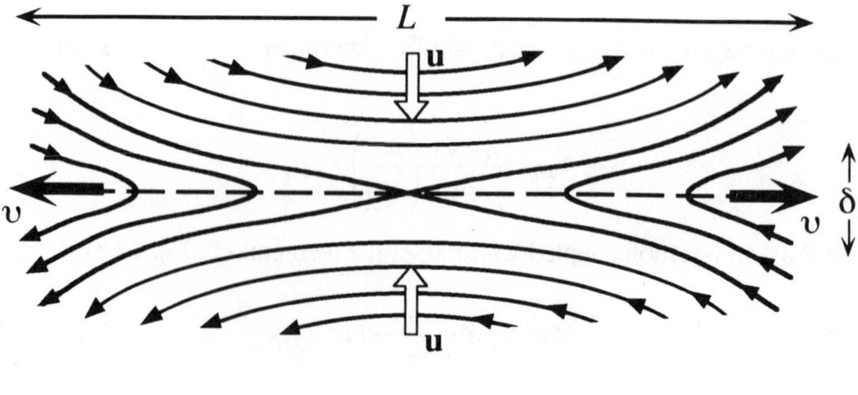

Figure 9.3 A schematic drawing of the magnetic lines of force of two opposite fields pressed together over a broad front L.

This empirical relation, apparently independent of the frequency of the classical Coulomb collisions, contradicts a standard theoretical formula $u = R_m^{-1/2} c_A$ derived below, or the result obtained from the growth rate of a tearing mode [Furth, Kilveen & Rosenbluth 1963]

$$u \simeq \frac{L}{\delta} R_m^{-3/5} c_A$$

where

$$R_m = \frac{4\pi c_A L}{\rho_E c^2} \tag{9.74}$$

is the *magnetic Reynolds number*. The latter two expressions depend on the frequency of Coulomb collisions through the electric resistivity ρ_E.

Hydromagnetic motion of the magnetic fields in a plasma may be described by the induction equation

$$\frac{\partial \mathbf{B}}{\partial t} = \nabla \times (\mathbf{v} \times \mathbf{B}) + \frac{\rho_E c^2}{4\pi} \nabla^2 \mathbf{B} \, . \tag{9.75}$$

This equation is obtained by substituting Ohm's law

$$\rho_E \mathbf{J} = \mathbf{E} + \frac{\mathbf{v}}{c} \times \mathbf{B} \tag{9.76}$$

and Eq. (3.82b) (neglecting the external and displacement currents) into Eq. (3.82a).

When two oppositely directed magnetic fields are pressed firmly together, as sketched in Fig. 9.3, the plasma squeezes out from between the two fields, letting the fields approach more closely, and continually steepening the field gradient until dissipation becomes important [Parker 1973]. Denote the thickness of the dissipation layer (the current sheet) between the two fields by δ, and the velocity with which fluid is ejected out of the dissipation layer by v. Assuming incompressibility, the continuity relation yields

$$uL \approx v\delta \, . \tag{9.77}$$

The characteristic diffusion velocity across the transition layer is given from the last term of Eq. (9.75) as

$$u \approx \frac{\rho_E c^2}{4\pi \delta} \, . \tag{9.78}$$

Hydrostatic equilibrium requires that the total pressure $P + B^2/8\pi$ be constant across the layer. Since B passes through zero on the central plane (broken line in Fig. 9.3), it follows that the fluid pressure P on the midplane is higher than the ambient plasma pressure by about $B^2/8\pi$, where B represents the strength of merging fields, and the ejection velocity is on the order of c_A. Therefore from Eq. (9.77) and (9.78) we have, in order of magnitude,

$$u \approx R_m^{-1/2} c_A \tag{9.79}$$

as quoted earlier. Since $R_m = 10^5$ to 10^{13} with the Spitzer resistivity around the Sun, Eq. (9.79) cannot be reconciled with Eq. (9.73); Parker's empirical law represents an anomalously fast process.

In a hydromagnetically turbulent plasma, the charged particles are scattered by the low-frequency magnetic-field fluctuations, $|\omega/k| < \sqrt{T/m}$; their drift motion is thereby dissipated. The electric resistivity arising from such processes can be calculated from Eq. (9.53), where Eq. (3.102) is used to convert $\delta\mathbf{E}(\mathrm{k})$ into $\delta\mathbf{B}(\mathrm{k})$; the result is

$$\rho_E = \sqrt{\frac{\pi m}{2T}}\frac{2\pi}{nmc^2}\int d^4\mathrm{k}\,\frac{k_{\parallel}^2}{k^3}\langle | B^2 |(\mathrm{k})\rangle \tag{9.80a}$$

$$\approx \sqrt{\frac{\pi m}{2T}}\frac{\langle \delta B^2\rangle}{3mc^2}\left\langle \frac{1}{k}\right\rangle . \tag{9.80b}$$

In Eq. (9.80b), $k_{\parallel}^2 \approx k^2/3$ is assumed, and the characteristic wavelength for the magnetic-field turbulence has been defined as

$$\left\langle \frac{1}{k}\right\rangle = \frac{2\pi}{\langle \delta B^2\rangle}\int d^4\mathrm{k}\,\frac{1}{k}\langle | B^2 |(\mathrm{k})\rangle \tag{9.81}$$

with Parseval's theorem

$$\langle \delta B^2\rangle = 2\pi \int d^4\mathrm{k}\langle | B^2 |(\mathrm{k})\rangle . \tag{9.82}$$

Diffusion velocity Eq. (9.78) with $P \approx B^2/8\pi$ is thus calculated as

$$u \approx \frac{\sqrt{\pi}}{3\delta}\left\langle \frac{1}{k}\right\rangle \frac{\langle \delta B^2\rangle}{B^2} c_A . \tag{9.83}$$

Since

$$\frac{1}{\delta}\left\langle \frac{1}{k}\right\rangle \lesssim 1 , \tag{9.84}$$

Eq. (9.83) becomes consistent with Parker's empirical law Eq. (9.73) if another equipartition relation

$$\frac{\langle \delta B^2 \rangle}{8\pi} \approx P \tag{9.85}$$

can be assumed for the magnetic-field turbulence [Ichimaru 1975b].

The process of magnetic-field-line reconnection was diagnosed in a large laboratory plasma by an experimental group at UCLA [Stenzel & Gekelman 1979; Wild, Gekelman & Stenzel 1981]. Detailed time- and space-resolved measurements of the electric fields and currents were made and the resistivity normalized to the classical Spitzer value was found to be spatially inhomogeneous. Its magnitude ranged $10 < \rho_E/\rho_E^0 < 250$, and did not maximize in regions of large currents.

The strength of the magnetic field in the third direction perpendicular to Fig. 9.3 in the transition layer depends on relative helicity between merging lines of force. The Tokyo-Princeton group [Yamada et al. 1990] measured such a helicity dependence of the reconnection velocity and thereby found that plasmas of antiparallel helicity (where the perpendicular field strength is approximately zero) merged much faster than those of parallel helicity where the field lines merge with finite angles).

9.3 MAGNETIC-FIELD TURBULENCE IN DISK GEOMETRIES

It has been recognized that the differential rotation and radial flow of plasmas in a disk geometry act to generate and amplify the magnetic-field fluctuations in it. The presence of such magnetohydrodynamic turbulence plays an essential part in the physical processes of various astrophysical objects, such as the Galaxy and the accretion model of compact X-ray sources [Prendergast & Burbidge 1968; Shakura & Sunyaev 1973; Eardley & Lightman 1975; Shapiro & Teukolsky 1983]. In the case of Galaxy the theory should be relevant directly to the question of the origin of its magnetic fields; the spectral distribution of magnetic-field fluctuations may then be correlated with observational data, such as cosmic rays [Ichimaru 1977b]. In the case of the accretion model the flux of angular momentum carried away by the stress tensor of the magnetic field enables the matter to flow toward an accreting star; the static and dynamic properties of such an accretion disk are vitally controlled by the rate of such an angular-momentum transfer in the plasma. In the cases of accretion onto a magnetic neutron star, the disk's inner boundary is determined by the pressure balance between the stellar magnetic field and the accreting plasma (Alfvén surface). The eventual fall of the plasma onto the stellar surface must be accounted for in terms of the theory of plasma diffusion across the Alfvén surface; the basic processes involved are the anomalous electric resistivity in the boundary domain caused by the presence of magnetic-field fluctuations in the disk.

In this section we thus consider a theory of magnetohydrodynamic turbulence appropriate to a plasma in such a disk geometry. Turbulence is generated mainly by differential rotation of the plasmas; it decays through current dissipation due to anomalous magnetic viscosity. The viscosity arising from such a turbulent mixing in fluctuating magnetic fields offers another example of those macroscopic transports that remain finite in the fluid limit. We derive a set of equations describing space-time evolution of the magnitudes and spectral distributions of both the diagonal and off-diagonal elements of the spectral tensor for the magnetic-field fluctuations. Its stationary solution yields explicit expressions for turbulence in terms of the macroscopic physical parameters of the system, such as the disk thickness and the rate of differential rotation. Finally the results are applied to the Galaxy, providing an essential account of the observed energy density of the magnetic fields in it.

A. Accretion-Disk Problems

Consider an axisymmetric disk configuration of a plasma (Fig. 9.4). Effective thickness $2H$ of an accretion disk can be calculated from the condition for equilibrium between the expanding force (P/H) due to the pressure P and the contracting force ($GM\rho_m H/r^3$) due to the gravitational attraction of the accreting star with mass M, i.e.,

$$2H \simeq 2r\sqrt{\frac{P}{\rho_m}}\sqrt{\frac{r}{GM}}\,. \tag{9.86}$$

Here $G = 6.6720 \times 10^{-8}\mathrm{erg} \cdot \mathrm{cm/g}^2$ is Newton's gravitational constant, and ρ_m is the mass density of the disk plasma.

We employ the cylindrical system of coordinates (r,φ,z) with the z axis chosen in the direction perpendicular to the disk. Along the z direction across the disk, the plasma is assumed to be uniform with an average number density n, temperature T, and average flow velocity,

$$\mathbf{v} = -u\hat{\mathbf{r}} + v\hat{\varphi}\ , \tag{9.87}$$

where $\hat{\mathbf{r}}$ and $\hat{\varphi}$ are the unit vectors in the $\mathbf{r}$ and φ directions.

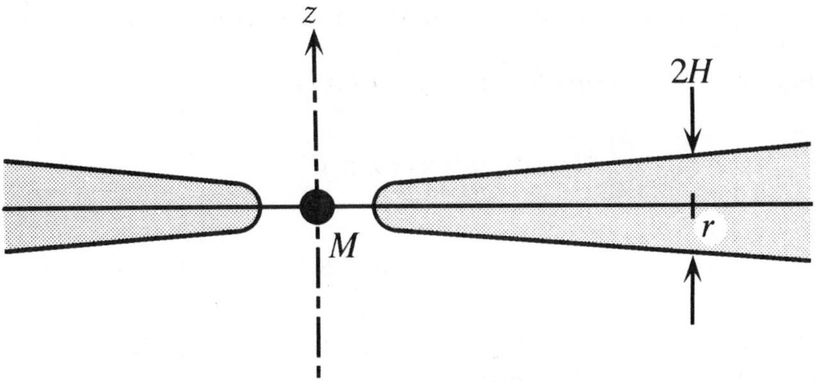

Figure 9.4 An accretion disk.

In an accretion disk, the plasma rotates with *Keplerian velocity*

$$v_K = \sqrt{\frac{GM}{r}} \tag{9.88}$$

in the azimuthal direction. The angular velocity, $\Omega = v/r$, depends on r, and this implies a differential rotation in the disk. A radial flow u then arises in the plasma, accompanied by radial transfer of angular momentum due to internal friction by turbulent mixing. We thus assume that u and v are functions of r alone.

The induction equation of the magnetic field is given by Eq. (9.75), where we now interpret the electric resistivity ρ_E as an operator in the sense of Eq. (3.1a). We expand the magnetic field as

$$\begin{aligned}\mathbf{B} &= \sum_{\mathbf{k}} \mathbf{B}_{\mathbf{k}} \exp\left[i(\mathbf{k} \cdot \mathbf{r} - \omega t)\right] \\ &= \sum_{\mathbf{k}} \left(B_k^r \hat{\mathbf{r}} + B_k^\varphi \hat{\varphi} + B_k^z \hat{\mathbf{z}}\right) \exp\left[i(\mathbf{k} \cdot \mathbf{r} - \omega t)\right].\end{aligned} \tag{9.89}$$

The expansion is carried out with the periodic boundary conditions appropriate to a cube of volume $(2H)^3$, so that the domain of the $\mathbf{k}$ summation is restricted to $k > \pi/H$. The spectral amplitudes, B_k^r, B_k^φ, and B_k^z, are slowly varying functions of r and t; the scales of those r and t variations are greater than $2\pi/k$ and $2\pi/|\omega|$.

Equation (9.89) is substituted in Eq. (9.75) to yield

$$\frac{\partial \mathbf{B_k}}{\partial t} - i\omega \mathbf{B_k} = \nabla \times (\mathbf{v} \times \mathbf{B_k})$$

$$+ i\mathbf{k} \times (\mathbf{v} \times \mathbf{B_k}) - \frac{\rho_{\mathbf{k}} c^2 k^2}{4\pi} \mathbf{B_k} + \mathbf{S}\,. \tag{9.90}$$

In this equation we have retained only the lowest-order contributions with respect to $|\partial \mathbf{B_k}/\partial r|/|k\mathbf{B_k}|$ arising from the last nonlinear term of Eq. (9.75). The quantity $\rho_{\mathbf{k}}$ denotes the electric resistivity for the current fluctuation with wave vector **k**, and gives rise to the magnetic viscosity term. The last term **S** has been included here to account for the rates of spontaneous fluctuations such as those in (7.26a).

The frequencies of fluctuations are determined from the equation,

$$\omega \mathbf{B_k} + \mathbf{k} \times (\mathbf{v} \times \mathbf{B_k}) = 0\,, \tag{9.91}$$

which reveals two (degenerate) transverse modes,

$$\omega = \mathbf{k} \cdot \mathbf{v} \text{ with } \mathbf{k} \cdot \mathbf{B_k} = 0\,. \tag{9.92}$$

The remaining mode, $\omega = 0$ with $\mathbf{v} \times \mathbf{B_k} = 0$, is devoid of coupling between the velocity field and the magnetic field, and will be neglected. We note that the frequencies of these modes vanish in the frame comoving with the plasma; their spectrum is broadened around $\omega = 0$ mainly through imaginary frequency arising from the magnetic viscosity in Eq. (9.90). Such a situation is therefore quite analogous to the case of ordinary hydrodynamic turbulence obeying the Navier-Stokes equation.

Equations describing space-time evolution of spectral amplitudes of the magnetic-field fluctuations may be derived from Eq. (9.90):

$$\frac{\partial B_k^r}{\partial t} = -\frac{\rho_k c^2 k^2}{4\pi} B_k^r + S^r\,, \tag{9.93a}$$

$$\frac{\partial B_k^\varphi}{\partial t} = \frac{\partial}{\partial r}\left(B_k^\varphi u + B_k^r v\right) - \frac{\rho_k c^2 k^2}{4\pi} B_k^\varphi + S^\varphi\,, \tag{9.93b}$$

$$\frac{\partial B_k^z}{\partial t} = \frac{1}{r}\frac{\partial}{\partial r}(B_k^z u r) - \frac{\rho_k c^2 k^2}{4\pi} B_k^z + S^z\,. \tag{9.93c}$$

The first terms on the right-hand sides of Eqs. (9.93b) and (9.93c) represent compressional effects on the magnetic field arising from the radial flow of the plasma. The second term on the right-hand side of Eq. (9.93b) describes a creation rate of B_k^φ out of B_k^z, owing to the differential rotation in the plasma. This effect may be clearly seen from an equation

$$\frac{\partial}{\partial r}(B_k^r v) = B_k^r\left(\frac{\partial v}{\partial r} - \frac{v}{r}\right) = B_k^r r\frac{\partial}{\partial r}\left(\frac{v}{r}\right) \tag{9.94}$$

stemming from $\nabla \cdot \mathbf{B} = 0$. The ρ_k-dependent terms in Eqs. (9.93) describe the decay of field fluctuations arising from the finite electric resistivity or the magnetic viscosity in the plasma.

To obtain a description for the spectral disributions of magnetic-field fluctuations, we multiply Eq. (9.90) by B_k^r, B_k^φ, and B_k^z, and carry out a statistical average denoted by $\langle \cdots \rangle$. Space-time evolution of the diagonal and off-diagonal elements in the spectral tensor of the magnetic-field fluctuations is thus described by the equations,

$$\begin{aligned}\frac{\partial}{\partial t}\langle \mathbf{B}_{-\mathbf{k}} \cdot \mathbf{B}_\mathbf{k}\rangle = {}& u\frac{\partial}{\partial r}\left[\langle B_{-k}^\varphi B_k^\varphi\rangle + \langle B_{-k}^z B_k^z\rangle\right] \\ &+ 2\frac{\partial u}{\partial r}\left[\langle B_{-k}^\varphi B_k^\varphi\rangle + \langle B_{-k}^z B_k^z\rangle\right] + 2\frac{u}{r}\langle B_{-k}^z B_k^z\rangle \\ &+ 2r\frac{\partial \Omega}{\partial r}\langle B_{-k}^r B_k^\varphi\rangle - \frac{\rho_k c^2 k^2}{2\pi}\langle \mathbf{B}_{-\mathbf{k}} \cdot \mathbf{B}_\mathbf{k}\rangle\,,\end{aligned} \tag{9.95}$$

$$\begin{aligned}\frac{\partial}{\partial t}\langle B_{-k}^r B_k^\varphi\rangle = {}& \frac{\partial}{\partial r}\left[u\langle B_{-k}^r B_k^\varphi\rangle\right] + \frac{u}{r}\langle B_{-k}^r B_k^\varphi\rangle \\ &+ r\frac{\partial \Omega}{\partial r}\langle B_{-k}^r B_k^r\rangle - \frac{\rho_k c^2 k^2}{2\pi}\langle B_{-k}^r B_k^\varphi\rangle\,.\end{aligned} \tag{9.96}$$

The contributions of spontaneous fluctuations have been ignored in the derivation of Eqs. (9.95) and (9.96). The mean square value of the fluctuations and the r–φ element $\chi/4\pi$ of the stress tensor are then calculated as

$$\langle \delta B^2\rangle = \sum_\mathbf{k}\langle \mathbf{B}_{-\mathbf{k}} \cdot \mathbf{B}_\mathbf{k}\rangle\,, \tag{9.97}$$

$$\chi = \sum_\mathbf{k}\langle B_{-k}^r B_k^\varphi\rangle\,. \tag{9.98}$$

B. Turbulent Magnetic Viscosity

For a complete formulation of a magnetic-field turbulence problem, one must establish a way to calculate the wave number-dependent electric resistivity ρ_k or the *magnetic viscosity* $\rho_k c^2/4\pi$ in Eqs. (9.95) and (9.96). A general expression for the dc electric resistivity in an electromagnetically turbulent plasma has been given by Eq. (9.53). In the presence of magnetohydrodynamic turbulence, the major mechanism for resistivity is provided by scattering of the current by the low-frequency magnetic-field fluctuations, resulting in Eq. (9.80a).

In the treatment of turbulent magnetic viscosity for a "test wave" magnetic field with wave number k, it is essential to evoke a hierarchy among the time scales involved in the fluctuations with different wavelengths. It then follows that the effective resistivity ρ_k for the test wave may be calculated as a summation of all the contributions from those scattering acts produced by field fluctuations with wave numbers greater than k. These fields fluctuate faster in space and time than those of the test wave; the scattering events may thus be looked upon as producing impulsive, collisional effects. Those field fluctuations with wave numbers smaller than k, on the other hand, vary slowly in space and time; their effects may thus be regarded as causing adiabatic deformation and diffusion of the test-wave field, with a net effect leading to an approach toward an isotropic spectrum. This is a hypothesis in the theory; a hypothesis similar to this has been implicit in the Kolmogorov-Heisenberg theory of turbulent viscosity [Kolmogorov 1941; Heisenberg 1948; Nakano 1972].

Adopting the hypothesis stated above, we express

$$\rho_k = \frac{1}{3nmc^2}\sqrt{\frac{\pi m}{2T}} \sum_{q(>k)} \frac{1}{q}\left\langle \mathbf{B}_{-\mathbf{q}} \cdot \mathbf{B}_{\mathbf{q}} \right\rangle. \tag{9.99}$$

Substituting this expression into Eqs. (9.95) and (9.96), we find that the spectral distributions

$$\left\langle \mathbf{B}_{-\mathbf{k}} \cdot \mathbf{B}_{-\mathbf{k}} \right\rangle = \left(\frac{\pi}{H}\right)^3 \frac{\langle \delta B^2 \rangle}{4Hk^4} \tag{9.100}$$

$$\left\langle B_{-k}^r B_k^\varphi \right\rangle = \left(\frac{\pi}{H}\right)^3 \frac{\chi}{4Hk^4} \tag{9.101}$$

fulfill those equations consistently. The amplitudes now satisfy the differential equations,

$$\frac{\partial}{\partial t}\langle\delta B^2\rangle = \frac{2}{3}u\frac{\partial}{\partial r}\langle\delta B^2\rangle + \frac{4}{3}\frac{\partial u}{\partial r}\langle\delta B^2\rangle$$

$$+\frac{2}{3}\frac{u}{r}\langle\delta B^2\rangle + 2r\frac{\partial\Omega}{\partial r}\chi - 2\nu\frac{\langle\delta B^2\rangle^2}{8\pi nT}, \qquad (9.102)$$

$$\frac{\partial}{\partial t}\chi = \frac{1}{r}\frac{\partial}{\partial r}(\chi ur) + \frac{1}{3}r\frac{\partial\Omega}{\partial r}\langle\delta B^2\rangle - 2\nu\frac{\langle\delta B^2\rangle}{8\pi nT}\chi, \qquad (9.103)$$

where

$$\nu \equiv \frac{\pi}{3H}\sqrt{\frac{\pi T}{2m}}. \qquad (9.104)$$

In the derivation of Eqs. (9.102) and (9.103), we have assumed isotropy in the diagonal elements of the spectral tensor, so that

$$\left\langle B^r_{-k}B^r_k\right\rangle \approx \left\langle B^\varphi_{-k}B^\varphi_k\right\rangle \approx \left\langle B^z_{-k}B^z_k\right\rangle \approx \frac{1}{3}\left\langle \mathbf{B}_{-\mathbf{k}}\cdot\mathbf{B}_{\mathbf{k}}\right\rangle \qquad (9.105)$$

as a consequence of the deformation effect by the long-wavelength field fluctuations.

The wavenumber-dependent magnetic viscosity is thus calculated from Eq. (9.99) as

$$\eta_M = \frac{\nu}{k^2}\frac{\langle\delta B^2\rangle}{8\pi nT}. \qquad (9.106)$$

The η_M is a decreasing function of k. We thus find an upper bound in k space for the applicability of this turbulence theory; the upper bound stems from a comparison between the anomalous magnetic viscosity Eq. (9.106) and the ordinary magnetic viscosity, $\eta_{M0} = \rho_E^0 c^2/4\pi$, arising from the Coulomb collisions between ions and electrons, where ρ_E^0 refers to the Spitzer resistivity.

In many practical cases including the accretion disk and the Galaxy, one may assume $|u| \ll |v|$ and $|\partial u/\partial r| \ll |\partial v/\partial r|$. In these circumstances the stationary solutions to Eqs. (9.102) and (9.103) are obtained from the last two terms as

$$\frac{\langle\delta B^2\rangle}{8\pi nT} = \frac{1}{\sqrt{6}\,\nu}\left|r\frac{\partial\Omega}{\partial r}\right|, \qquad (9.107)$$

$$\frac{\chi}{4\pi nT}=\frac{1}{3\nu}r\frac{\partial\Omega}{\partial r}. \tag{9.108}$$

Outward transport of angular momentum takes place when $\partial\Omega/\partial r<0$. Both the diagonal and off-diagonal elements of the stress tensor associated with the magnetic-field fluctuations are produced by the differential rotation of the plasmas.

It is instructive to consider the magnitude of the turbulent viscosity η_T defined through $|\chi/4\pi|=\eta_T(r/2)|\partial\Omega/\partial r|$; we find from Eq. (9.108)

$$\eta_T=\left(\frac{2}{\pi}\right)^{3/2}nm\sqrt{\frac{T}{m}}H. \tag{9.109}$$

This result may be interpreted in terms of a mixing-length theory as follows: Eq. (9.109) implies that momenta in the plasma are effectively mixed over a length comparable to H. Such an enhancement of the mixing length is a consequence of the k^{-4} spectrum obtained in Eqs. (9.100) and (9.101); the bulk of turbulence is contained in the long-wavelength regime, $k\gtrsim\pi/H$, so that mixing takes place over long distances.

The theory of magnetohydrodynamic turbulence described in the preceding subsection is equally applicable to the current fluctuations produced by ions or by electrons. Dependence on the particle species enters the formalism through the relaxation rate defined in Eq. (9.104). Since one can ordinarily assume $T_e/m_e>T_i/m_i$, the decay rate is larger for the electrons than for the ions. Hence one finds that the level of turbulence remains smaller for the electron-current fluctuations than for the ion-current fluctuations. We thus consider the intensity of magnetic-field fluctuations produced by ions and apply the result to estimate the energy density for magnetic fields in the Galaxy.

We first substitute Eq. (9.104) in Eq. (9.107) to obtain

$$\frac{\langle\delta B^2\rangle}{8\pi}=\frac{1}{\pi}\sqrt{\frac{3}{\pi}}\,nm\sqrt{\frac{T}{m}}H\left|r\frac{\partial\Omega}{\partial r}\right|. \tag{9.110}$$

Galactic plasmas consist mostly of protons (and electrons) with $n\approx 1\ \mathrm{cm}^{-3}$ and $T\approx 10^4$ K. We use the following values for parameters [Allen 1973]: $2H=2$ kpc and $|r\partial\Omega/\partial r|\approx 25$ km/sec kpc near the Sun.* Equation (9.110) is then computed as

$$\frac{\langle\delta B^2\rangle}{8\pi}\approx 0.74\ \mathrm{eV/cm^3}. \tag{9.111}$$

*1 pc ≈ 3.26 lt yr ≈ 3.09×10^{19} cm.

Observationally [e.g., Morrison 1961], the magnetic energy density in the Galaxy is said to be ~ 1 eV/cm^3. We thus find that the magnetohydrodynamic turbulence generated and maintained by the differential rotation of the plasma essentially accounts for the strength of the magnetic fields in the Galaxy.

The rate at which the field fluctuations are generated by the differential rotation may be estimated from Eqs. (9.102) and (9.103). Since $\langle \delta B^2 \rangle \approx \chi$, we find a characteristic time $\approx | r\partial\Omega/\partial r |^{-1}$ for the generation, which takes on a value $\sim 4 \times 10^7$ yr in the vicinity of the Sun. When the strength is given by its stationary value Eq. (9.110), its decay rate through the turbulent magnetic viscosity is naturally balanced by the generation rate. If the level of field fluctuations is greater than Eq. (9.110), the decay rate would be faster than the generation rate computed above. It thus appears that the "fossil" theory of the Galactic magnetic field may encounter a difficulty when the anomalous magnetic viscosity is taken into consideration.

Gas accretion onto a compact star that constitutes a member of a close binary system has been studied as a standard model for the compact such X-ray sources as the black-hole model of Cyg X-1 [e.g., Shakura & Sunyaev 1973, Novikov & Thorne 1973; Ichimaru 1977c]. Various plasma processes take place in the accretion disks so formed around compact stars. Turbulent viscosity such as that considered in Eq. (9.109) plays a leading role in accounting for a radial influx of plasmas sufficient to liberate gravitational energy and thereby to produce observed X-ray luminosity on the order of 10^{37} erg/s, some 10^4 times as large as the total luminosity of the Sun. Theoretical models have been constructed, explaining the static and dynamic properties of such an accretion disk around a black hole and consequently the observed features of Cyg X-1.

PROBLEMS

9.1 Show that the rate of spontaneous emission, Eq. (9.9), is a quantity of the first order in the discreteness parameters of Section 1.4C.

9.2 Show that the condition for the three-wave processes, Eq. (9.23), cannot be satisfied by the dispersion relation, Eq. (4.30), for the plasma oscillations, or by the dispersion relation, Eq. (4.53), for the ion-acoustic wave.

9.3 By symmetry considerations for interchanges between **k** and **q**, prove Eq. (9.34).

9.4 In the polarization potential approach of Section 3.2F, show that the function $\tilde{\varepsilon}$ (k) screens the field between a test charge and a plasma particle while ε(k) screens that between two test charges.

9.5 In the "α model"-model of Shakura and Sunyaev [1973], one assumes that the off-diagonal elements of the stress tensor are proportional to the static pressure in an accretion disk. In the treatment of Section 9.3B, χ is one of such off-diagonal elements. Consider if there is any relation between the α model and the magnetic-field turbulence theory in that section.

APPENDIX **A**

The Functional Derivatives

Formulation in terms of the functional derivatives is related closely to the linear response formalism in Chapter 3 and the multiparticle distributions in Vol. II. In this appendix, fundamental relations in the functional derivatives are summarized.

Let $\mathcal{F}[f(x)]$ be a functional of a function $f(x)$ with a variable x defined over a domain, $a \leq x \leq b$. The functional derivative, $\delta\mathcal{F}/\delta f(x)$, is given in terms of the increment $\delta\mathcal{F}$ produced by an infinitesimal variation $\delta f(x)$ of $f(x)$ as

$$\delta\mathcal{F} = \int_a^b dx \, \frac{\delta\mathcal{F}}{\delta f(x)} \, \delta f(x) \, . \tag{A.1}$$

It has the following properties:

(i) Identity relation:

$$\frac{\delta f(x')}{\delta f(x)} = \delta(x - x') \, . \tag{A.2}$$

(ii) Product rule: When a functional is expressed by a simple product of the functions $f(x_i)$ as

$$\mathcal{F} = \int_a^b dx_1 \cdots dx_N \prod_{i=1}^{N} f(x_i) \, ,$$

then

$$\frac{\delta \mathcal{F}}{\delta f(x)} = N \int_a^b dx_2 \cdots dx_N \prod_{i=2}^{N} f(x_i) \,. \tag{A.3}$$

(iii) Chain rule: When a functional $\mathcal{F}$ is a functional of $\mathcal{G}(x')$ which in turn is a functional of $f(x)$, then

$$\frac{\delta \mathcal{F}}{\delta f(x)} = \int_a^b dx' \frac{\delta \mathcal{F}}{\delta G(x')} \frac{\delta G(x')}{\delta f(x)} \,. \tag{A.4}$$

(iv) When the functional is given by

$$\mathcal{F} = \int_a^b dx F(f(x))$$

where $F(f(x))$ is a function of $f(x)$, then

$$\frac{\delta \mathcal{F}}{\delta f(x)} = \frac{dF(f(x))}{df(x)} \,. \tag{A.5}$$

APPENDIX **B**

Derivation of Eq. (3.13b)

Consider the application of an external disturbance field,

$$a(\mathbf{r},t) = a \exp\,[i(\mathbf{k}\cdot\mathbf{r} - \omega t) + t] + \mathrm{cc} \tag{B.1}$$

to the many-particle system which is assumed to be translationally invariant in space and time. The disturbance will create perturbations in the system; the increment $\delta B(\mathbf{r},t)$ of a physical quantity B may be calculated as the difference between its expectation values with and without the disturbance,

$$\delta B(\mathbf{r},t) = \langle B(\mathbf{r},t;a)\rangle - \langle B(\mathbf{r},t;0)\rangle\,. \tag{B.2}$$

Within the framework of the linear response theory elucidated in Section 3.1, Eq. (B.2) should be expressed as

$$\delta B(\mathbf{r},t) = B \exp\,[i(\mathbf{k}\cdot\mathbf{r} - \omega t) + t] + \mathrm{cc}\,. \tag{B.3}$$

The linear response function $K_{BA}(\mathbf{k},\omega)$ is then given by Eq. (3.13a).

The response of the system can be calculated with the aid of a standard perturbation theory of quantum mechanics. The coupling between the external disturbance and the physical variables of the system produces an extra Hamiltonian which acts to drive the system out of its unperturbed state. Such a situation can be analyzed through a quantum-mechanical equation of motion.

We thus begin with expressing the total Hamiltonian $\mathcal{H}_t$ as a summation between the exact Hamiltonian $\mathcal{H}$ of the system without the external disturbance and the extra Hamiltonian $\mathcal{H}'(t)$, that is,

$$\mathcal{H}_t = \mathcal{H} + \mathcal{H}'(t) . \tag{B.4}$$

The extra Hamiltonian may be explicitly written as

$$\mathcal{H}'(t) = -\int_{\text{unit volume}} d\mathbf{r}\mathcal{A}(\mathbf{r},t)a(\mathbf{r},t) . \tag{B.5}$$

This equation serves as the definition for the operator $\mathcal{A}(\mathbf{r})$ representing the physical variable that couples with the external disturbance.

The wave function $\Psi_S(t)$ of the system obeys the Schrödinger equation

$$i\hbar\frac{\partial\Psi_S(t)}{\partial t} = [\mathcal{H}_S + \mathcal{H}'_S(t)]\Psi_S(t) . \tag{B.6}$$

The subscript S is used here to imply that we are in the Schrödinger representation. One can formally solve Eq. (B.6) by going over to the interaction representation. In this representation, the wave function and the operator O are transformed from those in the Schrödinger representation according to

$$\Psi(t) = \exp\left(\frac{i\mathcal{H}t}{\hbar}\right)\Psi_S(t) , \tag{B.7}$$

$$O(t) = \exp\left(\frac{i\mathcal{H}t}{\hbar}\right)O_S \exp\left(\frac{-i\,\mathcal{H}t}{\hbar}\right) . \tag{B.8}$$

The equation for the wave function is then given by

$$i\hbar\,\frac{\partial\Psi(t)}{\partial t} = \mathcal{H}'(t)\Psi(t) \tag{B.9}$$

where

$$\mathcal{H}'(t) = -\int_{\text{unit volume}} d\mathbf{r}\mathcal{A}(\mathbf{r},t)a(\mathbf{r},t) . \tag{B.10}$$

is the extra Hamiltonian in the interaction representation. The differential equation (B.9) with an initial condition

$$\Psi(-\infty) = \Psi_0 \tag{B.11}$$

may be written in the form of the integral equation as

$$\Psi(t) = \Psi(0) - \frac{i}{\hbar}\int_{-\infty}^{t} dt' \mathcal{H}'(t)\Psi(t') . \tag{B.12}$$

This equation can be solved by interaction.

Since we are here interested in the linear response calculation, it is sufficient to retain only up to the first-order effects of the perturbation in (B.12); hence

$$\Psi(t) = \Psi_0 - \frac{i}{\hbar}\int_{-\infty}^{t} dt' \mathcal{H}'(t')\Psi_0 . \tag{B.13}$$

Assuming an appropriate normalization for the wave functions and adopting a standard bracket notation to express the quantum-mechanical expectation values, we calculate the induced perturbation Eq. (B.2) according to

$$\delta B(\mathbf{r},t) = \langle\Psi(t)|\, \mathcal{B}(\mathbf{r},t)\, |\Psi(t)\rangle - \langle\Psi_0|\, \mathcal{B}(\mathbf{r},t)\, |\Psi_0\rangle .$$

With the aid of Eqs. (B.10) and (B.13), we thus obtain

$$\delta B(\mathbf{r},t) = \frac{i}{\hbar}\int d\mathbf{r}' \int_{-\infty}^{t} dt' \,\langle\Psi_0|\, [\mathcal{B}(\mathbf{r},t),\mathcal{A}(\mathbf{r}',t')]\, |\Psi_0\rangle a(\mathbf{r}',t') \tag{B.14}$$

where the square brackets denote the commutator, $[\mathcal{B}(\mathbf{r},t),\mathcal{A}(\mathbf{r}',t')] = \mathcal{B}(\mathbf{r},t)\mathcal{A}(\mathbf{r}',t') - \mathcal{A}(\mathbf{r}',t')\mathcal{B}(\mathbf{r},t)$. Recalling the calculation in the derivation from Eqs. (3.1) to Eq. (3.3), we find the linear response function Eq. (3.13a) from Eq. (B.14) as

$$K_{BA}(\mathbf{k},\omega) = \frac{i}{\hbar}\int d(\mathbf{r}-\mathbf{r}') \int_{0}^{\infty} d(t-t') \,\langle\Psi_0|\, [\mathcal{B}(\mathbf{r},t),\mathcal{A}(\mathbf{r}',t')]\, |\Psi_0\rangle$$

$$\times \exp\{-i[\mathbf{k}\cdot(\mathbf{r}-r') - \omega(t-t')]\} . \tag{B.15}$$

The response function is thus expressed in terms of an expectation value of the commutator between two Hermitian operators, $\mathcal{B}(\mathbf{r},t)$ and $\mathcal{A}(\mathbf{r}',t')$.

Equation (B.15) is now averaged over the unperturbed states of the system with the aid of the statistical operator of the canonical ensemble

$$\mathcal{D} = \exp\left(\frac{F-\mathcal{H}}{T}\right). \tag{B.16}$$

The free energy F is to be determined from the normalization

$$\exp\left(-\frac{F}{T}\right) = \mathrm{Tr}\,\exp\left(-\frac{\mathcal{H}}{T}\right)$$

where Tr means the summation over the diagonal elements of the matrix. Finally we obtain

$$K_{BA}(\mathbf{k},\omega) = \frac{i}{\hbar}\int d(\mathbf{r}-\mathbf{r}')\int_0^\infty d(t-t')\mathrm{Tr}\,\mathcal{D}[\mathcal{B}(\mathbf{r},t),\mathcal{A}(\mathbf{r}',t')]$$

$$\times \exp\{-i[\mathbf{k}\cdot(\mathbf{r}-r') - \omega(t-t')]\}\,. \tag{B.17}$$

This coincides with Eq. (3.13b).

APPENDIX C

Derivation of Eq. (3.16)

To derive Eq. (3.16) out of the formalism presented in Appendix B, we introduce a set of eigenstates $|n\rangle$ of the unperturbed system with the energy eigenvalues E_n, such that

$$\mathcal{H}|n\rangle = E_n |n\rangle .$$

The explicit time dependence of such operators as $\mathcal{B}(\mathbf{r},t)$ and $\mathcal{A}(\mathbf{r}',t')$ may be recovered through Eq. (B.8). Their time-independent parts are decomposed into Fourier components as

$$\mathcal{A}(\mathbf{r}) = \sum_{\mathbf{k}} \mathcal{A}_k \exp(i\mathbf{k}\cdot\mathbf{r}) .$$

After a straightforward calculation, Eq. (B.17) is reexpressed as

$$K_{BA}(\mathbf{k},\omega) = -\frac{1}{\hbar}\sum_{m,n} D(E_n)\left[1 - \exp\left(-\frac{\hbar\omega_{mn}}{T}\right)\right]\frac{(B_k)_{nm}(A_{-k})_{mn}}{\omega - \omega_{mn} + i\eta} \tag{C.1}$$

where

$$D(E_n) \equiv \exp\left(\frac{F - E_n}{T}\right), \tag{C.2}$$

$$\omega_{mn} \equiv \frac{E_m - E_n}{\hbar}, \tag{C.3}$$

and the matrix elements are

$$(A_k)_{mn} = \langle m| A_k |n\rangle . \tag{C.4}$$

The spectral function defined by Eq. (3.15) can likewise be expressed with the aid of the statistical operator Eq. (B.16) as

$$S_{BA}(\mathbf{k},\omega) = \frac{1}{4\pi}\int d(\mathbf{r}-\mathbf{r}')\int_0^\infty d(t-t')\mathrm{Tr}\ \{\mathcal{D}[\mathcal{B}(\mathbf{r},t)\mathcal{A}(\mathbf{r}',t') + \mathcal{A}(\mathbf{r}',t')\mathcal{B}(\mathbf{r},t)]\ \}$$

$$\times \exp\{-i[\mathbf{k}\cdot(\mathbf{r}-\mathbf{r}') - \omega(t-t')]\ \} . \tag{C.5}$$

The calculation of Eq. (C.5) proceeds in much the same way as that of Eq. (C.1), so that

$$S_{BA}(\mathbf{k},\omega) = \frac{1}{2}\left[1+\exp\left(-\frac{\hbar\omega}{T}\right)\right]\sum_{m,n} D(E_n)[(B_k)_{nm}(A_{-k})_{mn}]\delta(\omega - \omega_{mn}) . \tag{C.6}$$

On the other hand, we derive from Eq. (C.1)

$$-\frac{i}{2}[K_{BA}(\mathbf{k},\omega) - K_{AB}(-\mathbf{k},-\omega)]$$

$$= \frac{\pi}{\hbar}\left[1-\exp\left(-\frac{\hbar\omega}{T}\right)\right]\sum_{m,n} D(E_n)[(B_k)_{nm}(A_{-k})_{mn}]\delta(\omega - \omega_{mn}) . \tag{C.7}$$

Comparison between Eqs. (C.6) and (C.7) yields Eq. (3.16).

Bibliography

Abe, R., 1959, *Prog. Theor. Phys.* **21**, 475.

Abella, I.D., N.A. Kurnit & S.R. Hartmann, 1966, *Phys. Rev.* **141**, 391.

Aigrain, P., 1961, in *Proc. Intern. Conf. on Semicond. Phys.*, Prague, 1960 (Academic, New York), p.224.

Akhiezer, A.I., I.A. Akhiezer, R.V. Polovin, A.G. Sitenko & K.N. Stpanov, 1967, *Collective Oscillations in a Plasma*, translated by H.S.H. Massey & R.J. Tayler (MIT Press, Cambridge, Mass.).

Alfvén, H., 1942, *Nature* **150**, 405.

Allen, C.W., 1973, *Astrophysical Quantities*, 3rd ed. (Athlone, London).

Allen, T.K., W.R. Baker, R.V. Pyle & J.M. Wilcox, 1959, *Phys. Rev. Letters* **2**, 383.

Al'tshul', L.M. & V.I. Karpman, 1965, *Zhur. Eksptl. Teoret. Fiz.* **49**, 515 [*Soviet Phys. JETP* **22**, 361 (1966)].

Ando, T., 1990, in *Strongly Coupled Plasma Physics*, edited by S. Ichimaru (North-Holland/Yamada Science Foundation, Amsterdam), p.263.

Ando, T., A.B. Fowler & F. Stern, 1982, *Rev. Mod. Phys.* **54**, 437.

Baker, D.R., N.R. Ahern & A.Y. Wong, 1968, *Phys. Rev. Letters* **20**, 318.

Balescu, R., 1960, *Phys. Fluids* **3**, 52.

Bernstein, I.B., 1958, *Phys. Rev.* **109**, 10.

Bernstein, I.B., E.A. Frieman, R.M. Kulsrud & M.N. Rosenbluth, 1960, *Phys. Fluids* **3**, 136.

Bernstein, I.B., J.M. Greene & M.D. Kruskal, 1957, *Phys. Rev.* **108**, 546.

Bethe, H., 1930, *Ann. Phys.* (Leipzig) **5**, 325.

Bloch, F., 1933, *Ann. Phys.* (Leipzig) **16**, 285.

Bogoliubov, N.N., 1946, *J. Phys. USSR* **10**, 257 & 265.

Bogoliubov, N.N., 1962, in *Studies in Statistical Mechanics*, Vol.1, translated by E.K. Gora, edited by J. de Boer & G.E. Uhlenbeck (North-Holland, Amsterdam), p.1.

Bohm, D., 1949, in *The Characteristics of Electrical Discharges in Magnetic Fields*, edited by A. Guthrie and R.K. Wakerling (McGraw-Hill, New York), Chap. 2, Sec. 5.

Bohm, D. & E.P. Gross, 1949a, *Phys. Rev.* **75**, 1851.

Bohm, D. & E.P. Gross, 1949b, *Phys. Rev.* **75**, 1864.

Bohm, D. & D. Pines, 1951, *Phys. Rev.* **82**, 625.

Bohm, D. & D. Pines, 1953, *Phys. Rev.* **92**, 609.

Bohr, N., 1948, Kgl. *Danske Videnskab. Selskab Mat.-Fys. Medd.* **28**, 8.

Bollinger, J.J., S.L. Gilbert, D.J. Heinzen, W.M. Itano & D.J. Wineland, 1990, in *Strongly Coupled Plasma Physics*, edited by S. Ichimaru (North-Holland/Yamada Science Foundation, Amsterdam), p.117.

Born, M. & H.S. Green, 1946, *Proc. Roy. Soc. Lond.* A **188**, 10.

Born, M. & H.S. Green, 1949, *A General Kinetic Theory of Liquids* (Cambridge U. P., Cambridge).

Bowers, R., C. Legendy & F. Rose, 1961, *Phys. Rev. Letters* **7**, 339.

Braginskii, S.I., 1957, *Zhur. Eksptl. Teoret. Fiz.* **33**, 459 [*Soviet Phys. JETP* **6**, 358 (1958)].

Braun, E., 1967, *Phys. Fluids* **10**, 731.

Briggs, R.J., 1964, *Electron-Stream Interaction with Plasmas* (MIT Press, Cambridge, Mass.).

Buchsbaum S.J. & J.K. Galt, 1961, *Phys. Fluids* **4**, 1514.

Buneman, O., 1959, *Phys. Rev.* **115**, 503.

Callen, H.B. & T.A. Welton, 1951, *Phys. Rev.* **83**, 34.

Ceperley, D.M. & B.J. Alder, 1980, *Phys. Rev. Letters* **45**, 566.

Chambers, R.G., 1956, *Phil. Mag.* **1**, 459.

Chandrasekhar, S., 1942, *Principles of Stellar Dynamics* (U. Chicago Press, Chicago).

Chandrasekhar, S., 1943, *Rev. Mod. Phys.* **15**, 1.

Chandrasekhar, S., 1960, *Plasma Physics*, compiled by S.K. Trehan (U. Chicago Press, Chicago).

Choquard, Ph., 1978, in *Strongly Coupled Plasmas*, edited by G. Kalman (Plenum, New York), p. 347.

Cole, M.W., 1974, *Rev. Mod. Phys.* **46**, 451.

Deacon, D.A.G., L.R. Elias, J.M.J. Madey, G.J. Ramian, H.A. Schwettman & T.I. Smith, 1977, *Phys. Rev. Letters* **38**, 892.

Debye, P. & E. Hückel, 1923, *Physik. Z.* **24**, 185.

Derfler, H. & T.C. Simonen, 1966, *Phys. Rev. Letters* **17**, 172.

Dessler, A.J., 1968a, J. *Geophys. Res.* **73**, 209.

Dessler, A.J., 1968b, J. *Geophys. Res.* **73**, 1961.

Dharma-wardana, M.W.C. & R. Taylor, 1981, *J. Phys. C* **14**, 629.

Drummond, J.E., 1958a, *Phys. Rev.* **110**, 293.

Drummond, J.E., 1958b, *Phys. Rev.* **112**, 1460.

Drummond, W.E. & D. Pines, 1962, *Nucl. Fusion Suppl.* **3**, 1049.

Dungey, J.W., 1953, *Phil. Mag.* **44**, 725.

Dungey, J.W., 1961, *Phys. Rev. Letters* **6**, 47.

Dupree, T.H., 1966, *Phys. Fluids* **9**, 1773.

Dupree, T.H., 1972, *Phys. Fluids* **15**, 334.

Eardley, D.M. & A.P. Lightman, 1975, *Ap. J.* **200**, 189.

Elias, L.R., W.M. Fairbank, J.M.J. Madey, H.A. Schwettman & T.I. Smith, 1976, *Phys. Rev. Letters* **36**, 717.

Englert, F. & R. Brout, 1960, *Phys. Rev.* **120**, 1085.

Fermi, E., 1924, *Z. Phys.* **29**, 315.

Fermi, E., 1949, *Phys. Rev.* **75**, 1169.

Fetter, A.L. & J.D. Walecka, 1971, *Quantum Theory of Many-Particle Systems*, (McGraw Hill, New York).

Forster, D., 1975, *Hydrodynamic Fluctuations, Broken Symmetry, and Correlation Functions* (W.A. Benjamin, Reading, Mass.).

Fortov, V.E., V.E. Bespalov, M.I. Kulish & S.I. Kuz, 1990, in *Strongly Coupled Plasma Physics*, edited by S. Ichimaru (North-Holland/Yamada Science Foundation, Amsterdam), p.571.

Fried, B.D. & R.W. Gould, 1961, *Phys. Fluids* **4**, 139.

Furth, H.P., J. Killeen & M.N. Rosenbluth, 1963, *Phys. Fluids* **6**, 459.

Galeev, A.A., V.N. Oraevskii & R.Z. Sagdeev, 1963, *Zhur. Eksptl. Teoret. Fiz.* **44**, 903 [*Soviet Phys. JETP* **17**, 615 (1963)].

Gell-Mann, M. & K.A. Brueckner, 1957, *Phys. Rev.* **106**, 364.

Ginzburg, V.L., 1970, *The Propagation of Electromagnetic Waves in Plasmas*, 2nd ed., translated by J.B. Sykes & R.J. Tayler (Pergamon, Oxford).

Glicksman, M., 1971, in *Solid State Physics*, Vol.26, edited by H. Ehrenreich, F. Seitz, & D.Turnbull, (Academic, New York), p.275.

Goldman, M.V., 1984, *Rev. Mod. Phys.* **56**, 709.

Gott, Yu.B., M.S. Ioffe & V.G. Telkovsky, 1962, *Nucl. Fusion Suppl.* **3**, 1045.

Gouedard, C. & C. Deutsch, 1978, *J. Math. Phys.* **19**, 32.

Gould, H.A. & H.E. DeWitt, 1967, *Phys. Rev.* **155**, 68.

Gould, R.W., T.M. O'Neil & J.H. Malmberg, 1967, *Phys. Rev Letters* **19**, 219.

Grimes, C.C., 1978, *Surf. Sci.* **73**, 397.

Gudmundsson, E.H., C.J. Pethick & R.I. Epstein, 1982, *Ap. J. Lett.* **259**, L19.

Guernsey, R., 1960, Ph. D. dissertation, U. Michigan, Ann Arbor.

Gupta, U. & A.K. Rajagopal, 1980, *Phys. Rev.* A **21**, 2064.

Hahn, E.L., 1950, *Phys. Rev.* **80**, 580.

Hamberger, S.M. & J. Jancarik, 1972, *Phys. Fluids* **15**, 823.

Hansen, J.-P. & I.R. McDonald, 1986, *Theory of Simple Liquids*, 2nd ed. (Academic, London).

Harris, E.G., 1959, *Phys. Rev. Letters* **2**, 34.

Harris, E.G., 1961, *J. Nucl. Energy* C **2**, 138.

Harrison, M.J., 1962, *J. Phys. Chem. Solids* **23**, 1079.

Hedin, L. & B.I. Lundqvist, 1971, *J. Phys.* C **4**, 2064.

Heisenberg, W., 1948, *Z. Physik*, **124**, 628.

Hill, R.M. & D.E. Kaplan, 1965, *Phys. Rev. Letters* **14**, 1062.

Hinton, F.L. & C. Oberman, 1968, *Phys. Fluids* **11**, 1982.

Hoh, F.C., 1962, *Rev. Mod. Phys.* **34**, 267.

Horton, W., 1984, in *Basic Plasma Physics* II, edited by A.A. Galeev & R.N. Sudan (North-Holland, Amsterdam), p.383.

Hubbard, J., 1961a, *Proc. Roy. Soc.* (London) A **260**, 114.

Hubbard, J., 1961b, *Proc. Roy. Soc.* (London) A **261**, 371.

Hubbard, W.B., 1980, *Rev. Geophys. Space Sci.* **18**, 1.

Ichimaru, S., 1962, *Ann. Phys.* (N.Y.) **20**, 78.

Ichimaru, S., 1965, *Phys. Rev.* **140**, B226.

Ichimaru, S., 1970, *Phys. Rev.* A **2**, 494.

Ichimaru, S., 1975a, *J. Phys. Soc.* Japan **39**, 261.

Ichimaru, S., 1975b, *Ap. J.* **202**, 528.

Ichimaru, S., 1977a, *Phys. Rev.* A **15**, 744.

Ichimaru, S., 1977b, in *Plasma Physics*, edited by H. Wilhelmsson (Plenum, New York), p.262.

Ichimaru, S., 1977c, *Ap. J.* **214**, 840.

Ichimaru, S., 1982, *Rev. Mod. Phys.* **54**, 1017.

Ichimaru, S., Ed., 1990, *Strongly Coupled Plasma Physics* (North-Holland/Yamada Science Foundation, Amsterdam).

Ichimaru, S. & H. Iyetomi, 1983, *J. Phys. Soc.* Japan **52**, 1730.

Ichimaru, S., H. Iyetomi & S. Tanaka, 1987, *Phys. Reports* **149**, 92.

Ichimaru, S., S. Mitake, S. Tanaka & X.-Z. Yan, 1985, *Phys. Rev.* A **32**, 1768.

Ichimaru, S., D. Pines & N. Rostoker, 1962, *Phys. Rev. Letters* **8**, 231.

Ichimaru, S. & M.N. Rosenbluth, 1970, *Phys. Fluids* **13**, 2778.

Ichimaru, S. & S. Tanaka, 1985, *Phys. Rev.* A **32**, 1790.

Ichimaru, S. & S. Tanaka, 1986, *Phys. Rev. Letters* **56**, 2815.

Ichimaru, S. & T. Tange, 1974, *J. Phys. Soc.* Japan **36**, 603.

Ikezi, H., 1973, *Phys. Fluids* **16**, 1668.

Ikezi, H. & N. Takahashi, 1968, *Phys. Rev. Letters* **20**, 140.

Ioffe, M.S., 1965, in *Plasma Physics* (IAEA, Vienna), p.421.

Iyetomi, H., S. Ogata & S. Ichimaru, 1989, *Phys. Rev.* B **40**, 309.

Jackson, E.A., 1960, *Phys. Fluids* **3**, 786.

Jackson, J.D., 1962, *Classical Electrodynamics* (Wiley, New York).

Jensen, T.H., J.H. Malmberg & T.M. O'Neil, 1969, *Phys. Fluids* **12**, 1728.

Jephcott, D.F., 1959, *Nature* **183**, 1652.

Kadomtsev, B.B., 1965, *Plasma Turbulence*, translated by L.C. Ronson (Academic, London).

Kaner, E.A. & V.G. Skobov, 1966, *Usp. Fiz. Nauk* **89**, 367 [*Soviet Phys. Usp.* **9**, 480 (1967)].

Kaner, E.A. & V.G. Skobov, 1968, *Advan. Phys.* **17**, 605.

Karpman, V.I., 1966, *Zhur. Eksptl. Teoret. Fiz.* **51**, 907 [*Soviet Phys. JETP* **24**, 603 (1963)].

Khanna, F.C. & H.R. Glyde, 1976, *Can. J. Phys.* **54**, 648.

Kihara, T. & O. Aono, 1963, *J. Phys. Soc. Japan* **18**, 837.

Kihara, T. & O. Aono, 1971, in *Kinetic Equations*, edited by R.L. Liboff & N. Rostoker (Gordon & Breach, New York), p.201.

Kim, H.C., R.L. Stenzel & A.Y. Wong, 1974, *Phys. Rev. Letters* **33**, 886.

Kirkwood, J.G., 1935, *J. Chem. Phys.* **3**, 300.

Kirkwood, J.G., 1946, *J. Chem. Phys.* **14**, 180.

Kivelson, M.G. & D.F. DuBois, 1964, *Phys. Fluids* **7**, 1578.

Klimontovich, Yu.L., 1957, *Zhur. Eksptl. Teoret. Fiz.* **33**, 982 [*Soviet Phys. JETP* **6**, 753 (1958)].

Klimontovich, Yu.L., 1967, *The Statistical Theory of Non-Equilibrium Processes in a Plasma*, edited by D. ter Haar, translated by H.S.H. Massey & O.M. Blunn (MIT Press, Cambridge, Mass.).

Kolmogorov, A.N., 1941, *Compt. Rend. Acad. Sci.* USSR **30**, 301.

Kono, M. & Y.-H. Ichikawa, 1973, *Prog. Theor. Phys.* **50**, 754.

Konstantinov, O.V. & V.I. Perel', 1960, *Zhur. Eksptl. Teoret. Fiz.* **38**, 161 [*Soviet Phys. JETP* **11**, 117 (1960)].

Krall, N.A. & M.N. Rosenbluth, 1963, *Phys. Fluids* **6**, 254.

Krall, N.A. & A.S. Trivelpiece, 1973, *Principles of Plasma Physics* (McGraw Hill, New York).

Kruskal, M.D. & M. Schwarzschild, 1954, *Proc. Roy. Soc.* (London) A **223**, 348.

Kubo, R., 1957, J. *Phys. Soc. Japan* **12**, 570.

Kukkonen, C.A. & A.W. Overhauser, 1979, *Phys. Rev.* B **20**, 550.

Landau, L.D., 1936, *Phys. Z. Sowjetunion* **10**, 154.

Landau, L.D., 1937, *Zhur. Eksptl. Teoret. Fiz.* **7**, 203.

Landau, L.D., 1946, *J. Phys.* (USSR) **10**, 25.

Landau, L.D. & E.M. Lifshitz, 1960, *Electrodynamics of Continuous Media*, translated by J.B. Sykes & J.S. Bell (Addison-Wesley, Reading, Mass.).

Landau, L.D. & E.M. Lifshitz, 1969, *Statistical Physics,* 2nd ed., translated by J.B. Sykes and M.J. Kearsley (Addison-Wesley, Reading, Mass.).

Lehnert, B., 1954, *Phys. Rev.* **94**, 815.

Lenard, A., 1960, *Ann. Phys.* (N. Y.) **10**, 390.

Lenard, A. & I.B. Bernstein, 1958, *Phys. Rev.* **112**, 1456.

Lindhard, J., 1954, *Kgl. Danske Videnskab, Selskab Mat.-Fys. Medd.* **28**, No.8.

Lindhard, J. & A. Winther, 1964, *Kgl. Danske Videnskab, Selskab Mat.-Fys. Medd.* **34**, 4.

Longmire, C.L. & M.N. Rosenbluth, 1956, *Phys. Rev.* **103**, 507.

Lundquist, S., 1949, *Phys. Rev.* **76**, 1805.

Malmberg, J.H. & C.B. Wharton, 1964, *Phys. Rev. Letters* **13**, 184.

Malmberg, J.H. & C.B. Wharton, 1966, *Phys. Rev. Letters* **17**, 175.

Malmberg, J.H. & C.B. Wharton, 1967, *Phys. Rev. Letters* **19**, 775.

Malmberg, J.H., C.B. Wharton, R.W. Gould & T.M. O'Neil, 1968, *Phys. Rev. Letters* **20**, 95.

Mayer, J.E. & M.G. Mayer, 1940, *Statistical Mechanics* (Wiley, New York).

Maynard, G. & C. Deutsch, 1982, *Phys. Rev.* A **26**, 665.

Mikhailovsky, A.B., 1983, in *Basic Plasma Physics* I, edited by A.A. Galeev & R.N. Sudan (North-Holland, Amsterdam), p.587.

Mikhailovsky, A.B. & L.I. Rudakov, 1963, *Zhur. Eksptl. Teoret. Fiz.* **44**, 912 [*Soviet Phys. JETP* **17**, 621 (1963)].

Mikhailovsky, A.B. & A.V. Timofeev, 1963, Zhur. Eksptl. Teoret. Fiz. **44**, 919 [*Soviet Phys. JETP* **17**, 626 (1963)].

Miller, P.B. & K.K. Haering, 1962, *Phys. Rev.* **128**, 126.

Montroll, E.W. & J.C. Ward, 1958, *Phys. Fluids* **1**, 55.

Morrison, P., 1961, *Handbuch der Phys.* **4611**, 1.

Mostovych, A.N., K.J. Kearney & J.A. Stamper, 1990, in *Strongly Coupled Plasma Physics*, edited by S. Ichimaru (North-Holland/Yamada Science Foundation, Amsterdam), p.589.

Nakano, T., 1972, *Ann. Phys.* (N.Y.) **73**, 326.

Novikov, I.D. & K.S. Thorne, 1973, in *Black Holes*, edited by C. DeWitt and B. DeWitt (Gordon and Breach, New York), p.344.

Nozières, P. & D. Pines, 1958a, *Phys. Rev.* **109**, 762.

Nozières, P. & D. Pines, 1958b, *Nuovo Cimento* **9**, 470.

O'Neil, T.M., 1965, *Phys. Fluids* **8**, 2255.

O'Neil, T.M., 1968, *Phys. Fluids* **11**, 2420.

O'Neil, T.M. & R.W. Gould, 1968, *Phys. Fluids* **11**, 134.

O'Neil, T.M., P.G. Hjorth, B. Beck, J. Fajans & J.H. Malmberg, 1990, in *Strongly Coupled Plasma Physics*, edited by S. Ichimaru (North-Holland/Yamada Science Foundation, Amsterdam), p.313.

O'Neil, T.M. & N. Rostoker, 1965, *Phys. Fluids* **8**, 1109.

Parker, E.N., 1973, *Ap. J.* **180**, 247.

Pathria, R.K., 1972, *Statistical Mechanics* (Pergamon, Oxford).

Penrose, O., 1960, *Phys. Fluids* **3**, 258.

Perrot, F. & M.W.C. Dharma-wardana, 1984, *Phys. Rev.* A **30**, 2619.

Pines, D., 1963, *Elementary Excitations in Solids* (W.A. Benjamin, New York).

Pines, D., 1966, in *Quantum Liquids*, edited by D.F. Brewer (North-Holland, Amsterdam) p.257.

Pines, D. & D. Bohm, 1952, *Phys. Rev.* **85**, 338.

Pines, D. & P. Nozières, 1966, *The Theory of Quantum Liquids* (W.A. Benjamin, New York).

Pines, D. & J.R. Schrieffer, 1961, *Phys. Rev.* **124**, 1387.

Pines, D. & J.R. Schrieffer, 1962, *Phys. Rev.* **125**, 804.

Post, R.F. & M.N. Rosenbluth, 1966, *Phys. Fluids* **9**, 730.

Prendergast, K.H. & G.R. Burbidge, 1968, *Ap. J. (Lett.)* **151**, L83.

Raether, H., 1980, *Excitations of Plasmons and Interband Transitions by Electrons* (Springer, Berlin).

Rosenbluth, M.N., 1965, in *Plasma Physics* (IAEA, Vienna), p.485.

Rosenbluth, M.N., N.A. Krall & N. Rostoker, 1962, *Nucl. Fusion Suppl.* **1**, 143.

Rosenbluth, M.N. & C.L. Longmire, 1957, *Ann. Phys.* (N.Y.) **1**, 120.

Rosenbluth, M.N., W.M. MacDonald & D.L. Judd, 1957, *Phys. Rev.* **107**, 1.

Rosenbluth, M.N. & R.F. Post, 1965, *Phys. Fluids* **8**, 547.

Rosenbluth, M.N. & N. Rostoker, 1962, *Phys. Fluids* **5**, 776.

Rostoker, N., 1964a, *Phys. Fluids* **7**, 479.

Rostoker, N., 1964b, *Phys. Fluids* **7**, 491.

Rostoker, N. & M.N. Rosenbluth, 1960, *Phys. Fluids* **3**, 1.

Rudakov, L.I. & R.A. Sagdeev, 1959, *Zhur. Eksptl. Teoret. Fiz.* **37**, 1337 [*Soviet Phys. JETP* **10**, 952].

Rukhadze, A.A. & V.P. Silin, 1961, *Usp. Fiz. Nauk* **74**, 223 [*Soviet Phys. Usp.* **4**, 459(1961)].

Rukhadze, A.A. & V.P. Silin, 1962, *Usp. Fiz. Nauk* **76**, 79 [*Soviet Phys. Usp.* **5**, 37 (1962)].

Rutherford, P.H. & E.A. Frieman, 1963, *Phys. Fluids* **6**, 1139.

Sagdeev, R.Z., 1967, *Proc. Symp. Appl. Math.* **18**, 281.

Salpeter, E.E., 1954, *Aust. J. Phys.* **7**, 373.

Salpeter, E.E., 1960, *Phys. Rev.* **120**, 1528.

Sato, N., H. Ikezi, Y. Yamashita & N. Takahashi, 1968, *Phys. Rev. Letters* **20**, 837.

Schatzman, E., 1958, *White Dwarfs* (North-Holland, Amsterdam).

Scheuer, P.A.G., 1967, in *Plasma Astrophysics*, edited by P.A. Sturrock (Academic, New York), p. 289.

Schülke, W., U. Bonse, H. Nagasawa, S. Mourikis & A. Kaprolat, 1987, *Phys. Rev. Letters* **59**, 1361.

Shakura, N.I. & R.A. Sunyaev, 1973, *Astr. and Ap.* **24**, 337.

Shapiro, S.L. & S.A. Teukolsky, 1983, *Black Holes, White Dwarfs, and Neutron Stars* (Wiley, New York).

Singwi, K.S. & M.P. Tosi, 1981, in *Solid State Physics*, Vol. 36, edited by H. Ehrenreich, F. Seitz & D. Turnbull (Academic, New York), p. 177.

Singwi, K.S., M.P. Tosi, R.H. Land & A. Sjölander, 1968, *Phys. Rev.* **176**, 589.

Sitenko, A.G. & K.N. Stepanov, 1956, *Zhur. Eksptl. Teoret. Fiz.* **31**, 642 [*Soviet Phys. JETP* **4**, 512 (1957)].

Soper, G.K. & E.G. Harris, 1965, *Phys. Fluids* **8**, 984.

Spitzer, L., 1960, *Phys. Fluids* **3**, 659.

Spitzer, L., 1962, *Physics of Fully Ionized Gases*, 2nd ed. (Wiley Interscience, New York).

Starrfield, S., J.W. Truran, W.M. Sparks & G.S. Kutter, 1972, *Ap. J.* **176**, 169.

Stenzel, R.L. & W. Gekelman, 1979, *Phys. Rev. Letters* **42**, 1055.

Stern, E.A., 1963, *Phys. Rev. Letters* **10**, 91.

Stix, T.H., 1962, *The Theory of Plasma Waves* (McGraw-Hill, New York).

Su, C.H. & C. Oberman, 1968, *Phys. Rev. Letters* **20**, 427.

Sweet, P.A., 1969, *Ann. Rev. Astr. and Ap.* **7**, 149.

Tago, K., K. Utsumi & S. Ichimaru, 1981, *Prog. Theor. Phys.* **65**, 54.

Tanaka, S. & S. Ichimaru, 1985, *J. Phys. Soc.* Japan **54**, 2537.

Tanaka, S. & S. Ichimaru, 1986a, *J. Phys. Soc.* Japan **55**, 2278.

Tanaka, S. & S. Ichimaru, 1986b, *Phys. Rev.* A **34**, 4163.

Tanaka, S. & S. Ichimaru, 1987, *Phys. Rev.* A **35**, 4743.

Tanaka, S. & S. Ichimaru, 1989, *Phys. Rev.* B **39**, 1036.

Tanaka, S., X.-Z. Yan & S. Ichimaru, 1990, *Phys. Rev.* A **41**, 5616.

Tange, T., 1974, J. *Phys. Soc. Japan* **36**, 596.

Tange, T. & S. Ichimaru, 1974, *J. Phys. Soc. Japan* **36**, 1437.

Taylor, J.B., 1961, *Phys. Fluids* **4**, 1142.

Taylor, J.B., 1963, *Phys. Fluids* **6**, 1529.

Taylor, J.B., 1965, in *Plasma Physics* (IAEA, Vienna), p.449.

ter Haar, D., 1960, *Elements of Statistical Mechanics* (Holt, Rinehart & Winston, New York).

Thompson, W.B. & J. Hubbard, 1960, *Rev. Mod. Phys.* **32**, 714.

Tonks, L. & I. Langmuir, 1929, *Phys. Rev.* **33**, 195.

Totsuji, H. & S. Ichimaru, 1973, *Prog. Theor. Phys.* **50**, 753.

Totsuji, H. & S. Ichimaru, 1974, *Prog. Theor. Phys.* **52**, 42.

Van Horn, H.M., 1971, in *White Dwarfs*, edited by W.J. Luyten (Reidel, Dordrecht), p.136.

Van Horn, H.M., 1990, in *Strongly Coupled Plasma Physics*, edited by S. Ichimaru (North-Holland/Yamada Science Foundation, Amsterdam), p.3.

Van Hoven, G., 1966, *Phys. Rev. Letters* **17**, 169.

Vashishta, P., P. Bhattacharyya & K.S. Singwi, 1974, *Phys. Rev.* B **10**, 5108.

Vedenov, A.A., E.P. Velikhov & R.Z. Sagdeev, 1962, *Nucl. Fusion Suppl.* **2**, 465.

Vieillefosse, P. & J.-P. Hansen, 1975, *Phys. Rev.* A **12**, 1106.

Vlasov, A.A., 1938, *Zhur. Eksptl. Teoret. Fiz.* **8**, 291.

Vlasov, A.A., 1967, *Usp. Fiz. Nauk* **93**, 444 [*Soviet Phys. Usp.* **10**, 721 (1968)].

Wallenborn, J. & M. Baus, 1978, *Phys. Rev.* A **18**, 1737.

Washimi, H. & T. Taniuti, 1966, *Phys. Rev. Letters* **17**, 996.

Wharton, C.B., J.H. Malmberg & T.M. O'Neil, 1968, *Phys. Fluids* **11**, 1761.

Wigner, E.P., 1934, *Phys. Rev.* **46**, 1002.

Wigner, E.P., 1938, *Trans. Faraday Soc.* **34**, 678.

Wild, N.W., W. Gekelman & R.L. Stenzel, 1981, *Phys. Rev. Letters* **46**, 339.

Wilcox, J.M., F.I. Boley & A.W. DeSilva, 1960, *Phys. Fluids* **3**, 15.

Williams, G.A. & G.E. Smith, 1964, *IBM Journal* **8**, 276.

Wong, A.Y. & P.Y. Cheung, 1984, *Phys. Rev. Letters* **52**, 1222.

Wong, A.Y., N. D'Angelo & R.W. Motley, 1962, *Phys. Rev. Letters* **9**, 415.

Wong, A.Y., R.W. Motley & N. D'Angelo, 1964, *Phys. Rev.* **133**, A436.

Wyld, H.W. & D. Pines, 1962, *Phys. Rev.* **127**, 1851.

Yamada, M., Y. Ono, A. Hayakawa, M. Katsurai & F.W. Perkins, 1990, *Phys. Rev. Letters* **65**, 721.

Yamada, M. & M. Raether, 1975, *Phys. Fluids* **18**, 361.

Yan, X.-Z. & S. Ichimaru, 1987, *J. Phys. Soc. Japan* **56**, 3853.

Yvon, J., 1935, *Act. Scient. Ind.* **203** (Hermann, Paris).

Yvon, J., 1966, *Les Corrélations et l'Entropie sn Mécanique Statistique Classique* (Dunod, Paris).

Zakharov, V.E., 1967, *Zhur. Eksptl. Teoret. Fiz.* **53**, 1734 [*Soviet Phys. JETP* **26**, 994 (1968)].

Zakharov, V.E., 1972, *Zhur. Eksptl. Teoret. Fiz.* **62**, 1745 [*Soviet Phys. JETP* **35**, 908 (1972)].

Ziman, J.M., 1961, *Philos. Mag.* **6**, 1013.

Ziman, J.M., 1972, *Theory of Solids* (Cambridge University, London).

Index